Telecommunications Department Management

For a complete listing of the *Artech House Telecommunications Library*, turn to the back of this book.

Telecommunications Department Management

Robert A. Gable

Artech House
Boston • London

Library of Congress Cataloging-in-Publication Data
Gable, Robert A.
Telecommunications department managment / Robert A. Gable.
p. cm. — (Artech House intelligent transportation systems library)
Includes bibliographical references and index.
ISBN 0-89006-650-7 (alk. paper)
1. Telecommunications—Management. I. Title II. Series.
HE7661.G33 1999
384'.068—DC21 99-23715
CIP

British Library Cataloguing in Publication Data
Gable, Robert A.
Telecommunications department management.—(Artech House telecommunications library)
1. Telecommunications systems—Management 2. Business enterprises—Communication systems—Management
I. Title
651.7'068
ISBN 0-89006-650-7

Cover and text design by Darrell Judd

685 Canton Street
Norwood, MA 02062

International Standard Book Number: 0-89006-650-7
Library of Congress Catalog Card Number: 99-23715

10 9 8 7 6 5 4 3 2 1

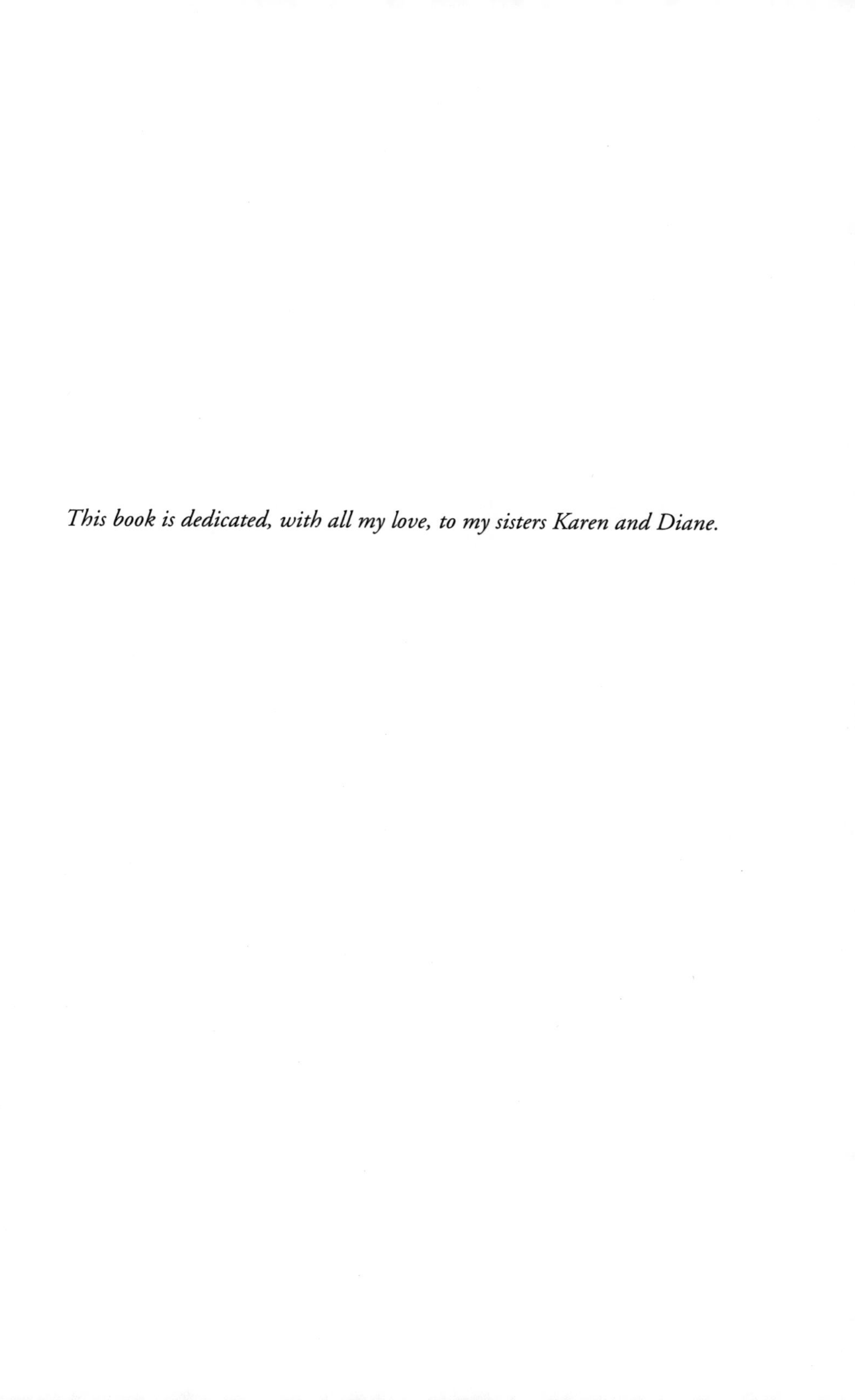

This book is dedicated, with all my love, to my sisters Karen and Diane.

Contents

Introduction

I came to the field of telecommunications quite by accident. In 1983, I was working at an insurance company, laboring away at what could only be described as a boring, clerical position. Often, while in the midst of the agonizing, paper-pushing drudgery, I thought that there had to be a better way to make a living. I longed for something more exciting, certainly more challenging, and I often felt that continued employment at that insurance company would seriously affect my sanity. There was also a fear that I might never find the type of challenging and rewarding career that I desired, and would languish in a boring desk job for the rest of my life.

As I pondered what my next move would be, my dreary job ended sooner than I had anticipated. A recession had struck the American economy, and I soon found myself out on the street, the victim of a corporate downsizing. While I wanted desperately to leave my job at the insurance company, I certainly did not want to leave on those terms.

Employment opportunities were somewhat abysmal in 1983. A recession was raging and my resume was weak, having only worked at an entry level clerical position since I had graduated from college. Given my qualifications, and the fact that I had just recently married, I was not seeking employment in any particular field, I was simply looking for a job. My modus operandi was: Money first, career second!

Jobs were scarce in 1983, but I did notice that there seemed to be a plethora of opportunities in the field of telecommunications, a field that was on the verge of a major upheaval. I was working through an employment agency and the recruiter kept trying to convince me that a career in long distance sales would provide a promising future.

Concerning the field of telecommunications, I must admit I was ambivalent. I had no technical training (my degree was in journalism), and knew little of communications matters except that a large monopoly seemed to control everything.

In addition to my lack of technical expertise, I also had grave concerns about a career in sales. I knew that I did not have a forceful personality, and also wondered if my cynical nature might not be a tremendous deterrent in a field where unbridled enthusiasm is a plus, if not a necessity. But as the bills began to accumulate, and there seemed to be no other opportunities, I reluctantly took employment as a sales rep with a new long distance company, one that was competing directly with the large monopoly that controlled everything: AT&T.

I suppose I can only describe the year 1983 as being tumultuous. It was a year of great excitement, but also of great anxiety. For AT&T's competitors, the world was to be their oyster. On January 1, 1984, the world's largest monopoly would be dismantled, and customers would no longer be "victims" of the large, bureaucratic monolith known as "Ma Bell." The breakup which came to be known as the divestiture would make available the billions of dollars in revenue that had once been the exclusive domain of "Ma Bell."

Competitors of AT&T were salivating over the prospects. Companies such as MCI, Sprint, and a myriad of small resellers felt as though they had infinite growth potential. Equipment providers were also optimistic. While it was certainly permissible to sell equipment (other than AT&T's) prior to divestiture, the fact that "Ma Bell" was breaking up changed the way Americans viewed the field of telecommunications. End-users—both large and small—now had more choices and more opportunities. There was no longer the one-stop-shop option of "Ma Bell."

My foray into telecommunications and sales proved to be a turning point in my life. I had found a career that was exciting, challenging, and constantly changing. Communications, I soon learned, was an essential component of economic life, for both the nation and each individual company. I came to realize that a nation that did not invest in its telecommunications infrastructure was a nation in economic decline. The same could be said of a company that did not develop its communications infrastructure. And with the divestiture of the Bell system, end-users inherited more responsibility than ever before to develop and administer a telecommunications infrastructure.

During those early years, I found the reactions of potential customers to be interesting and sometimes puzzling. People were upset and disoriented about divestiture. I remember one sales call where I was proposing thousands of dollars in savings. The perspective customer interrupted me in the middle of my sales presentation and asked me what I was doing about my home telephone. Of course, I was dumbfounded. Why did this man care about his home telephone so much when I was proposing thousands of dollars in savings? This example demonstrates the anxiety and uncertainty that people felt when their comfort level was removed. Before divestiture, nobody thought much about telecommunications. It was an accepted fact that it worked and there was little that the average person needed to know about this technology. If you did know about this technology, you worked for the telephone company.

The monopoly of "Ma Bell" also created many paradigms. It was interesting to note that some customers thought that if they used an alternative long distance service, they would be committing an illegal act. "What can AT&T do about this?" they would often ask me.

These examples demonstrate how many Americans viewed telecommunications in 1983. In many companies, there was no need for telecommunications staffs. In fact, only large companies often considered it necessary to retain telecommunications expertise. After all, there was only one place you could order telecommunications services or products, so why incur that additional expense? Divestiture changed this.

There were two concepts that struck me in 1983. First, I absolutely abhorred sales, and the second was that most companies lacked, and yet needed, telecommunications expertise. Given the fact that I despised sales, but loved the technology of telecommunications, it was only natural for me to apply my efforts toward telecommunications management.

Before I made my move from sales to that of end-user, I often took a keen interest in who was responsible for a given company's telecommunications services. The people you would call upon were a disparate group indeed, ranging from the telecommunications manager to the maintenance man. However diverse the group seemed, across the board there seemed to be a general lack of knowledge about telecommunications services and products. Not only was the group diverse, there was also no standard method by which companies administered telecommunications. Even if a company did happen to develop a telecommunications staff, it was often done on an ad hoc basis, and most staff members learned through on-the-job training.

Throughout the years, as a salesman, consultant, and ultimately as an end-user, I have spent a great deal of time observing how companies develop and administer their telecommunications infrastructures. Some sixteen years later, the results of these observations are often as disparate as when I began my career in 1983.

I have often thought of the field of telecommunications as an odd profession. Unlike accounting, law, medicine, or other professions, there is no mandatory certification process or degree that one must obtain in order to enter the field. In fact, many people find that they come to the field completely by chance. And while many people eventually take a smattering of courses, it is rare that you find a telecommunications professional who has a bachelor's degree in the field, let alone an advanced degree.

Since most telecommunications professionals achieve their knowledge via on-the-job-training, there are also no standard guidelines for the structure or specific duties of the telecommunications department. The support of telecommunications functions within a given company might range from a formally structured department to an office manager who administers the PBX, cable, and a few servers connected to a local area network (LAN). I have even seen this responsibility given to secretaries and maintenance personnel.

Even for companies that have dedicated personnel, there are no official guidelines for the structure or duties of the telecommunications department. There are companies that maintain strict divisions between voice and data, while other companies have combined these functions. The degree of support within each company also varies widely. Companies may outsource a number of functions and only offer support from a high level, or they may be completely hands-on, maintaining a staff of qualified technicians, controlling every aspect of telecommunications products and services.

Throughout my years of observation and experience, about the only standard thing I can say about telecommunications departments and staffs is that there are no standards. This statement is somewhat ironic since, in the modern postdivestiture era, most companies, both large and small, are often dependent upon telecommunications expertise. Increasingly, as we approach the new millennium, telecommunications services and products are taking on an increasingly more important role in the business world. It is no longer simply a necessary business expense. Telecommunications services and products give companies a competitive advantage, and they also support a major portion of a company's infrastructure.

Consider that most functions within a company have been computerized. Whether it is inventory control, payroll, accounts receivable, accounts payable, or order entry, in the modern business world these functions have usually been computerized. Then consider that computing has been migrating from a mainframe environment to one that is client-server based. Such an infrastructure, where each department may have their own computer, mandates that quality communications are established between the various computers. While each computer might be maintained by the individual department, there is still a need for the rest of the corporation to have access to the data.

Distributed computing is only one part of the problem. Computers are also becoming increasingly smaller, less expensive, and more powerful. PCs now have multimedia capabilities and have storage capacity in the gigabit range. Now that computing power has migrated to the desktop, there is a greater need for more bandwidth. Hence the proliferation of LANs, and the development of high-speed wide area network (WAN) services such as frame relay. LAN and WAN technologies are both complex and expensive and without a capable technical staff, the net result will be poor network performance and high operating expenses.

Beyond intracompany communications, there is also a need to connect computers publicly. The Internet is not only accessed by many residential computer users for their own use, many companies now use it as a powerful business tool. With Internet access comes more responsibility and liability. Opening a corporate network to the public can be dangerous, inviting industrial espionage or malicious hackers.

The work environment has also affected corporate telecommunications requirements. Pending environmental legislation, many states may require companies to offer work-at-home capabilities. Hence, there is the concept of telecommuting, whereby an employee has access to voice and data capabilities from the home. As many states or cities attempt to reduce pollution or traffic congestion, one viable option will be telecommuting.

Beyond data communications requirements, all companies still have to support good old fashioned POTS (plain old telephone service). This includes local service (perhaps a variety of local calling plans), long distance service (with countless number of providers), and a myriad of equipment that provides automatic call distribution (ACD) and voice mail, to name but a few features.

Just by citing these few examples, it is obvious that the field of telecommunications is in a state of turmoil. Companies face many challenges, not only in recruiting top flight technical talent, but also in structuring the telecommunications department so that it provides optimum performance for the company. Many questions face businesses as we approach the twenty-first century. While there are many questions, the answers are not always straightforward. For instance, how many people should be hired to service the work load? How does one determine what is a sufficient head count? Should voice communications be separated from data communications? Should one manager oversee all group activities, and would it be better to have assistant managers within the department? What division should the telecommunications department be in, and what person should it ultimately report to? What positions would be better served by specialists and what positions would be better served by generalists? When is it necessary or advantageous to outsource? How does a telecommunications manager maintain morale, while still pushing a weary, understaffed department to perform under pressure? Can a telecommunications manager prevent staff from leaving? How can a telecommunications manager maintain a heavy workload, but still find the time and budget for continuing education?

These are but a few of the questions that may plague the modern corporation and the pressure to support technologies that are mission-critical, increasingly complex, and changing by the day. In my professional roles as a salesman, consultant, and end-user I often observed what type of people were responsible for telecommunications support and what their assigned duties were. Throughout those years I asked many questions and increasingly began to think that the field of telecommunications was one that was in dire need of the same criteria as other professions. For instance, is there an organization that sets standards for telecommunications department structure, job descriptions, certification processes, and continuing education? What if a company wanted to reorganize their telecommunications department or perhaps start one from scratch? Is there an organization or a resource available that could provide such information?

The sad truth is that most companies do not approach telecommunications management in such a logical manner. This is a pity, since telecommunications is such a critical part of a business's capability. Companies that do not establish a telecommunications infrastructure and retain telecommunications expertise will find themselves at a distinct disadvantage.

This presents many challenges to the modern company. Consider first the rapid pace of technological change. Existing technologies continue to be refined, while new technologies continue to surface. What direction should a company take? Should they consider a new technology that appears to be superior to any existing technology? But is that new technology a deliverable technology? Even if it is, will it survive in the marketplace? The best technology in the world will not serve a company's needs if they cannot buy parts or find trained service people. Companies that commit to leading edge technology can find themselves an island in the business world, and may need to retire expensive assets prematurely if the technology is abandoned by both vendors and end-users.

Conversely, there are also dangers to relying too much on existing technology. If companies are frightened of new technology they may, once again, become an island in the business world, and perhaps be at a distinct disadvantage with their competitors.

Technology aside, there are constant changes in the providers of telecommunications services and products. Just a cursory glance at the many telecommunications newspapers and magazines will reveal a myriad of up-start companies, mergers, takeovers, spin-offs, and bankruptcies.

To further complicate the situation, regulatory change did not cease in 1984. Local service now has competition (which it did not in 1984). Now end-users will begin to see competition on a local level.

The domestic market is not the only place that challenges the modern telecommunications department. Increasingly, many companies are now making forays into foreign markets, thereby necessitating the need for international connectivity.

It is with these issues in mind that I developed the concept for this book. The modern telecommunications department has become a critical component of any major business. Without telecommunications services, very few modern companies could survive. Considering how critical telecommunications services are, it is only logical then for any company to administer the department so that the company will gain full advantage of telecommunications services. The telecommunications department is more than simply corporate overhead. In many cases the services and products that these people provide generate and maintain revenue.

This book examines the concept of the telecommunications department and its role in corporate America. It is not meant to be a rigid sets of rules, but rather an educational vehicle. This book is also not the final word in telecommunications department structure and management. No book could be since the business world constantly changes and new ideas are always emerging. It does, however, offer guidelines and advice for structuring and administering the department. My hope in writing this book is not that people will take it literally. That would smack of dogma and there is certainly too much of that in the business world today. I actually have two primary goals in writing this book. First, I hope that it will have some impact in influencing the business community relative to the importance of telecommunications in the modern business world. Then perhaps more companies will view the telecommunications department as a critical part of their infrastructure and future, instead of merely overhead. Second, I hope that it will stimulate questions and thought, even above and beyond the ones that have been posed in this book. After all, no book can be the final word. But hopefully, this book will stimulate activity that will lead to companies examining the importance of telecommunications and how a well-structured department can stimulate revenue in addition to maintaining a healthy business infrastructure.

1

The Telecommunications Department and the Business World

How important are telecommunications technologies to the modern business world? Many statistics could be cited to support the importance of these technologies, but perhaps the question could best be answered by some casual observations. If you were to stand on the corner of an average city street, you would notice many drivers and pedestrians engaged in conversation using cellular telephones. Many of these people are probably using a company-provided telephone, and conducting business while they are commuting to work, or traveling on business. Probably none of the cell phone users would garner much attention, indicating the commonality of the technology. Just twenty years ago, a person walking down the street using a cell phone would probably have attracted many curious stares, and commute time to a job was simply time where an employee was inaccessible. Business travel time was considered the same. In this world of the past, people were inaccessible while they were driving or walking, and relied on public telephones (if available) for their remote communications needs. This caused circumstances that would be considered difficult by today's standards. For instance, if a salesman were late for an appointment in today's world, he might worry about how his business prospect would perceive him, perhaps not believing there was such a large traffic jam that caused his tardiness. In today's world, the salesman calls from his car, letting his prospect know that he will be late and the reason why. Moreover, if the salesman did not call from a cellular phone, the customer might ask why he does not own one. In today's business world, people expect connectivity, regardless of where you are.

As you continue to observe the activity from this same street corner, you will probably see a number of people carrying portable computers. Many of these laptop computers will be equipped with modems, allowing access to their company's local area networks (LANs), servers, mainframe computers, or the Internet. These people might dial into the Internet, or to remote computers to check e-mail, inventory, enter an order, or simply obtain news that might be critical to their job. They could do this via telephone cable, plugging into a publicly available telephone line, or perhaps use a wireless modem. Once again, just twenty years ago, the sight of somebody sitting on a bus, pecking away at a keyboard on a small computer would have attracted many curious stares. Today, a personal computer is a common and accepted technology, as are the use of modems and the Internet.

If you casually walked into an office on this city street, you would see workers sitting at their desks, engaged in conversation on the company telephone system, or accessing data on their personal computers. The telephone they are using could very well be a digital telephone, digitizing the employee's voice, and connecting them through the public network, perhaps transmitted through high capacity fiber optic lines, over trans-oceanic cable, or through satellite links. The conversation might go through international gateways and perhaps the party they are speaking to is walking down the street of some foreign city, also speaking on a cellular telephone.

You might then walk into a conference room of this office and see a group of people engaged in a videoconference, viewing fellow employees who are located on the other side of the world. These employees may be writing information on an electronic white board or viewing a PC-based program, each party being able to view this information on their video units, just as if everybody were sitting in the same room. Just a few short years ago these people might have struggled through this meeting over the telephone, or may have even traveled to the distant office, spending thousands of dollars in travel expenses, not to mention the potential productive time lost to travel.

Such technologies a few decades ago would have been the musings of science fiction writers. Today these are common and accepted technologies, very often taken for granted by the average employee. In most companies, a new employee will not only be assigned a telephone, but he will also be issued a personal computer. The employee might also be issued a portable computer equipped with a modem, a cellular telephone, and a pager depending on his duties. The employee will also probably be given a voice mail box so that messages can be recorded in his absence. When the employee is issued a business card, there will be listings for a telephone, a fax machine, perhaps a pager, and an Internet e-mail address. All of these technologies will usually be provided or coordinated through the efforts of individuals who have specialized knowledge in the field of telecommunications. In smaller to mid-sized companies these could be individuals who have been assigned telecommunications duties as only a part of their overall responsibilities. In larger companies, these technologies will have been provided by a telecommunications department, a group of professionals who maintain a high level of technical expertise and business skills.

As these common examples illustrate, the technologies of communication have invaded almost every aspect of the business world. In a nutshell, business is now conducted on networks. But the presence and proliferation of these technologies does not stop there. Advanced communications technologies are now readily available to the average consumer. Children are now assigned homework on personal computers and people surf the Internet to seek information, shop, or communicate with other Internet users around the world. People use cellular telephones for their own personal use and subscribe to voice mail for their home telephone service. Because these technologies have proliferated in the business world and for personal use, new economic opportunities have opened for both businesses and consumers. Yet, for as many advantages as these technologies provide, they also offer many challenges.

One of the first problems that any company faces is the rapid pace of technological change. It has been estimated that the rate of technological change doubles every ten years. This may be a difficult fact to quantify, but nobody (neither novice nor professional) can deny that technology advances at a furious rate. This means that the telecommunications technologies are changing too rapidly for the "part time or novice telecommunications professional" to grasp. Even experienced telecommunications professionals will complain that they find it difficult to keep up with the changes in the technologies. Nevertheless, companies cannot shy away from the rapidly evolving technologies because they are fundamental to our modern economic infrastructure. Would a company dare to not offer toll-free numbers to its customer base to order product? In a free market economy, where competition is plentiful, this would be economic suicide. Should a company hesitate to offer a digital catalog via a web site on the Internet? It may not be economic suicide, but it would be safe to assume that a company's competitor has already done it and is generating additional revenue because of it. Should a company hesitate to provide cellular telephones to its sales force, thinking the service to be too expensive? Once again, it is probably safe to assume that a competitor has already employed the technology and is gaining efficiencies and additional revenue. For most companies in the information age, the question is not whether to employ the technology. Rather, it is to aggressively employ the technologies and to keep the company at the forefront of technological change. This can only be accomplished with dedicated expertise that must be consolidated into a single department: the corporate telecommunications department.

1.1 The Evolution of the Telecommunications Department

January 1, 1984 was a pivotal date for the world of telecommunications. On this date, the world's largest monopoly, AT&T, was broken up. Commonly known as divestiture, the government set a calculated course for total deregulation of the telecommunications environment in the United States. AT&T was permitted to retain Bell Laboratories, Western Electric (the equipment division), and its long distance network. The former monopoly was no longer allowed to provide local service which was to be provided by seven regional bell operating companies (RBOC) and a myriad of independent telephone companies (ITC). A primary reason for divestiture was to introduce more competition into the telecommunications marketplace. Even though competition existed prior to divestiture, it was certainly not an even playing field. The government wanted to provide opportunities for AT&T's competitors while keeping a tight reign on the telecommunications giant which could, given its sheer size, quickly overcome smaller companies should regulatory restraints not be imposed.

With the advent of divestiture, there were immediate and dramatic changes for the users of telecommunications services, both residential and business. The benefits of these changes to the end-user were to be more choices and lower prices, the natural by-products of a competitive environment. The irony of the situation was that many end-users were both confused and wary of the situation. For as much as people used to complain about "Ma Bell," residential and business users alike found a certain degree of comfort in the one-stop-shop offered by the controlled monopoly of the old Bell system. What fueled the wariness and discomfort of many end-users was the fact that AT&T had always maintained a high degree of quality in their products and services. Had AT&T provided inferior products or transmission quality, it might have been a different story, but in the early eighties, the most prevalent complaints about "Ma Bell" was that she was expensive, bureaucratic, and there were no other choices. Seldom did anyone hear the complaint that "Ma Bell's" products did not work.

During the monopoly era, telecommunications departments were usually only prevalent in the largest of companies. In smaller or mid-sized companies there was seldom a telecommunications department or even a dedicated telecommunications professional. The thinking was, there is only one company that supplies everything, what is the need for dedicated expertise? In such situations, an employee was assigned the telecommunications duties, but it was usually one of many duties. Sales representatives of the old Bell system found that a sales call could place them in front of almost any variety of employee, from janitors and secretaries on up to senior managers. During these sales calls, the salesman probably encountered a variety of attitudes. Some employees inevitably were resentful that they had been assigned the telephone duties while others might have been indifferent, thinking that telecommunications was simply another necessary business expense. It is doubtful that many salespeople encountered outright enthusiasm or people who were very knowledgeable about telecommunications products and services. The prevailing thought was not that these technologies provided a competitive edge for their company, it was simply a necessary cost of doing business, the same as copiers or typewriters.

The lack of telecommunications expertise in many of these smaller companies was not necessarily a bad business decision. Consider that many of these companies might have had but a single location. If there was more than one location, there usually wasn't a need for dedicated voice tie lines and data (probably transmitted at a very low speed) was transmitted via a dedicated analog data line, or perhaps batched over a low speed modem at night. The overall communications requirements for smaller companies were simple and manageable without the added expense of dedicated professionals. Moreover, most people probably didn't even know that there was such a thing as a telecommunications professional.

Large companies faced different issues. For a company that had multiple locations geographically dispersed throughout the country and perhaps the world, telecommunications products and services took on a more demanding role. First, the sheer volume of communications devices and services required full-time employees to manage them. At each site there might have been a private branch exchange (PBX) that supported thousands of telephones. This equated to thousands of extensions, numerous PBX features, and trunk lines that interfaced to the PBX from the public network. A large company could very well pay AT&T personnel to manage their telephone system, but many companies found there was no alternative but to dedicate personnel to this function. There were simply too many devices and services to hand over to a third party, and some decisions plainly had to be made by company personnel.

In addition to the sheer number of devices and services, large companies incurred enormous telecommunications bills. As is the case with any business expense, senior management demanded that these costs be brought under control. In an environment where a monopoly controls the entire industry, it would seem that large businesses had no choice except to pay the bill. This was not the case and large businesses actually had a number of options that were open to high volume users.

Depending on the volume of long distance business and calling patterns, companies could install and manage their own networks such as the electronic tandem network (ETN), share in high volume networks that were used by other large businesses such as the enhanced private switched communications services (EPSCS), or subscribe to long distance discount services such as wide area telecommunications services (WATS). However, the decision to use such capabilities was not simple. For instance, an ETN was a private network comprised of dedicated tie lines that might have saved a company thousands of dollars in long distance charges. In order to install and maintain an ETN, telephone switches required additional software and hardware. The telephone switches would interface to both ETN private voice grade tie lines and local central office (CO) lines. A company would have to determine how many tie lines were required, how much they cost, and how much telephone traffic could be carried on these lines at peak periods of usage. If the analysis to justify an ETN was not properly done, the system could actually increase telecommunications costs. In addition, as telephone traffic patterns changed, the design of the ETN would change. The ETN required constant monitoring, and corporate staff had to develop specific skills in order to keep the network financially and operationally efficient. This entailed a number of duties. Telecommunications professionals needed to apply complex mathematical formulas to telephone traffic patterns in order to determine how many tie lines were required for each site. This discipline, known as traffic engineering, was based on statistical formulas. Based on the results of these formulas, staff members would order additional tie lines or cancel existing ones. The orders would have to be tracked and the tie lines, both the number of lines and the cost per line, would have to be maintained in a department inventory. In addition, many companies would charge back the cost of network services and administration to the end-users.

As this one example illustrates, the decision to use such services was complex. A company needed personnel who could comprehend technical subject matter; oversee large, complicated projects; perform high level financial and statistical analysis; and continually maintain such an environment. There was no choice but to develop an in-house staff. Inevitably, as companies grew or changed, and as the size of such staffs increased, a manager was required to oversee this department. The department was typically known as the voice communications (or telecommunications) department which oversaw local services (toll calls and CO trunk lines), telephone equipment (key systems and PBXs), and long distance services (direct dial, dedicated tie lines, and WATS—inbound and outbound). Once again, all of these services and equipment were obtained from the monopoly known as the Bell system.

During the time prior to divestiture there also came a need for data communications capabilities. The mainframe computer was adopted by American business on a large scale basis during the 1960s. During this time, the computer was located under lock and key in a computer room; a closed, sterile environment that was the sole domain of computer professionals. In this first paradigm of computing, all interaction with the computer took place in the computer room and end-users were completely dependent upon computer professionals for their computing needs. If, for instance, a manager desired a report, he would submit a request to the computer department and wait for the results. This was the era of batch computing where punch cards were used to input information. The era of batch computing did not necessarily mandate that a company maintained a data communications staff: end-users waited for their information and did not expect immediate results. In spite of the limitations of the batch era, computing made rapid advances and American business became more reliant on computers to support their business infrastructure. Inventory control, payroll, order entry, and cost accounting were just a few of the functions that migrated from a manual function to the computer.

The 1970s ushered in a new paradigm of computing: time-sharing. During this era, terminals began to appear on the desks of selected workers. The first terminals were often only used by trained specialists, and there was still the perception by the general public that computers were cold, complex machines only meant to be understood by a select few. Still, the appearance of the first terminals introduced the concept of response time. Under the batch paradigm, the user would submit a job and come back at a later time to obtain the results. It was not mandatory that the results were provided immediately, nor was it expected. A terminal, however, did introduce interaction between the user and the computer. It was now necessary to ensure that the user received information from the computer quickly. Whether the end-user was situated in the same building as the computer, or at a remote site, the appearance of terminals mandated real-time and reliable data transmission. For terminals that were located at remote sites, computer professionals were challenged to provide quick response time. The stage was set for the emergence of the data communications specialist, a role that would evolve as a discipline separate from other computer professionals.

The third paradigm of computing was known as desktop computing which occurred in the 1980s. This era generally introduced computing to the end-user. At first, the end-users were issued dumb terminals and all computing took place on the mainframe. As the decade progressed, the dumb terminal was replaced by the PC and computing entered the new era of client-server. This was a pivotal era for telecommunications professionals. During this time, several trends emerged that would have a tremendous impact on the discipline of telecommunications management. For the first time in the history of computing, end-users (other than computer professionals) were performing their work on the computer. Consequently, many functions were soon automated and the computer became a tool for the average worker.

Most mission-critical business functions were performed on mainframe computers in the eighties and companies with remote offices needed to provide access to the mainframe. Consequently, elaborate data networks began to evolve, known as the wide area network (WAN). The WANs (see Figure 1.1) of this era were comprised of dedicated analog lines. But the design and management of such a network was more than simply ordering a circuit between two locations. Depending on the type of data that was transmitted over the circuit, companies could purchase varying degrees of line conditioning (for more reliable transmission) or opt for multi-point circuits as a cost saving measure. If a circuit failed, many companies (especially financial institutions) employed dial backup capabilities which was the temporary use of a dial-up modem until the main circuit could be restored. In this era of slave-host computing (i.e., a dumb terminal accessing a mainframe computer), terminals were connected to a cluster controller at a remote site. The controller required programming and provided management reports. The programming for the controller was constantly being tweaked as data traffic changed with business trends. The controller interfaced to a dedicated modem which was physically connected to the data circuit. The data circuit was then terminated on a modem at the host site which was connected to a front end processor that interfaced to the mainframe. As in the case of voice communications, many companies developed a separate department comprised of staff and manager known as the data communications department.

As illustrated in the example of the ETN, a variety of business and technical skills had to be employed in order to manage a corporate data communications environment. Unlike the ETN, the WAN did not require staff members to balance dial-up costs versus dedicated data lines. Dial-up via modems offered poor response time and unreliable transmission. The only option was to use dedicated data circuits. Each location required a data circuit, two modems (for each end of the circuit), dial-up modems for back-up, and a telephone line for modem access. There was also a need for a cluster controller at each remote site. All of these items needed to be purchased, programmed, inventoried, and budgeted. There were financial decisions that needed to be made. Should the modems be leased or purchased? What sort of maintenance contract should be used for the modems? Would the data communications department assume full responsibility for the data budget or would the budget be charged back to the end-users? There were also many technical skills that were required in order to keep response time acceptable. Was the data circuit adequate to transmit the amount of data transmitted from a given location? If response time was slow, where was the true source of the problem? Line speed or a poorly written application? This very question introduced many inter-departmental conflicts. Was the cluster controller properly programmed or was the problem in the front-end processor? Data communications specialists needed to monitor network

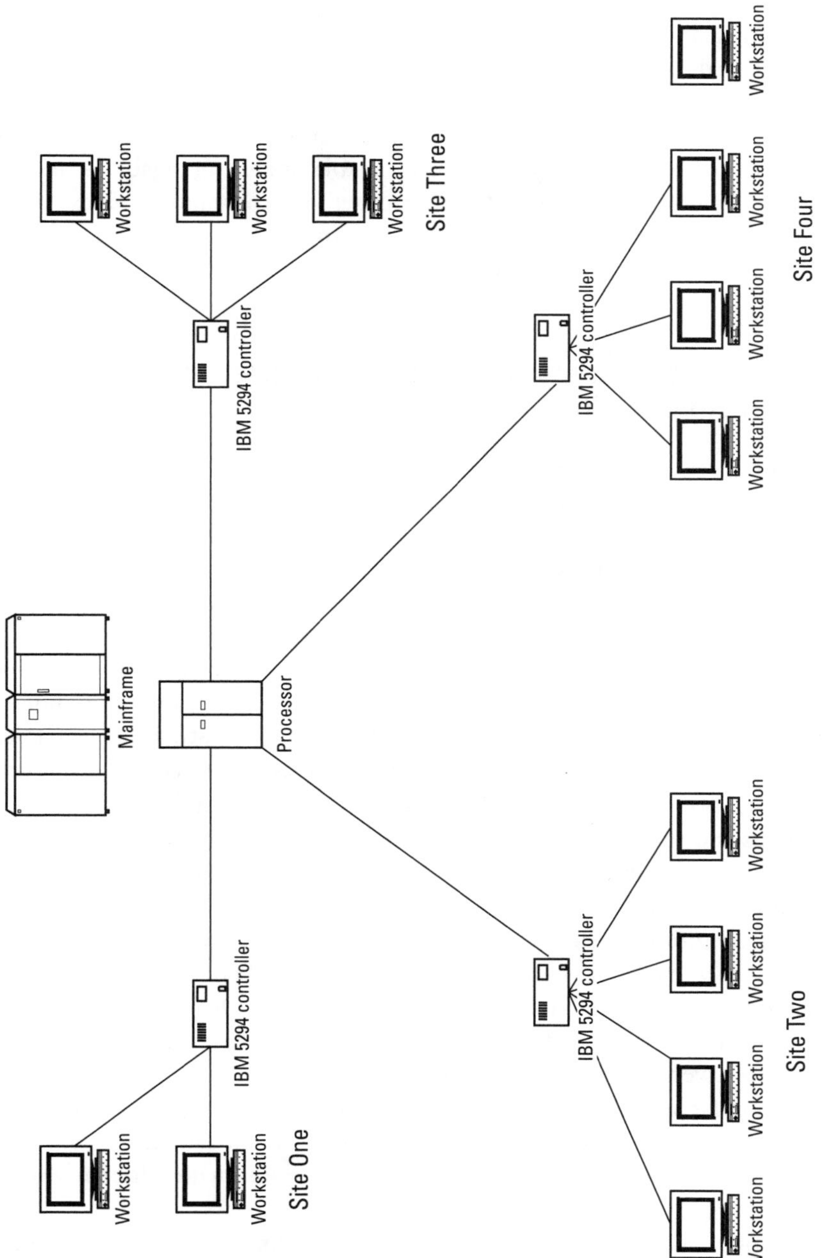

Figure 1.1 Slave-host computer environment with dedicated data circuits.

activity in order to keep response time acceptable. Slow response time was not only a nuisance, it cost revenue.

The fourth paradigm of computing introduced the era of networked computing. In the 1990s, the PC had come of age and computing became distributed. In the era of networked computing, the complexity of data communications increased tremendously. PCs were capable of storing gigabytes of data and local area networks became pervasive throughout corporations. New capabilities emerged in WAN technologies and the Internet emerged as a publicly available worldwide computer network.

1.2 The Telecommunications Department, Divestiture, and Technological Change

As stated in the previous section, the era of predivestiture saw two types of communications departments and two seemingly different disciplines. Even when both departments reported to the same division, the two departments were often managed separately, with very little cross-over of duties. A number of events and trends began to bring the two departments under one umbrella, several of which began prior to the breakup of the world's largest monopoly.

Once competition was introduced into the field of telecommunications, it was no longer a matter of ordering product from a limited menu of choices provided by AT&T. The decision process now encompassed more variables and more analysis. Telecommunications professionals had to consider the stability and quality of new telecommunications companies and the viability of the products and technologies they were offering. The groundwork for competition was laid with three decisions by the federal government.

The Carterfone Decision of 1968 was a landmark ruling that set a platform for competition in the telecommunications equipment market. The Carterfone was actually a two-way radio system that had the capability of interconnecting with the public telephone network. Only 4,000 of these devices were ever installed and yet legislation raged over the right of a person, or company, to have the ability to interconnect any device to the Bell system that was not manufactured by AT&T. The FCC ruled that devices could be connected to the public network if they were, "privately beneficial, but not publicly harmful." This decision set the foundation for equipment alternatives to AT&T.

The long distance business was also opened to competition prior to divestiture. The MCI Decision of 1969 expanded competition in the inter-city private line market. It allowed Microwave Communications Institute (MCI) to become the first specialized common carrier to provide competitive services to

AT&T. The MCI Decision, combined with the SCC (Specialized Common Carrier) Decision of 1971 allowed carriers to compete in the long distance market. By the time divestiture was instituted, end-users had a variety of choices regarding communications equipment and long distance services. The decision- making process became more complex and, as the one-stop-shop advantage became a thing of the past, telecommunications departments began to find themselves coordinating a number of disparate vendors, trying to make them act as one.

Divestiture changed the telecommunications environment dramatically in the United States and paved the way for deregulation in many foreign markets. The concept was not without its critics and there were those who wondered if a deregulated environment would yield chaos in lieu of the highly touted benefits of lower prices, better service, and more advanced technologies. The environment was also not completely deregulated nor completely open to competition. The RBOCs and ITCs were divested from AT&T, but they had virtually no competition in the local arena. In essence, the local telephone companies still held a monopoly over their areas of coverage. While many of the aforementioned benefits were realized, there were still many challenges for the modern telecommunications department. Many of the challenges that resulted from divestiture still exist today. Decisions are more complex and many carry a high degree of risk.

Once competition entered the picture, any telecommunications project usually entailed the selection of a minimum of three different vendors: long distance, local service, and equipment. Consider the selection of a basic PBX system. Under the Bell system, all three categories were the sole domain of AT&T. After divestiture, long distance could be provided by AT&T, MCI, or Sprint. The end-user could also consider using smaller carriers or resellers. Local service could only be provided by the local exchange carrier (LEC). The equipment would be supplied by a PBX vendor which could have been a manufacturer direct (such as AT&T or Rolm) or a distributor. To further complicate the situation, cable could be purchased from the LEC, the PBX vendor, or an independent cable company.

This simple example demonstrates how the role of the telecommunications department became more complicated and carried higher risks. A closer examination of these three types of choices will demonstrate some of the issues faced by the telecommunications department. Consider the issue of long distance. In 1984, carriers such as MCI and Sprint were known as other common carriers (OCC). While these companies offered many of the same features and services that AT&T did, the OCC networks were still in a state of development. Consequently, the quality of transmission was far inferior to that of AT&T. Today this is not a problem, but in 1984 there was a price to pay for the savings realized with alternate long distance services. Many telecommunications managers faced hostile end-users who had experienced cross-talk, distortion, background noise, and an occasional unintentional disconnection. While the telecommunications manager might have realized a substantial discount over his old AT&T bill, senior management and end-users alike often felt that alternate long distance services were unusable. Consider the quandary the telecommunications manager faced; under pressure to reduce operating costs, but forced to resort back to the original (and more expensive) service. Telecommunications managers also faced scrutiny and skepticism when they chose services and products that were not recognized names. Just as IBM stood as the most recognizable name in the computer field, AT&T was recognized as a company that provided quality and reliability in the field of telecommunications. Of course, not all calls placed on an OCC network were of poor quality, and some companies were willing to accept the lower quality as a trade-off for the savings. But this one example demonstrates that telecommunications management in the brave new world of deregulation would be both complex and risky.

The choice of local service and a PBX incurred less risk than long distance. Local service was normally via an LEC that had been part of the old Bell system. The telecommunications manager simply had to evaluate the menu of available services and make the appropriate choice. This choice was often financial, based on a list of options that offered discounts based on usage.

Choosing a PBX was not nearly as straightforward. It was a complex piece of equipment that offered a myriad of features. The purchase decision was also complicated by the bevy of vendors selling PBXs which were marketed directly by manufacturers or sold through distributors. Unlike alternate long distance services, most major PBX manufacturers offered a high quality product. But a high quality product did not guarantee a successful installation, nor did it mean that the system would be accepted by the end-users. One of the first problems that telecommunications managers faced was the paradigm left by the old Bell system. Telephones are very familiar technology to people. They are so used to using telephones that they seldom think about the technology or features that can be associated with them. Consequently, when people face change, they often resist it, however positive the change might be. If the PBX vendor did not do a good job of training the end-users, the first day of the installation was usually a nightmare and the telecommunications department was inundated with complaints about the "terrible new phone system." Senior managers and end-users alike would ask why the telecommunications department had selected these new, strange systems. Who were these Rolms, Northern Telecoms, Mitels, or NECs? And what was really wrong with our old Bell Dimension system?

If the PBX installation was perceived to have been poor, the situation was often exacerbated by poor service from the chosen vendor. As new manufacturers and distributors began to emerge, they were often faced with the same growing pains that any new business faces. There were limited funds and too few people. Technicians were often new to the field or behind in training. This was especially true of distributors that were in a start-up mode. When problems surfaced, technicians would arrive late or would not be able to solve the problem quickly. Although this was a service-related issue, end-users often blamed the telephone system itself.

Of course, not all installations went poorly. Many telecommunications managers and their staffs were able to provide good products and services and still save money. But there were still new responsibilities that further complicated the telecommunications department's job. Prior to deregulation, when there was one choice, problems were owned completely by the Bell system. When the customer had a choice of network services and customer premise equipment (CPE), telecommunications staffs suddenly found themselves thrust into large, complex problems. Vendors were often quick to point a critical finger at each other. "The problem is not in the tie line, it appears to be in your PBX... the problem is not in our PBX, it appears to be in the network." Such scenarios drew telecommunications staff members further into the problem solving process. Telecommunications managers and staff members had to develop strong problem solving skills in addition to enhancing their technical knowledge. Very often it was necessary to prove to a vendor that a problem was indeed theirs.

The selection of new systems was only one aspect of the changes that deregulation imposed on telecommunications departments. Equipment needs to be installed, inventoried, upgraded, replaced, or repaired. In a regulated environment, there was only one provider of this equipment. But for a large company with multiple locations, this could still entail thousands of devices that needed to be managed. This in itself was a daunting task. In a deregulated environment, the task was exacerbated by multiple vendors. Databases were complex and new questions began to emerge. Should the company standardize on a single vendor? Such a move might offer volume discounts and simplify management. Standardization, however, also had it drawbacks. What if the chosen vendor suddenly became financially insolvent, had to declare bankruptcy, or was the victim of a hostile takeover? Even if the situation was not this dire, financial problems often cause vendors to cut costs internally, which often includes laying off quality technicians. The net result of such a move is that service often suffers. What if the chosen vendor suddenly found their technology to be obsolete? A large, corporate agreement might mandate that "X" number of systems be purchased in a specified period of time, or the negotiated discounts would not be available. Moreover, what if the telecommunications requirements of the company changed and the vendor's product was no longer viable? Such scenarios in the tumultuous world of telecommunications are common.

Cable also became a new technology that required support. When the Bell system was first divested, there was a question mark regarding the issue of "inside" or "house" cable plant. Cable that connected telephones had once been the sole domain of the Bell system. Not long after divestiture, the responsibility for in-house cable was relinquished to the end-user. Many companies, for the first time, had to manage their own cable. This entailed detailed record keeping and a new world of knowledge that was given little notice when the telephone company had this responsibility.

Technological developments in the field of cable also began to have an impact. Coaxial cable was used in many companies to connect data devices while twisted pair was used for telephone connectivity. Coaxial cable was bulky and expensive, but it did offer high transmission capacity. Twisted pair was inexpensive and easier to install and manage. However, it was limited in its ability to carry high speed data. Innovations in this field led to the development of twisted pair cable that could carry high capacity data transmission. The concept of a universal structured cable system developed. Such a system was capable of carrying any type of transmission. The introduction of the LAN also demanded high speed data capabilities. In the case of a campus environment, this often mandated the use of a high capacity data backbone provided via fiber optic cable. When high capacity cable was installed, it did not always make sense to lay additional copper for voice communications. The sensible, cost effective solution was often to place all telecommunications services on the same backbone. The emergence of unified cable systems was one more factor that eroded the lines that separated the disciplines of voice and data communications.

The deregulation of the U.S. telecommunications environment was not complete in 1984. Local phone service was still a virtual monopoly. The Telecommunications Act of 1996 was a federal bill signed into law on February 8, 1996 "to promote competition and reduce regulation in order to secure lower prices and higher quality services for American telecommunications consumers and encourage rapid deployment of new telecommunications technologies." The act covered a number of areas. Key among them was the opportunity for new local telephone companies to offer local service. As of this writing, it would be safe to say that the local exchange carriers (LECs) still have a monopoly on local service. Statistics show that alternate local carriers have gained less than three percent of the available local market. There is a degree of competition for private cable and certain data services that are provided by competitive access providers (CAPs). But for all intents and purposes, the incumbent LECs still maintain a stranglehold on local telephone services. While a number of CAPs have attempted to offer alternative local telephone service, a principle roadblock has been local number portability (NP). That is, the ability of a customer to retain his telephone number after changing local service providers (SP). A system has been designed and is currently being deployed in large metropolitan markets that will allow subscribers to change their SP (see Chapter 2). While this capability signals progress, there are only 93 markets slated for the next few years.

The changes in the regulatory environment offered by the Telecommunications Act of 1996 have, once again, introduced a number of choices and new responsibilities to the telecommunications department. Interexchange carriers (IXCs) such as AT&T, MCI, and Sprint now have the capability to offer local service. At the time of this writing, many of the IXCs are aggressively building local, metropolitan networks or purchasing CAPs or CLECs. The distinctions set forth by divestiture in 1984 are blurring even further. Local service can now be purchased through an IXC and long distance service can be obtained through a LEC.

The United States telecommunications infrastructure does not stop with the blurring definitions of IXCs and LECs. Television cable companies have entered the picture. Some of these companies have merged with traditional carriers (both LEC and IXC) while others will compete directly with these traditional carriers. If the reader perceives this scenario to be chaotic, the diagnosis is correct. There is constant lobbying in Washington, DC for regulatory change, in addition to the incessant acquisitions, mergers, investing, and divesting of telecommunications companies which paints a picture that is forever changing. These companies are jockeying their positions to gain new revenues in previously unavailable markets and, in some instances, position themselves so that they can provide all types of communications services and products. The reader may ascertain that we are evolving back to the one-stop-shop of the old Bell system, only via a number choices. This is, in part, true. It may be possible one day to order all services and products from a single vendor, anywhere in the country. In the interim, the telecommunications department will have to closely monitor these changes, cautiously evaluating new offerings as they unfold. The deregulation of the United States telecommunications industry, however, was not the only factor affecting the role of the telecommunications department. Competition also fosters new services and products. Because new players entered the market, technology began to take on dramatic new forms.

Up until the late seventies virtually all communications, whether voice or data, were transmitted over analog lines. Gradually, during the early 80s, digital technology began to appear in both CPE and in the public network. Digital technology offered higher transmission capabilities and more flexibility. For instance, an end-user could combine both voice and data over a T1 line that connected two locations. Data communications could also be transmitted through the PBX which now had the potential to become the office communications controller. This generally did not happen for a number of reasons, but the example illustrates that the lines that once divided voice and data communications began to blur. Because these lines began to blur, and because transmission technologies began to become more ubiquitous, voice and data communications responsibilities began to overlap more. In some companies, the two units actually merged and if they did not, there was still a need to cooperate on many projects.

The capability to digitize information spawned many technologies that affected the field of telecommunications. In fact, it can be safely said that the field was revolutionized. Speech could now be digitized and stored so that it could be retrieved by an end-user or forwarded to a fellow employee, hence the birth of voice mail. Video could also be digitized, stored, or transmitted. The enabling technology was the coder/decoder (Codec), a device that could translate an analog signal into a digital format. Once this technology was applied to telephone systems and video systems, virtually any form of human communications could be stored or transmitted in a digital format. Digital transmission technologies were also more reliable and offered clearer transmission than their analog counterparts. The development of digital technologies has paved the way for the concept of convergence; that is, the fusion of various types of communications media: voice, video, or data. Today, this is not a concept that is always universally achievable across a company's network or via domestic or international public networks. There are still analog segments of public networks and transmission technologies that have not been accepted as universal standards. However, convergence is possible under limited circumstances, normally guided by economics. The capability to merge analog and digital transmissions under one uniform digital medium has drawn the disciplines of voice and data communications closer together. This trend will continue and much progress has been made in technologies that merge voice, video, and data. Asynchronous transfer mode (ATM) is just such a technology that has found its way into public and private networks that allow for multiple types of transmissions.

Beyond the emergence of digital transmission technologies, a technological advance that was to have a major impact on the role of corporate telecommunications management was the emergence of the personal computer. In 1981, IBM announced the personal computer (PC), a small, self-contained unit that actually placed computing power in the hands of the average worker.

When the first PCs began to appear they were considered to be nothing more than toys by computing professionals. However, in a short period of time the PC began to infiltrate the office environment. It was then only a matter of time before there was a need to connect PCs to corporate networks. The PC changed the nature of data communications management dramatically. When PCs first began to appear in offices, they were connected as dumb terminals to the mainframe computer. This was done with a circuit card and special software that emulated the type of terminal that was exclusive to the mainframe manufacturer (e.g., 3270 for IBM, VT100 for DEC). The PC was still a standalone unit, the computing power inaccessible to all but the person using it. Eventually there was a need to connect to other PCs (known as peer-to-peer computing) or to access data on a server (known as client-server computing).

Novell introduced the first commercially available local area network in 1983. Primitive by today's standards, Netware initially was designed to allow networked PCs access to each other's hard drives. Soon, the capabilities of the LAN were expanded to include printer services, e-mail, and file servers.

When the PC was used as a dumb terminal, there was little impact on the network. A dumb terminal actually sends very little data and all of the actual computing takes place on the mainframe. When a PC accesses data from either another PC or a server, new demands are placed on the network. Large volumes of data can be stored on a PC and it became necessary to provide high capacity data transmission capabilities and to impose controls on what data the end-users could (or should) access.

As hard as many computing professionals fought the invasion of the PC, they fought a losing battle. PCs were user-friendly and inexpensive. They enhanced and stimulated productivity by placing computing power in the hands of the average worker. It was no longer a matter of having all computing power controlled by a department of skilled individuals. Computing power was now at the desktop and desktops needed to be connected.

Once the PC became established in the business environment, a myriad of innovative capabilities began to emerge. In today's environment, PCs can store Gigabits of data, provide telephone functionality, e-mail, videoconferencing, and multimedia capabilities, as well as perform basic computational functions. Many of these functions require special types of transmission capabilities, in addition to high bandwidth. PCs stimulate innovation and new inventions are constantly being marketed. Some find a place in the marketplace and others quickly fade from the business world, either being impractical or too expensive. But new applications are always surfacing while proven ones are refined.

The PC not only brought computing capability to the average person, it also brought an increased knowledge of technology. Because of this fact, technological knowledge is no longer the sole domain of the computing or the telecommunications professional. This has had a dramatic effect on the relationship between the telecommunications professional and the corporate end-user. There was a time when the average American worker possessed little or no knowledge of computing or communications technologies. This was very common well into the late seventies and early eighties. The technologies of computing and communications were viewed as complex and inaccessible to the average person. From an executive management standpoint, decisions regarding these technologies—whether financial or strategic—were often based on information provided by technology professionals. Many executive managers were ill-equipped to challenge what had been proposed by these people.

The closed and secret world of computing and telecommunications professionals changed with the advent of the PC. Suddenly, scores of people became amateur computer (and communications) wizards. Magazines began to appear on newsstands with titles such as *Byte*, *PC Magazine*, and *PC World* offering simple, user-friendly articles about the technologies of computing and communications. Suddenly the average person knew about local area networks, wide area networks, modems, e-mail, and videoconferencing. In the late 70s and early 80s, the telecommunications professional was the person who pushed the technological envelope of a company. In the era of the PC, that same person often found themselves fielding questions from fellow employees about new and developing technologies. These questions often came from people who were completely removed from the profession of computing or telecommunications. In fact, many telecommunications managers often found themselves defending their decisions to people who once blindly accepted them. Ironically, many telecommunications managers often found themselves defending older, proven technologies over emerging technologies that were gaining attention in the popular media. This paradigm shift continues to this day.

Once the PC began to become firmly entrenched in the world, there came a need for universal connectivity. People could buy off-the-shelf PC communications packages, but these were proprietary packages that were used to establish a connection with a PC equipped with the same package. They did not address e-mail or provide a uniform data communications standard that would allow universal connectivity. The solution to this problem was a network that was developed exclusively for connecting different computer systems: the Internet. The Internet was originally developed by the Department of Defense in 1969 and was then known as Arpanet. It was a network that was developed with robust data communications protocols that were developed to connect disparate computer systems. During the time that Arpanet was developed, data communications protocols were exclusive to manufacturers. For example, IBM computers could only communicate with IBM computers. In order for disparate computers to communicate, data protocol conversions had to be performed, a complex and cumbersome process. With Arpanet, there was a need to connect the computers of various universities with each other and to the computers used by the federal government. Initially, Arpanet was used for research that related to military applications. In the later part of the twentieth century, it is more appropriate to define the Internet as a commercial and publishing medium.

The Internet is one of the most hyped technologies of the twentieth century, but in this case the hype may be justified. It is often difficult to assess the importance of a technology and what its impact will be on society. Many technologies have been hyped and many have passed, only to be replaced by better (or more appropriate) technologies. The Internet, however, is one of those rare concepts, such as the telephone and television, that has the potential to dramatically change society. There are even some who contend that the Internet will have the same impact on society as did the Gutenberg Press in 1455. Regardless of what the final verdict will be, the Internet has certainly made a great impact in a very short period of time. Consider that as of 1995 the Internet linked 59,000 networks, 2.2 million computers, and had 15 million users in 92 countries. Today, this explosive growth shows no signs of slowing down. In today's world, telecommunications departments must provide connectivity to the Internet, in addition to their own proprietary corporate networks. This entails the administration of a robust data communications protocol known as transport control protocol/internet protocol (TCP/IP), the protocol of the Internet. It also involves the administration of domain name service (DNS), the protocol that masks the complexity of TCP/IP addressing from the average end-user and allows them to use English names. Also, the telecommunications department needs to be concerned with security issues because the Internet is a public network. Because of the evolving technologies of PCs, LANs and the Internet, specialized roles have developed in the modern telecommunications department. Telecommunications professionals who specialize in the Internet have some of the most in-demand sets of skills. These people are often difficult to find and keep.

The computing revolution was not exclusive to PCs. Computer technology has invaded virtually every type of device imaginable. It is no different for telecommunications devices. PBXs are now essentially computers. It was not always that way. The first PBXs were electromechanical in nature. Today's PBXs are software driven and must be programmed. The programs are stored on hard disks to be retrieved for various functions such as least cost routing (LCR), automatic call distribution (ACD), or voice mail. Routers, hubs, and other data communications devices are also computers. Consequently, telecommunications professionals, both voice and data, had to develop a high degree of computer literacy. They needed to understand the fundamental components of a computer such as processing power, storage capacity, input/output devices, throughput, and memory. They also had to have fundamental programming skills. These skills were necessary so that telecommunications professionals could understand what they were evaluating and buying, and how to support these devices on an ongoing basis. All modern telecommunications devices are computers and most features and capabilities are software driven. It is a matter of understanding what they had to support and also what they connected. PCs are also being developed as telecommunications devices. For instance, the Internet can be used as a voice network (known as voice over IP) or for video-conferencing. A PC can also be set up as a small telephone system with voice mail. This trend shows no sign of slowing down, and computing and communications technologies continue to blend.

The complexity of telecommunications devices and services demanded that specialized positions be created in the telecommunications department. Devices became so feature rich, and networks became so complex, that telecommunications departments needed to develop special groups within the department. This led to a number of new positions that continually evolve to this day. For instance, network design was now far more complicated with multiple vendors and emerging technologies. Hence, the position of network designer or network engineer began to surface. There was also a need for people who were able to oversee large-scale, complicated projects, a need that yielded the position of project leader. Administrators were needed to program software on PBXs, routers, or hubs. In certain cases, somebody needed to perform physical work, such as the installation and termination of cable or telephones. Hence, the positions of installers or technicians were developed. There were also people who served more of a generalist role, such as the communications analyst, who was responsible for understanding, evaluating, and recommending new technologies. In the larger corporations, there were many positions and departments within a telecommunications department. In smaller companies, staff members took on many (or all) of these duties. It is interesting to note that when I began conducting research for this book, I asked several AT&T employees, veterans of the old Bell system, if they knew of any standards that were used for telecommunications departmental structure. They all stated there were none that they knew of, but they also stated that most telecommunications organizations had similar structures. Given the fact that telecommunications management has not traditionally been a profession with a national or universally recognized certification processes, it is this author's theory that the telecom- munications department structure evolved for two reasons. First and foremost, the business environment dictated this delineation of duties within the department. Second, telecommunications managers obtained ideas from their colleagues and vendors.

The development of specialized roles also introduced a new dilemma. Specialists began to emerge but there was also the issue of the workload versus the available personnel resources. A single specialist within the department meant that there was a liability. What if the specialist quit or became sick? Conversely, the department budget almost never seemed to allow managers to hire the people that were needed to address the workload. This problem continues today and is an issue that telecommunications managers will continually have to address. Many companies found temporary relief of this dilemma by hiring consultants, contract workers, or outsourcing. While these alternatives offered relief, they were not always panaceas. Outside sources do not always understand the inner workings of an organization as well as the actual employees. Traditionally, they also provide service that is of lesser quality than a company's own employees. Also, funding is not always available to procure outside sources.

The advance of technology also added new challenges. Staff members needed to be educated on general concepts and on vendor specific products. There were so many technologies and vendors that telecommunications managers had no choice but to develop educational programs for their departments. This entailed vendor-specific certification processes, university sponsored courses and curriculums, independent courses, and self-teaching methods. Most staff members found that it was an uphill battle to keep up with the technology.

In today's business environment, the telecommunications department performs a critical role for business. But as critical as that role is, there are still no standards employed across the country, and the structure of telecommunications departments across companies and industry segments remains diverse. Voice and data technologies are beginning to overlap and yet many companies maintain these as separate disciplines. The complexity and plethora of technologies demand specialization while the business environment demands generalization. These issues will be examined in more depth in later chapters. The modern telecommunications department carries a role that is critical to conducting business. The business environment is changing as is the workplace. The telecommunications department plays a major role in how those changes are being executed.

1.3 The Changing Workplace

The American workplace has changed dramatically over the past two decades. Many of the changes have been due to the development of new managerial techniques, however, technology has also been a major influence. Consider the impact of the PC. Before the PC, the average worker did not type his own correspondence or reports. The worker would either write something by hand or dictate to a secretary who knew shorthand. The secretary would then type the work, probably one of many documents placed in a large queue. In many businesses, there were secretarial pools to process the large volumes of correspondence and reports that an office generated. Secretaries needed to type quickly and accurately, but a single mistake often meant that an entire page would have to be retyped. Regardless of how efficiently the organization was run, there was still a long wait to receive the finished product.

In the modern business environment, many workers now type their own work. Word processing programs offer the electronic page, where work can be visually examined, manually changed, and automatically proofed for spelling and grammatical errors before printing. Documents that are stored in software can also be shared and reviewed by a number of people before the final product is approved and printed. A number of employees might examine the document, adding changes or corrections, perhaps making dozens of revisions without printing a single page. The people who work on the document might be located in the same room, the same building, across the country, or around the globe.

This simple example demonstrates several basic themes that have affected business in the latter part of the twentieth century. First, sophisticated technological tools are now in the hands of virtually every worker. From the CEO to the maintenance person, it is becoming rare when a worker does not have some form of computing or communications device at their disposal. Second, job roles are changing. Because of technology, workers have been enabled and empowered to do more, and more productivity is often expected of them. Computer proficiency is no longer a desired skill, it is now often mandatory. Finally, the technologies of computing and communications have speeded up the pace of business. Information in readily available and more abundant. Tasks are completed faster and decisions are not only made faster, but more accurately. Consider some of the following examples of computing and communications technology that has affected the workplace today.

Example One

A special assignment has been given to an employee that will require an extraordinary amount of research. In years past, an employee would have spent long hours pouring through thick volumes of books and papers in corporate and public libraries. Perhaps the employee would have trudged with a stack of thick, heavy reference books to his table, only to find that he had selected the wrong volumes or there was scant information. He or she would have to collect the books and trudge back to the shelves to replace them. They would then look through banks of cards that were cataloged in drawers in order to find more suitable materials. In today's environment, this very same person might access information through a LAN attached to a CD tower (capable of holding multiple CDs). Each CD is capable of holding volumes of data; 650 megabytes which is equivalent to approximately 250,000 printed pages. The employee simply types in the desired subject and the PC will perform a search, scanning thousands of articles and reports within seconds. They can then scan the titles offered from the search and determine what articles will be of value. Continuing with the same assignment, the employee might dial into the Internet, accessing the databases of thousands of computers. Once again, a search is executed and the employee will probably have a multitude of sources for his or her project. Perhaps the biggest dilemma will not be finding information, but rather eliminating redundant or unnecessary information.

Example Two

A person has just retired. He or she is financially stable, but would still like to work, perhaps on a part-time basis, just to remain active. They do not, however, wish to fight the morning traffic anymore, or even bother to dress up and deal with annoying office politics. He or she finds just the right position. They now get up in the late morning, boot their PC, and dial into a corporate network. He or she checks for voice mail messages, accepts and makes calls on a company provided telephone, exchanges e-mail messages and documents via remote LAN access, and receives and sends faxes on a PC. This person is telecommuting, a work-at-home concept made possible by telecommunications technologies. Telecommuting is bringing a number of possibilities to modern business. The example of the retiree is only one possibility for modern business. House-bound parents, the handicapped, and rural residents are just a few of a previously unavailable pool of potential workers that are becoming available to modern business because of this concept.

Example Three

An employee walks into the office one morning, a portable PC hanging from his shoulder, a wireless telephone in his pocket. He or she is not sure where they will work today, because they have no office, even though he reports to the same building every day, and requires a PC and telephone to do the job. It is a beautiful fall day and there is a courtyard outside one of the buildings on the company campus, where several picnic tables are shaded by large trees. He or she decides that it would a nice day to spend outside and enjoy the outdoors.

They go outside with a coffee, boot their PC, and turn on the telephone. He or she spends the morning typing a report, exchanging e-mail, and answering and making calls. The employee is taking advantage of the floating workpoint, a concept made possible by wireless technologies. The employee's PC is equipped with a wireless modem and the company campus has been equipped with a wireless LAN capability. Workers are free to walk to any of a number of designated work spaces to spend their day. They are empowered to be responsible for their work and are given the tools to accomplish their duties, regardless of their location. The campus has also been equipped with a wireless PBX that will work anywhere within the campus and for several hundred yards beyond. The worker decides to eat lunch at the same spot and then, growing bored with his or her surroundings, decides to find another spot for the remainder of the working day.

Example Four

Several workers are discussing a concept via telephone, and they soon realize that they are having difficulty understanding each other. The difficulty is the result of trying to describe to each other a three dimensional design for a new product. They send graphics files through the Internet to one another, but also find that they need to draw designs on a whiteboard and do a bit of brainstorming in order to finish the design. They use videoconferencing units and an electronic whiteboard to illustrate their ideas. They draw designs, make corrections, and then print the results before ending the meeting. By using the concept of videoconferencing, the workers have taken advantage of the age old concept, "A picture is worth a thousand words." These workers might have traveled in years past, and the concept of videoconferencing certainly allows for savings in travel expenses. However, the larger benefit they have gained is efficiency. By using the technology of videoconferencing, these workers are able to bring a product to market quicker than was ever possible in the past.

Each of these examples demonstrates how the workplace has changed because of advancing computing and communications technologies. These are not wishful extrapolations of the future, but technologies that are deliverable today. Not all of these technologies are used by every business, but these common examples demonstrate how many workers are reliant upon computing and communications technology in order to accomplish their daily tasks. They also demonstrate how the work environment can be improved, not only to increase productivity, but also to provide a more attractive work environment. The employee who spent the morning outside would hesitate to consider another job. It would indeed be a big change for such a person to have to sit at the same desk everyday in the same drab cubicle.

There are three different concepts demonstrated by these scenarios that exemplify the benefits of communications technologies: efficiency, time, and space. Efficiencies are gained by performing work more quickly and more accurately. Once again, consider the example of the document that was typed on a word processing program. Mistakes were automatically corrected and there was no wait for a secretary to complete the task. Efficiencies are also gained by having access to more information. The document can be sent through a corporate computer network or through the Internet for review by superiors or peers. Project members have instant access to the information so that decisions can be made more quickly. Efficiencies are also gained because more information is accessible. Continuing with the example of the research project, the employee was able to access vast quantities of information instantly. Once again, decisions can be made faster and more accurately.

Time has also been affected. Workers now make better use of their time because they are more efficient and more accessible. Unproductive time has been reduced. Time has also been affected because the limitations of the standard eight hour workday are beginning to erode. Technologies such as e-mail and voice mail allow workers to communicate, regardless of work shift, location, or time zone differences.

The space that has traditionally been occupied by a worker is also changing. The office can now be located at home, at some remote location, or in a virtual sense, as illustrated in Example Three. The walls that once confined workers are now being torn down. This allows workers to be more mobile or, as illustrated in Example Three, used as a motivational tool.

In the later part of the twentieth century, it is not just an issue of gaining efficiencies through technology; work is performed with technology. Simply put, if the technologies fail, the work cannot be performed. This dictates that new skills are demanded of workers. It has also changed how business is done, and has placed new demands on the telecommunications department. Technologies are adopted and implemented within a company for a variety of reasons. Regardless of the reason, the telecommunications department is responsible for connecting people and computers to support the modern work environment. The members of this department must be a bridge, supporting legacy technologies and looking towards emerging technologies to maintain the modern worker's needs.

The technologies of communications are "enabling technologies" for the modern worker and the modern corporation. On a very basic level, the telecommunications department supplies two fundamental capabilities to a company and its workers: access to computer stored data, and communications capabilities between people. Computer stored data might be functions such as order entry, sales figures, inventory checks, pricing checks, budget control, e-mail, or research data obtained via the Internet. The enabling technologies are modems, LANs, WANs, and the Internet.

People-to-people communications is any technology that allows a person to communicate with another person. Such technologies are telephone systems and services, voice mail, e-mail, and videoconferencing. In both categories, the information can be transmitted over a public or private network, via tethered or wireless communications.

Reporting structures have also changed in the workplace. Managers not only expect more out of their employees, they must also adjust to employees who have more power (because they are more efficient and have access to more information) and are more mobile. The concept of the 9 to 5 job where an employee spends his entire day at a desk is changing. Telecommunications technologies are allowing this.

Business is now conducted on networks and, as the reader can see, the technologies of communications stimulate new ideas for the modern work environment and promote collaboration. In the information age, the workplace will continue to evolve, and the telecommunications department will be responsible for supporting the workplace of the information age: wireless or tethered; national or global; voice, data, or video.

1.4 The Changing Corporation

We live in the Information Age. It is an age where information and people need to be accessible: anywhere, anytime, anyplace. People and businesses need information quickly. They expect an avenue to this information and a method to retrieve it quickly. Computers and communications technologies provide the infrastructure that fuels the economy of the modern era. Just as the agricultural age was based on plows and the animals that pulled them, or the industrial era was based on machines and fuel, the information age is based on computers and the technologies that connect them. Computer and communications technologies are not in the process of invading the business world, they are a part of it. Business has taken advantage of the price and performance improvements of computing and communications technologies which have been increasing at a rate of 25 percent per year for the past 20 years. The trends have been set: They are inexorable.

Executives make decisions based on data fed from subsidiaries located all over the world. Many industries, in fact, make decisions based on information that arrives instantaneously. Financial markets are a prime example of this. Telecommunications technologies have become an essential tool of this process. In today's world, when a person picks up a telephone, they expect it to work, and they expect to be connected. If a consumer dials a toll-free number, they expect the call to be instantly answered and the order to be processed. If a credit card is used for a purchase, it is expected that the sale will be instantaneous. If the card holder's credit limit has been reached, additional credit should be available within minutes. The modern economy demands speed and accuracy.

There is virtually no aspect of the modern business world that has not been touched by these technologies. Computers, due to their size, power, and cost efficiency, are omnipresent. Computerized robots perform precision manufacturing processes, supercomputers analyze complex medical images, people retrieve information from computers via spoken words, and complex three dimensional models are designed, analyzed, and perfected prior to the manufacturing process. Computers accept, store, process, and present information. Networks connect the computers and provide the information that drives business. The telecommunications department is the force that connects all of these processes. The employees who need access to this information are located throughout the country and around the globe.

Consider when a shirt is sold in a department store. The sales clerk scans a bar-code which registers the price on the cash register, but also sends data through the network. One shirt sold; replenish stock by a factor of one. There is no paperwork to fill out, the process is fully automated. Inventory control is more accurate and less expensive. Computers and networks provide this function.

A customer calls a toll-free number because he or she has been browsing through a catalog and wants to order a shirt. The call is routed to an automatic call distributor (ACD) located in St. Louis. The system "senses" that there are no agents available in St. Louis, but there are several agents available in Denver. The call is automatically routed to the pool of agents located in the Denver call center. In this example, the telecommunications department has designed the ACD systems so that all resources will be available when a customer calls. They also know the cost of an average order and what the potential lost revenue is if a call should be lost. When the call arrives at the call center, the customer's computer record is presented along with the call to the agent. When the call arrived at the ACD, the customer's telephone number was downloaded to a computer, cross-referenced against a database, and the call was matched to the computer record. Known as a computer telephony integration (CTI), the order entry function of this catalog operation has gained tremendous efficiencies over their competition. They have improved accuracy and response time to their customer base.

Just as computing and communications technologies have affected the way that business is conducted, these technologies have also affected the very organizational structure of the modern corporation. Corporate structures have become flatter as information technology enables more decisions to be made at the operating level. When decisions are all made at the top, the decision making process is often slowed. Executive managers will certainly not become powerless. Rather, they are free to take a more global approach to the decision making process, while lower levels of management are being empowered to make more decisions and gain efficiencies for the entire organization.

Corporate cultures have now become more open and the concepts of teamwork, empowerment, and collaboration are being adopted as more productive ways to use the talent that resides within the employee workforce. Moreover, it is actually more difficult to maintain a strict dictatorial style of management, where the vast majority of decision are being made at the higher levels. This is because employees are naturally empowered to make more decisions when they have more information. This theory was proven when applied to political structures. One factor that contributed strongly to the collapse of communism was the accessibility to information via telephone, fax, modem, and the Internet. With so much information available as a result of communications technology, it simply became impossible to control all of it.

The open culture has given rise to new terms for specific types of software. Groupware is software written for the specific purpose of allowing employees to collaborate over networks. Within seconds, employees distributed around the globe can share ideas and information, coming to quick consensus on a project or idea. Because of this capability, the product or service may get to market faster and have a higher degree of quality.

The American economy has also made major shifts in the latter half of the twentieth century. The industrial age was based on manufacturing and the later part of this century has seen a shift to a service-based economy. A service-based economy demands that business finds a way to service customers quickly and efficiently, because the service (not the product) is often the difference. Because of these requirements, technological concepts based on customer service have infiltrated businesses. Toll-free services invite a customer to call free of charge. An ACD distributes the calls to the least-busy agent. ACD reports tell the business how well they are servicing the customer. Customers order product over the Internet or have a direct line to a vendor's computer using electronic data interchange (EDI). Businesses scratch and claw to find the technology that will help them gain an edge over their competition. This trend will continue and will increasingly place more pressure on telecommunications departments to provide these capabilities. They must also be efficient and reliable.

In addition to becoming more service oriented, the economy has also become more global. Canadian sociologist Marshall McLuhan once coined the phrase, "the global village." McLuhan theorized that modern transportation and communications technologies had woven the post-war world into a single community. While there are still distinct economic, political, and cultural boundaries dividing various countries, nobody can deny that the world has become smaller. Some reasons are the advances made in transportation. Telecommunications technologies, however, have probably had the largest impact. U.S. companies aggressively penetrate foreign markets and expect the same technological support that they have in the domestic market. Foreign countries have also recognized that an advanced telecommunications infrastructure is paramount to economic growth. Consequently, advanced voice, video, and data capabilities are becoming available in foreign markets. Because the economy is becoming more global, the telecommunications department faces new issues that relate to foreign telecommunications capabilities and regulatory issues. While the telecommunications infrastructure of many foreign countries is still controlled by the government, many foreign countries are aggressively deregulating and investing in their infrastructures. Telecommunications professionals are faced with the task of understanding the capabilities within the foreign markets and also keeping an eye on future developments.

1.5 Future Developments

We live in an age that is constantly transforming itself. It is a dynamic age, one where ideas are becoming reality so quickly, we who are living it cannot comprehend the complexity and scope of what is transpiring. This incredible age even boggles the minds of the people who are at the forefront of the technological innovations. Computers are omnipresent; communications technologies are becoming universal. It would seem that every aspect of our lives has been touched by these technologies. We perform our work on a networked PC and when we go home, we find that our children are doing their homework on a PC, perhaps even surfing the Internet. We browse through a magazine and see that every advertisement offers a toll-free number, an Internet address, or both. People spend hours on the Internet visiting chat rooms. They make friends from around the globe without leaving their homes. Shopping is conducted via a computer and millions of dollars worth of product are sold without a single person leaving their home. The presence of computers and communications has changed our social and working lives. We have the ability to compute; we have the ability to connect. What we are witnessing, in the later half of the twentieth century, is the fusion of computer and communications technology. Devices are becoming smaller and more intelligent, and connectivity is becoming more ubiquitous. This is changing our concept of time, space, and distance. It is changing the way we work, it is changing the way we play. We have experienced awe-inspiring technological development in the latter part of this century. What can we expect to experience five, ten, or twenty years into the future? Can we even grasp how technology will evolve and how this will affect society?

Predicting the future is a risky and inaccurate business. Books of quotations are often filled with bold, critical statements about new technology from seemingly intelligent people, only to be proven wrong when the technology is quickly adopted by society. There was the case of a London businessman who proclaimed that at the turn of the century there would be no need for telephones in London since they had such a sophisticated network of messenger boys. In my own experience, I once attended a lecture given by the MIS director of a large company who said, "I do not understand why PCs exist. They are fine for doing grocery lists, but other than that, I do not understand why they exist!" These were people who were actually able to see and experience the technology and they still failed to grasp its value. Given this fact, can we truly envision what will transpire in the future? Will we see continued improvements of existing technologies, or will there be innovations that are completely different from anything we could have imagined? Only time will tell, but telecommunications professionals will need to be keenly aware of changes and developing technologies. Their role in the business world of the future will be critical, because they will be a crucial part of any business's infrastructure.

It is not my intent to speculate wildly about the future. Science fiction writers have done this for years. Their extrapolations range from the wildly unimaginable to the plausible, whether they base their predictions upon existing or developing technologies. Indeed, many of these predictions can be quite intuitive. In his book *Paris in the Twentieth Century* Jules Verne predicted the use of the fax machine back in the nineteenth century. Issac Asimov correctly predicted the hand-held calculator in his famous *Foundation* series. In spite of these accurate predictions, there have been many extrapolations that were quite adventuresome and made many wild assumptions. But as wild as some of these predictions may seem, many are not completely outside the realm of possibility. Many telecommunications professionals would be surprised if they were to tour a research facility where emerging telecommunications technologies are being developed. Telecommunications professionals must be able to stretch their imaginations in order to understand new technologies. As new ideas are brought forth by telecommunications companies, they must evaluate each idea for its value to business and not judge those technologies against established paradigms.

This book is directed to the telecommunications professional who needs to practically apply communications technologies to a business environment. Technologies must be provided that are deliverable, reliable, manageable, and financially feasible. But the reality of the Information Age is that technology is in a constant state of flux. New technologies are always waiting in the wings, ready to replace the ones with which we are familiar and comfortable. The telecommunications professional needs to keep one foot in the past and one foot in the future. He must also be open to change, always ready for the next paradigm shift. There is a danger in placing either foot too firmly in either the past or the future.

New technology cannot be blindly accepted, as it may not gain acceptance in the business world, or it may not be mature. Such a move would leave a company with expensive equipment that cannot be supported. Many quality technologies have not been accepted over the years for a variety of reasons. Perhaps they were too expensive, too complex, the economy was bad, the product was not mature, or the time simply was not right. Regardless, telecommunications managers who blindly adopt new technologies often find their reputations and jobs in jeopardy. They are perceived as managers who have no practical business knowledge and spend money frivolously. In my own personal experience, I knew of such a person who was dubbed, both by colleagues and vendors, as "Buck Rogers."

Conversely, there is a danger in staying in the comfort zone. I have heard of telecommunications professionals who have been referred to as Luddites. In nineteenth-century England, there was a band of textile workers who earned fame by destroying machinery that had been purchased to replace their jobs. In modern times, the name is often used in a derogatory sense for people who cannot grasp or oppose technological change. Telecommunications professionals who are slow to adopt change are seen as roadblocks to progress. Fellow employees will visit other companies or executive management will sit on the board of directors of other companies. They will see and experience new technologies that enhance productivity, reduce operating costs, or produce revenue. It is also just as probable that one of these people may read of a new technology in a magazine they picked up at the local newsstand. Neither circumstance bodes well for the telecommunications professional who has no knowledge of the subject or arbitrarily dismisses it, like the MIS director who stated that PCs were only good for doing grocery lists.

As the millennium comes to an end, telecommunications professionals, both managers and staff members, have no choice but to extrapolate into the future and make the best judgments possible about emerging technologies. Telecommunications technologies are enabling technologies. They make modern business work. In the corporate environment, one of the primary ways that a company stays competitive is by employing computing and communications technologies. Companies cannot make frivolous decisions regarding technology, but neither can they always take a wait and see attitude. This means that all telecommunications professionals must, at one time or another, assume the role of visionary. It also means that the job carries an inherent degree of risk. Technological change is highly visible in the modern corporation. The rapid pace of technological development is difficult to monitor, but there are general trends that will continue to affect and redefine the telecommunications environment.

Miniaturization of computer components continues to be a major trend. The first computers filled an entire room and were capable of 500,000 instructions per second. Today one computer chip is one square centimeter and is twenty times faster. Small devices continue to be developed and connected. Once again, the trend is toward the fusion of computing and communications devices. Wireless telephones that possess crude computing capabilities have already been marketed on a limited basis. The user can send or receive a call, use the telephone as a pager, or receive and send simple e-mail or text via the touch tone pad that also doubles as a keyboard. Although the screen is small and crude, this device certainly is usable, albeit on a limited basis. This small telephone/computer may be a portent of the future. Imagine a device, small enough to fit into a purse or pocket. It will have a small handset that folds, much in the same way that a cellular flip phone does today. The handset components will be thin so that it will be no larger than a business card when completely folded. The handset will attach to a main computing device. The device will be a small computer, also with thin foldable parts. The parts will unfold to unveil a keyboard and screen. A mouse will be built into the keyboard. The unfolded computer will be large enough to use comfortably, but certainly not as large as a modern desktop. The quality will be very good, perhaps erasing the mindset that many users have that they are using a sub-par device where functional concessions were made for the sake of size. The device will provide an all-in-one functionality. It will be a powerful computer, capable of the basic business software applications such as word processing, spreadsheets, presentations, database functions, and e-mail. The color monitor will have high quality resolution. There will be a small camera and scanning device so that desktop videoconferencing can be performed, and there will be a utility that will offer voice mail from the hard drive, if the end-user chooses not to answer the telephone. The device would also be equipped with a jack for plugging into a standard telephone port and a wireless modem. There might also be a pen device so that the user can write on the screen. A handwriting utility will translate the scribble into readable fonts. The microphone that augments the videoconferencing function will also serve as a dictation tool, turning the user's speech into readable text. This device will be connected either via cable or wireless, as situations dictate. It will be powered via battery or electrical outlet. Is it demanding too much of our imaginations to dream of such a device? Small enough to fit in a purse or perhaps your suit coat pocket, and yet provide such an array of functionality? Remember that every time industry experts feel that the "minimum" size for a computer chip has been achieved, the barrier has been broken. The same can be said for every time the "maximum" speed of a computer chip has been achieved. Based on current technologies, this prediction is not really very outrageous. This example also demonstrates how telecommunications technologies and computing are fusing, and how that relates to telecommunications department management. Telecommunications professionals cannot rely solely on their knowledge of communications technologies. They must understand computers, software, and the technological trends that are affecting them.

In addition to miniaturization, speed—in computers and in networks—has been improved immensely. As electronic devices shrink, signals traverse them faster, more operations can be performed, and the cost of computing inevitably falls. More devices are packed on a chip and manufactured in larger batches, reducing the cost of manufacturing. Lower costs lead to increased sales of computers, for both businesses and computers. More computers means that more connectivity will be needed. Telecommunications departments need to provide connectivity to the business units within their own companies, or to the consumers who will connect to their company to order product or obtain information. Because of this trend, devices such as the one illustrated in the example of the small, folding computing/communications device are entirely possible, and probably will become plentiful in the not-too-distant future. The stumbling blocks are factors such as speed, weight, size, ruggedness, cost, and consumption of power. This may seem like a large list of obstacles, but many of these obstacles have been refined in today's technology. Modern laptop computers have color displays, weigh only a few pounds, are capable of storing gigabytes of data, and have multimedia capabilities. Wireless modems and cellular telephones exist today as do voice and handwriting recognition technologies. Battery technology continues to be refined, providing lighter batteries and longer life.

Television is widely viewed as a technology geared toward entertainment and the consumer, and has not traditionally been a responsibility of the telecommunications department (with the exception of videoconferencing which is still not ubiquitous in the modern corporate environment). American television is still, for the most part, analog technology. Television manufacturers, however, have steadily been incorporating computer technology into their products, and it will only be a natural progression to incorporate television into the computing/communications environment once digital television is introduced into the American marketplace on a wide scale basis. This integration may bring about the concept of the "telecomputer" which would have dramatic implications for both businesses and consumers. A "telecomputer" would not only offer tremendous potential for the entertainment industry, it would also offer new business and educational opportunities. Analog television offers poor picture quality compared to that of a PC monitor. Digital television would resolve this issue and allow for input devices that would transform the basic television into a computer. If this is combined with the high bandwidth capabilities to the doorstep, the foundation has been laid for what has been labeled the information superhighway.

Regardless of how fast the television will migrate to the computing environment, current computer trends continue with a heavy emphasis on communications, including software development. Software has an impact on how people use computers and the GUI made computers user-friendly and second nature. Software is now developed with networks in mind, which dovetails with current business trends. For instance, groupware is a software concept that promotes and allows collaboration and communications between workers. Operating systems are now developed to include internet access. It is only logical to assume that voice and video communications capabilities will soon become integral parts of commercial software packages.

Software will also continue to become more intelligent and user-friendly. End-users will simply plug in the appropriate device, the software will then "learn" of its presence and automatically enable it. Voice and video communications will become immediate. Eventually, we can expect computers to be so user-friendly, that "computer literacy" will be a term that has become virtually obsolete. The most profound technologies are those that disappear. They weave themselves into the fabric of everyday life. Whenever people learn something sufficiently well, they cease to be aware of it. That is the direction where software development is going and there will eventually be a seamless integration of computers to networks.

Multimedia for current PC technology will continue to be refined. However, multimedia has not found a common place in the average business environment. While it is common in PCs, it is still not very practical for many business applications. Telecommunications experts often talk of the "killer application," a computer application that will demand robust communications capabilities and provide justification for some of the advanced communications capabilities such as ATM. If multimedia begins to become common in the general business environment, it will, like the concept of the telecomputer, drive network providers to bring fiber optics to the doorstep and provide robust communications capabilities in the public network.

Virtual reality may also provide some interesting applications. Virtual reality simulates experience through stereo-graphic visual displays, tactile sensation, and response to gestures. This may have interesting applications for business. It could be used in meetings in much the same way that videoconferencing is used today or has been valuable for training applications. This technology is still in its infancy and, for most business applications, lacks practical value. There are many more technologies that are beyond the scope of this book. Concepts such as molecular computing, nanomechanical logic gates, or reversible logic gates are being investigated at the theoretical level. There are also many theories that have been ventured regarding computing and communications. Only time will tell what technologies will be adopted, but an open mind is critical and telecommunications professionals must always be ready for change.

Network evolution will have a critical impact on how business is conducted in the future. Primary areas of development will be higher bandwidth capability, the ability to carry voice, video, and data communications on a real-time basis, wireless capability (on a domestic and international basis), Internet enhancements, and the further development of network intelligence.

Cabled communications will certainly continue to expand, but copper will continue to be replaced by fiber optics, especially in the local loop (the cable segment that connects homes or businesses with the LEC CO). There are already tremendous amounts of fiber optic cable that have been installed to support the backbones of various carriers. Many of these carriers have gigabit capacities, but this is not generally available to end-users. Fiber optic lines have tremendous potential for high capacity transmissions. The transmission capabilities of copper have been improved, but still do not present the capacity offered by fiber optics. If the average residential home begins to demand service such as high capacity Internet access, video on demand, hundreds of cable television stations, or multimedia-based computer services, fiber optics will be the only physical medium by which they can be delivered. While wireless communications are limited because of the electromagnetic spectrum, fiber is only limited by the amount that is installed.

Wireless technologies will continue to be refined and will allow for handoffs from private wireless networks to public wireless networks. For instance, a worker may be engaged in conversation, walking from a building to his car. While the employee is within the building, he will be using the wireless PBX. Once he is outside the building and the range of the wireless PBX antennas, the signal would be handed off to a public network, and his company would begin to incur usage charges for the call. Public wireless networks will continue to be developed, but there are limitations. Frequency bands are licensed and the electromagnetic spectrum is limited. Coverage will be improved to more rural areas of the country but as high bandwidth applications become prevalent, only fiber- based land lines will offer this type of capacity.

Wireless capability will not be limited to land-based cellular systems. Powerful satellite systems are being launched that will provide global telephone and data services. These services will be independent of traditional terrestrial networks. For instance, a satellite telephone can now provide direct access to a satellite network, completely bypassing all other traditional networks. The initial foray into this service was COMSAT Handheld Telephone Service, which was made available in 1997. The service was not inexpensive, but it did offer many advantages. The service utilizes a spot beam, which requires that the satellite be directly overhead. Unfortunately, with a satellite in geosynchronous orbit 22,500 miles above the earth, users experience a delay in conversations that many people find annoying. Still, in situations where no communications capabilities are available, this is a viable application.

Low earth orbiting satellite (LEOS) systems and medium earth orbiting systems (MEOS) are now being constructed. LEOS and MEOS will be a series of satellites that encompass the earth in lower orbits (2,000 and 10,000 km) than the standard geosynchronous satellites (36,000 km). The advantages will be that the user does not have to be directly under the satellite and that the lower orbit reduces the amount of delay that users have found so annoying in the past. A number of companies have applied for (or have already received) licenses to provide LEOS services. The most publicized of these is Motorola's Iridium network. Both voice grade and high capacity data services will be offered. Wireless capabilities, both terrestrial and cellular, will aid in the expansion of global economies. This is because many foreign telecommunications infrastructures cannot keep up with growth by installing cable. Wireless technologies have proven to be a faster methodology to providing communications capabilities. In addition to wireless, the Internet is also proving to be a stimulus to global economic expansion.

The Internet has grown so fast, nobody really knows how big it is or how many people use it. Statistics show that usage on the Internet's heaviest links increases by 20 percent each month. Rapid growth and capacity limitations have led to performance problems on the Internet. Still, the growth of this network shows no signs of slowing. The Internet will continue to experience growing pains, but the internet service providers (ISP) will aggressively upgrade their networks. Video will become more common and voice communications is already a highly touted technology. Performance will eventually improve but network intelligence could be a major enhancement. The concept of a "Knowbot" has been theorized for a number of years and may prove to be the type of "personal planner" that people will need to sift through the overwhelming mountains of information available on the Internet. A Knowbot is a software agent that would serve as an electronic secretary. It would know your habits and wishes. It would act without instructions, organize information, hunt for data, and protect confidential information. People spend hours on the Internet, often frustrated and unable to find what they are looking for. An intelligent agent, the Knowbot, will save much time and frustration.

This brief overview is of technological enhancements that may occur in the immediate future. As the reader may ascertain, things change quickly in the field of telecommunications. But there is no turning back. The technological advance is unstoppable and business is reliant upon it.

1.6 Summary

The telecommunications department developed due to deregulation of the telecommunications environment and technological advances. Two separate tracks developed, voice and data, but the lines that distinguished these disciplines are blurring. The workplace in American business is changing due to advanced computing and telecommunications technologies. The business world is changing also. The economy is becoming more service oriented and global. The rate of technological change continues at an inexorable rate. Telecommunications professionals must monitor and assess those changes so that their companies can maintain a competitive edge.

2

Basic Technological Overview: Products, Services, and Vendors

The profession of telecommunications is highly technical in nature and all members of the telecommunications department must maintain a high level of technical knowledge. This knowledge must be kept current, and telecommunications professionals must always position themselves so that they are in a mode of continuous development; learning new technologies and strengthening weaknesses. In my experience, there is no way to avoid this, and it is rare that any telecommunications professional has the luxury of resting on his laurels, or assuming that his existing knowledge base will carry him for an entire career. For as soon as one project is completed, another more complex project looms on the horizon. The situation is also complicated by more educated end-users who are computer literate and have easy access to publicly available technical information. Today's telecommunications professionals are constantly challenged and need to support a demanding environment and demanding end-users.

The array of telecommunications technologies is overwhelming and in a constant state of change. It behooves the telecommunications manager to not only encourage continuous education, but to also set an official educational policy. This will be covered in more depth in Chapter 10. The intent of this chapter is provide an overview of products and services supported by the telecommunications department. It is by no means the last word and readers are encouraged to seek more in-depth knowledge via the many books, periodicals, seminars, and classes available today. This chapter will address telecommunications products and services at a high level so that readers may gain an understanding of what technologies, services, and products need to be supported by the modern telecommunications department. It will also address issues that are important to the management and support of each individual technology. It is my intention to provide an overview while still conveying some of the details that are important in understanding the technologies and how they interconnect.

2.1 Local Carriers

The local carrier is commonly known as the local exchange carrier (LEC), which services both business and residential subscribers. The LEC provides physical connectivity to the public network via cable and offers various telecommunications capabilities through the physical connection. The concept of an LEC is becoming more difficult to define. At the time of divestiture, the LEC picture was much clearer. There were seven regional bell operating companies (RBOC), also known as regional holding companies (RHC). The original seven were Ameritech, Bell Atlantic, Bell South, NYNEX, Pacific Telesis, Southwestern Bell, and US West. Each of these RBOCs provided local service in a specified geographical area, which could actually be very large, often encompassing a number of states. The original seven were forbidden to manufacture telecommunications equipment, offer cable television services, or market long distance services. They were, however, allowed to sell equipment manufactured by other companies, and to offer a variety of local services, both voice and data. At the time of divestiture, there were also many independent telephone companies (ITC). These were not nearly as large as the seven RBOCs, but they offered local coverage in areas where the old Bell System was not able to establish service. The ITCs, however, were fully integrated with the old Bell system, adhered to standards established by local and federal regulatory bodies, and offered service of a similar quality to the Bell system.

Quite a bit has changed since 1984. As outlined in Chapter 1, the Telecommunications Act of 1996 set the stage for redefining the local telecommunications market. Since divestiture and the establishment of the original seven RBOCs, there have been a number of mergers, acquisitions, and expansions between RBOCs, ITCs, cable companies, and interexchange carriers (IXCs). One of the largest and most notable is the recent merger of Bell Atlantic and NYNEX. In addition to the business transactions that have been occurring, new companies have been formed to service potentially lucrative local markets.

Before the Telecommunications Act of 1996, a number of companies began to emerge in local markets. Known as competitive access providers (CAP), these companies offered alternative access services such as dedicated cable and T-carrier services (described later in this chapter). Subscribers often found CAP-provided services to be more cost effective than those offered by the traditional LEC. In addition, CAP-provided services could also add a level of insurance in the event of a failure on the part of the incumbent LEC. For instance, a subscriber might have purchased a form of access from the LEC and a similar service from a CAP. If the primary service fails, the subscriber could switch to the CAP service (see Figure 2.1). Because the CAP service is provided through separate cable, a separate entranceway, and a separate switching center, the subscriber continues service and is not affected by a disaster that may have affected the LEC. When the CAP only offers cabling and various data access services, they are known as a CAP. When the CAP begins to offer local telephone service, they become a competitive local exchange carrier (CLEC). Established LECs that were once part of the old Bell System have become known as incumbent local exchange carriers (ILEC). CLECs may offer local telephone service; however, it is often necessary for the subscriber to change telephone numbers if they are changing from existing service. This is not an easy decision when a business has held the same telephone numbers for a number of years. As stated in Chapter 1, a system has been developed to provide local number portability (NP). As of this writing, it is not widely available; however, 93 local markets have been targeted for the next several years and competition is beginning to emerge in the local telephone service market.

Long distance carriers, also known as interexchange carriers (IXC), are now permitted to market local services. This has further blurred the traditional definition of an LEC. For example, a company may be an IXC on a national

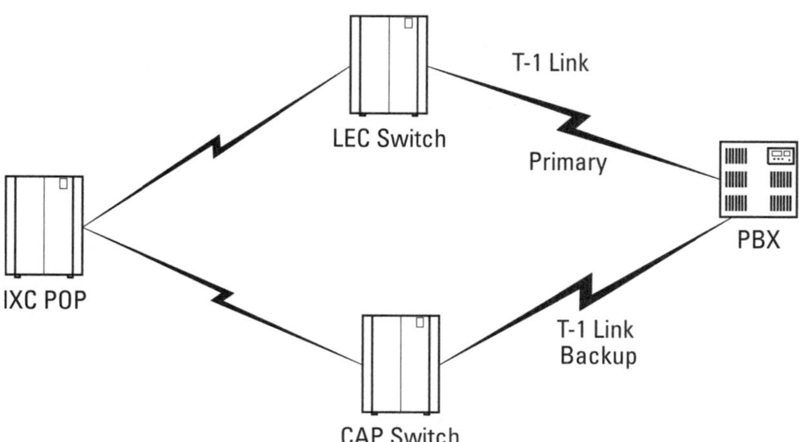

Figure 2.1 Redundant access through CAP switch.

basis, a CAP in a limited number of cities, and a CLEC in yet other cities. Cable companies are also continually vying for a slice of the local pie and some are aggressively installing telephone switches. The eventual goal of many of these companies is to offer a vast array of communications services to business and residential customers. Included in this list of services would be entertainment media such as cable television, video on demand, and Internet access.

The local carrier situation has become very confusing for the telecommunications professional. New choices offer new opportunities, but they also entail risk. For example, traditional LECs offer superior reliability for local telephone service. Consider the last time your electricity went out. Chances are the telephone still worked. In fact, most people have probably never experienced a local carrier outage. The same cannot be said for local cable television service. It is not unusual for cable television companies to experience outages during electrical storms. If a cable company offered local telephone service, even at a substantial discount, this is one of the factors that would have to be considered. However, as the new players gain more experience, they will begin to approach the quality and reliability of the established LECs, just as long distance carriers did after the market was open to competition. As the local environment continues to change, telecommunications professionals will have to weigh the quality of the new carriers against the established quality of the existing LECs. This, in turn, will have to be weighed against the potential savings offered by the new local carriers. New local carriers should not be dismissed out of hand, but they should be examined carefully before a decision is made.

The LEC provides local subscribers with a building that serves as a cable and communications switching hub. The central office (CO) is a building that houses a sophisticated telephone switch and various peripheral equipment. The CO switch provides voice communications services; either plain old telephone service (POTS), a type of business telephone service called Centrex, or trunk line services that interface to various types of telephone systems. The CO also provides data communications capabilities, either through the CO switch, or through peripheral data communications equipment. The CO switches are manufactured by a number of companies, major manufacturers being Alcatel, Lucent (formerly AT&T), Nortel (formerly Northern Telecom), and Siemens. LECs are free to purchase their CO switches from any manufacturer, as long they are approved for the public network interconnection by the Federal Communications Commission (FCC). They must also provide a high degree of reliability so that service continues in the event of a disaster. Generator power provides continuous service during power outages and redundant cable paths prevent busy signals if there is a cable cut, or if the network becomes excessively busy. LECs are regulated by federal and state agencies. Public utility commissions (PUC) oversee issues such as local rate structures and business practices.

One of the primary functions of an LEC is to provide access to the public network. LECs serve residential and business customers, in addition to providing services for long distance carriers. For example, if a company orders a dedicated data communications circuit between Pittsburgh, Pa. and Chicago, Ill., the access lines from the buildings in both Pittsburgh and Chicago that connect to the IXC's point of presence (POP) is normally provided via an LEC. In the case of basic long distance, when a customer dials a long distance number, the call is sent to the cable that connects the business with the LEC. The LEC then sends the call to the long distance carrier chosen by the customer. LECs provide other services and features, but one of the primary roles they play in the telecommunications infrastructure is access.

It should be noted that subscribers are not legally bound to use the dedicated access provided by LECs. In fact, many telecommunications departments are able to save money by installing bypass capabilities. This can be done via microwave, satellite, infrared, or private cable. In some foreign countries, where the telecommunications infrastructure is poor, bypass may actually be a necessity. Bypass designs can also be expensive, complex, and difficult to manage. Telecommunications departments that utilize bypass on a regular basis usually need to maintain a high degree of in-house expertise in the form of engineers and technicians. Utilities such as gas and electric are prime examples of companies that often bypass the LEC access. Because these companies have right-of-way access to most residences and businesses, they can install their own cable in order to gain the services they need.

Access begins with the physical connection to the CO, which is typically via copper cable (see Figure 2.2). The cable that connects a subscriber is known as the local loop. Most local loops that exist today are copper that was originally installed by the old Bell System. Fiber optic connections are possible for the local loop, but this is not used for residential subscribers because, traditionally, it has not been economically feasible. In addition, there has generally been no need to provide such high capacity capabilities to the average residence. Business subscribers use fiber optics from the LEC on an as-needed basis. Typically, these are businesses with high capacity voice or data requirements. As the local arena changes, and as the price to install fiber optics reduces, fiber may gradually replace copper as the standard for the local loop. Demands by consumers and businesses for high capacity telecommunications services may mandate this. However, most local loops in place today are copper with fiber being used only for special applications by the largest users.

The LEC terminates their cable at a point within the subscriber's facility, a physical point located somewhere inside a home or business. This is known as the point of demarcation. The responsibility of the LEC ends at the point of demarcation. Whether a business or residential application, the subscriber is

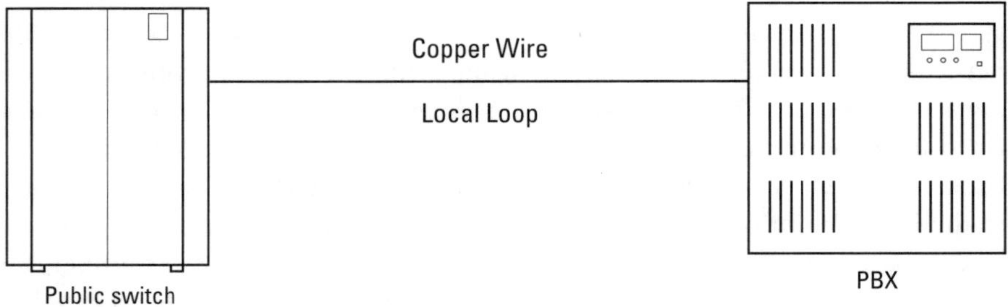

Figure 2.2 Local loop.

responsible for extending cabling within the facility and providing their own communications equipment (e.g., telephone systems, data communications devices, etc.). The LEC can provide inside cable, but this is an additional charge above the installation cost of the service ordered. Inside cable is a primary responsibility of the telecommunications department. Whether the application is an entire building or an employee's residence, the department must provide the physical connectivity from the point of demarcation to the communications devices. This responsibility can entail anything from a simple cable that is terminated on a telephone jack to a complex, structured wiring system that requires detailed documentation and continuous management. Cable will be covered in more detail later in this chapter.

In spite of the proliferation of new wireless services, bypass capabilities, and new emerging local carriers, the traditional LECs still play a fundamental role in the national telecommunications infrastructure. The services they provide are a major responsibility of the corporate telecommunications department.

2.2 Local Communications Services and Providers

The most fundamental of services provided by an LEC is telephone service. This includes access (the local loop), dial tone, the assignment of a telephone number, billing, long distance access, and various services (e.g., call waiting, voice mail, or caller ID). The common designation for basic telephone service is POTS, and a closer examination of the billing components demonstrates three fundamental financial components of telecommunications services: one-time charges (installation), fixed charges (monthly fees), and usage-based charges (message units or long distance). All telecommunications services are subject to at least one of these components.

All network-based services usually incur an installation charge of some sort. There may also be other fees, but this is a primary budgetary consideration for telecommunications planning. In the case of a basic POTS line, there is a one time fee to establish service. Regardless of how the LEC provides the service (whether through new or existing cable), the installation fee is fixed. More advanced telecommunications services may entail charges for such things as network engineering fees, but these charges are still one-time.

Once the POTS line has been installed, the subscriber is charged a monthly fee for service(s). These are the fixed charges. A bill provided for a telecommunications service, whether voice or data, can be complex and difficult to read. There may be many different components, but the fixed costs should be itemized and will not vary from month to month. When operating budgets are developed, fixed costs are the easiest to predict and administer.

Usage-based charges introduce an element of uncertainty into the budgeting process. Since charges are based on usage, how can they be accurately budgeted? Telecommunications professionals address this issue in several ways. First, the rate becomes a budgetary factor. In the voice communications arena, this is typically known as cost per minute. By negotiating the best possible rate, the telecommunications department is able to help control costs that are not fixed. A second factor is to forecast usage based on past history. If a department within a business typically spends $2,000 per month on long distance charges, a large variance of this will indicate either abuse or an increase in business. If it is the later, there may be options to change service, renegotiate rates, or look to high volume services that offer better rates.

LECs are also responsible for administering local numbering plans. When a subscriber orders telephone service, telephone numbers are assigned by the LEC. The LEC administers a database of local numbers, assigning, disconnecting, and reassigning as the subscriber base changes. The numbers assigned by the LEC are part of a standard national plan known as the North American numbering plan (NANP). The NANP employs a standard 10-digit format. The first three digits are known as the area code. The more technical designation is the numbering plan area (NPA). Within an area code there may be dozens of COs. Each CO is assigned unique central office codes that are also known as exchanges or NXXs. Area codes and exchanges are also referred to as NPA NXX. The N represents any digit from 2 to 9, and X can be any number. Theoretically, the central office code can support up to 10,000 individual telephone numbers. The remaining four digits of a 10-digit telephone number are known as the subscriber code. Whereas the area code and exchange services specific geographical areas, the four digit subscriber code is assigned to a specific business or individual subscriber.

As previously stated, the LEC provides the first point of access to the public network. Beyond the physical connection and simple dial tone, the LEC provides local telephone service, or interconnection with other subscribers in the same area. Local calls may be billable at a flat rate per call (known as message units), or may be "free" through various service plans where the monthly service charge is higher. With such plans, the subscriber may receive "X" number of free message units before charges are incurred. Long distance charges are incurred beyond the local calling area known as the LATA.

LECs offer "local" service within the local access and transport area (LATA). In this case, the definition of "local" bears some explanation. LATAs are 161 geographical areas defined in the United States. These are areas that have traditionally been service areas that were once the exclusive domain of the LECs. When a subscriber called another subscriber within the LATA, the call was terminated through the LEC network, and toll (or local) charges were generated through the LEC. When a subscriber dialed a number that terminated outside the LEC, the call was passed to an IXC. Billing for inter-LATA calls can be through the LEC or the IXC.

When local telephone service is ordered, subscribers must choose a long distance carrier. If they do not, the LEC may actually block long distance calls. Once a long distance carrier is selected, a software parameter is entered by the LEC, known as the primary interexchange carrier (PIC). This determines which IXC will transport the call. Telecommunications staff members must indicate to the LEC who the long distance carrier will be. This simple task often becomes an exercise in futility for many telecommunications departments. LECs may assign the wrong PIC, and even if they select the correct IXC, problems may ensue. For example, a large company may sign a contract with an IXC with special negotiated rates. If the company was AT&T, the PIC would be 288 for their basic service. The special rate requires a different PIC. It is not simply a matter of telling the LEC AT&T; the instruction must be specific to the PIC that relates to the contract.

Beyond POTS, LECs provide business services such as CO trunk lines and direct inward dialing (DID) service. These services interface to customer provided telephone systems, either a private branch exchange (PBX) or a key telephone system. CO trunk lines are normally provided in hunt groups. That is, a number of lines are provided and connected to the telephone system, but only one number is published. A call arrives at the first trunk line and is answered. While the first trunk line is engaged, a second call arrives and, sensing that the first trunk line is busy, the call hunts over to the second open line. Trunk lines can be two-way (inbound and outbound) or one-way, depending on how the subscriber would like them to be configured. Ordering the correct amount of trunk lines is a critical function of the telecommunications department. For a new application, where no telephone traffic information is available, telecommunications professionals must use estimates based on industry standards. Once service has been installed, usage on trunk lines must be monitored in order to assure sufficient capacity.

DID is a type of trunk line service offered by the LEC that typically interfaces to larger telephone system installations. The basic principle is to provide a direct dial number to most (or all) employees. In larger installations, this helps to reduce the number of calls directed to the main listed number and reduces the work load of the operator. It would not be economical to provide a trunk line with a unique telephone number to each employee, therefore a larger number of telephone numbers are directed to a smaller group of trunks (see Figure 2.3).

LECs also offer point-to-point trunk line services. Point-to-point trunk lines are also called tie lines. In a PBX environment, these circuits are used to connect two telephone systems. In effect, this is a method of bypassing the public network (see Figure 2.4) and saving on direct dial or toll costs. Point-to-point services are also offered for data connectivity. Analog point-to-point lines that are ordered specifically for data are conditioned to provide more error free service.

Trunk line services (voice or data) can be provided in either analog or digital format. The aforementioned trunk line services have traditionally been offered in analog format, but with the advent of digital services, the use of

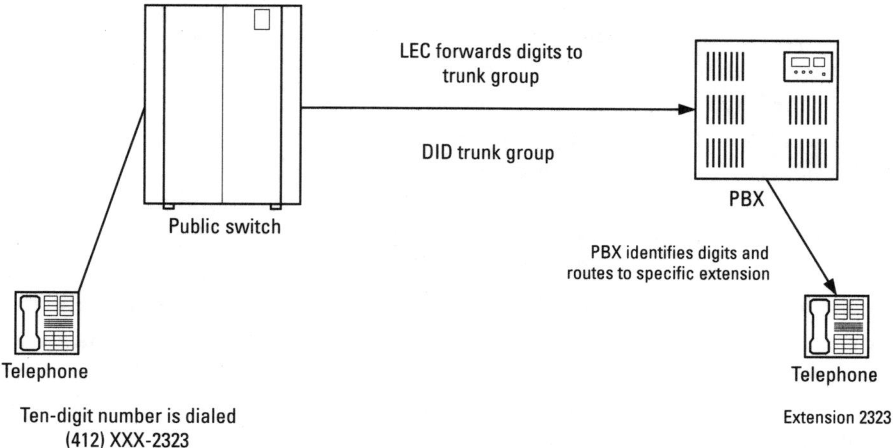

Figure 2.3 Direct inward dialing.

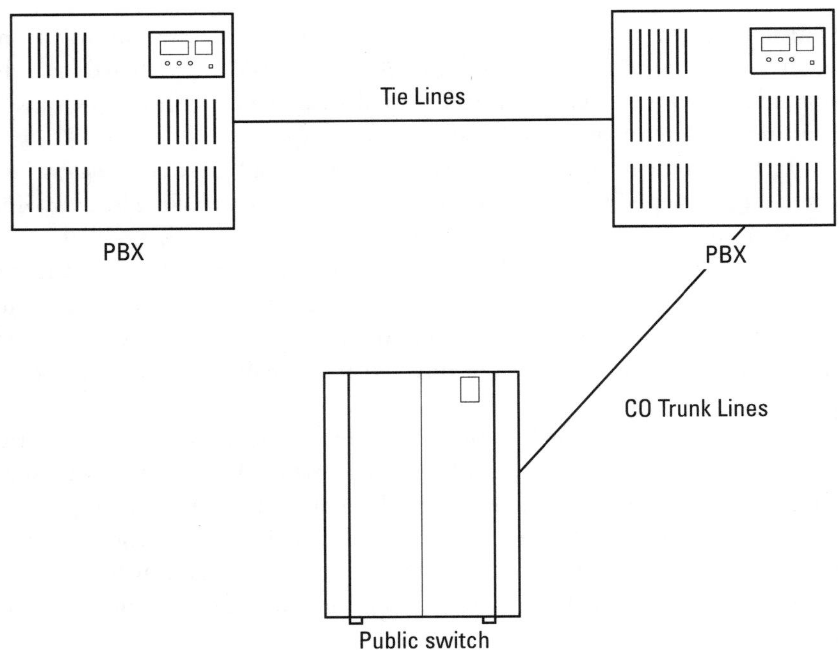

Figure 2.4 PBXs connect via tie lines.

analog facilities is diminishing. An analog trunk line carries one channel of capacity. A digital trunk line can be provided in a variety of ways; either single or multiple channel capacities, dedicated or dial-up. A single channel digital trunk line is typically installed to support data communications and is provided at base rates of 56 or 64 Kbps. Single channel digital lines are sold directly to subscribers as point-to-point circuits (see Figure 2.5) if both termination points are within the LEC coverage area. If one of the termination points is outside of the LEC coverage area, then the circuit is usually provided by an IXC. Under these circumstances, the LEC will often still supply the local loop portion of the point-to-point circuit to the IXC. The IXC will pass this charge onto the customer, however, this is not a binding requirement. Customers can negotiate their own local loops. This does, however, introduce a new level of complexity, and most telecommunications professionals will opt for a one-stop-shop, rather than incurring three bills for a single circuit.

The point-to-point trunk line falls within the category of private lines. There are other types of private lines that are not commonly used anymore,

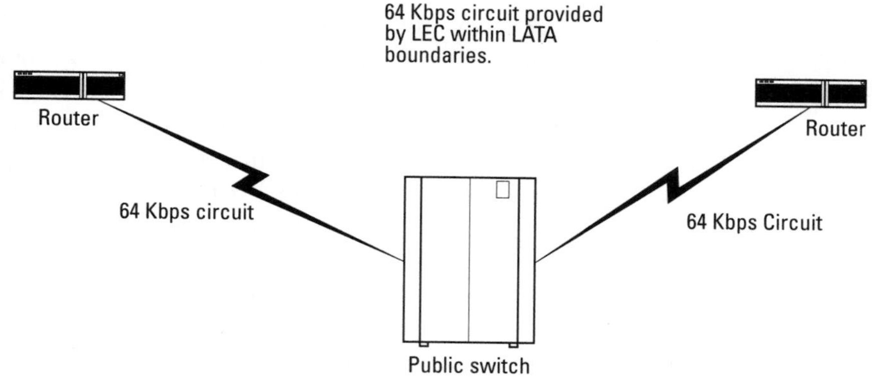

Figure 2.5 64 Kbps circuit provided by an LEC within LATA boundaries.

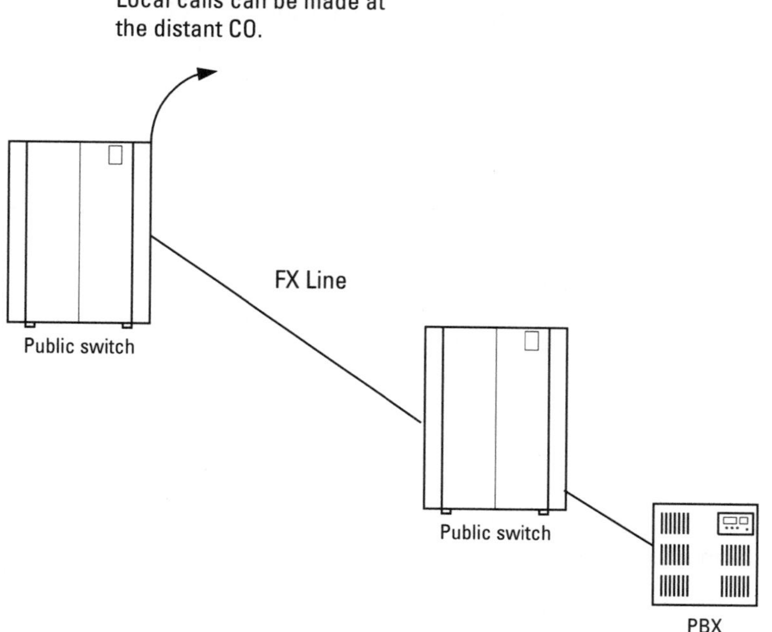

Figure 2.6 Foreign exchange line.

however they do merit a brief overview. A foreign exchange (FX) line is a dedicated line connected to a PBX from the distant CO (see Figure 2.6).

For this example assume that the distant CO is located within the LATA, but far enough away to incur long distance charges if dialed directly. By comparing the long distance charges to the fixed price of the FX line, a savings is realized. A second type of dedicated private line is the off premise extension (OPX). An OPX is simply an extension of a PBX connected via a private line. The end-user gains all features and functionality associated with the PBX. The final type of private line is the shout-down circuit, also known as a hoot and holler line. A shout-down circuit is simply a private line that terminates on a speaker. As strange as the application may seem, I once knew of a company that owned a number of junk yards that were geographically dispersed. When a customer needed a particular part, a manager would dial each location and ask if the part was available, via speakers that covered each junk yard. Because the employees were always somewhere in the yard, the speaker system via the shout-down circuit was the only method of communication. It should be noted that each of these private line services are available via LECs or IXCs, depending on whether the private line terminates outside the LATA.

Single channel dial-up digital service is offered in the form of switched 56 Kbps. Switched 56 (as it is commonly referred) is essentially dial-up data. However, unlike modem access through an analog connection, switched 56 offers the advantages of clear digital transmission and higher bandwidth than most modems (although analog 56 Kbps modems are now being marketed). Switched 56 is often used as a backup capability to point-to-point circuits. That is because it is a dial-up service and usage charges are incurred, either local or long distance. There is a monthly service charge for Switched 56 plus usage charges when a distant location is dialed. Switched 56 is becoming less common because ISDN offers more bandwidth and is a more versatile service, capable of carrying voice, data, or video.

Multichannel digital trunk lines are offered via T carrier services, also known as digital signal (DS) carrier services. T carrier refers to a copper-based land system that is exclusive to AT&T. The proper term is DS, however, the most commonly used name is T carrier and will be used in this book. The most basic form of a T carrier system is the T-1 (also known as DS-1). A T-1 line offers 24 channels of capacity. Each channel is 64 Kbps. The total bandwidth of a T-1 line is 1.54 million bits per second (Mbps). Readers who are new to the technology may notice that 24 times 64 does not equal 1.54 Mbps. That is because an extra bit is used for synchronization of the line. T carrier lines use a system of transmission known as time division multiplexing (TDM). For instance, in a digital PBX, a device called a codec (short for coder/decoder) samples an analog signal at a rate of 8,000 times per second. This sampling is coded into an eight bit word (8 times 8,000 = 64 Kbps). Since there are 24 channels in a T-1 line, a sample of each channel is combined into a frame. Each frame is 24 channels times 8 bits plus a synchronization bit, for a total of 193 bits. 193 bits times the sampling rate of 8,000 equals 1.54 Mbps. A single channel of a T-1 is called a DS-0. It may seem odd, but the entire transmission system is commonly called a T-1 while the single channel is almost always called a DS-0. T carrier systems are offered in increasing levels of capacity (see Table 2.1). The most common is T-1, however T-3 is finding more acceptance for larger companies with high capacity requirements. Both IXCs and LECs offer fractional T-1 service. In this case, the end-user purchases a number of channels less than 24, realizing a savings on the cost of a full T-1.

T carrier systems are used for a variety of applications. They can be used to combine voice and data transmissions as a means of saving money, or simply as a cost effective solution to a basic trunking requirement. For example, a company may require a dozen trunk lines to interface to their PBX. The cost of 12 analog trunk lines might be equal to a single T-1 with 24 channels. The comparison, however, is not that simple. A T-1 does require a specific type of multiplexing equipment, which may be purchased or leased. If the required number of trunk lines exceeds 12, then the purchase of the multiplexing equipment is probably justified. T carriers can also be used to connect PBXs and bypass the public network, in the same way that analog tie lines can be utilized. Moreover, a channel may be dedicated to data communications, thereby increasing the potential savings. T carrier services are point-to-point services that incur a one-time installation charge and a fixed monthly cost.

LECs also provide integrated services digital network (ISDN), a digital dial-up service. This service is offered in low and high capacity versions. The basic rate interface (BRI) of ISDN is a digital service consisting of three channels (see Figure 2.7).

BRI has two bearer (or "B")channels and one data (or "D") channel. The B channels are 64 Kbps and the D channel is 16 Kbps. The B channels can be

Table 2.1
T-Carrier Hierarchy

Type	Bandwidth	Channels
T-1(DS-1)	1.544 Mbps	24
T-1C	3.152 Mbps	48
T-2(DS-2)	6.312 Mbps	96
T-3(DS-3)	44.736 Mbps	672
T-4(DS-4)	274.176 Mbps	4032

used for voice, video, or data. The D channel is used for signaling or some data applications. ISDN has gained acceptance over switched 56 and is becoming a more widespread offering from most LECs. The proliferation of ISDN has been aided by national standards being accepted by most carriers. Consequently, most new ISDN installations are national ISDN-1. Before the acceptance of a national standard, ISDN was proprietary to specific manufacturers. For instance, a Lucent (formerly AT&T) 5ESS CO switch required ISDN Custom. This required equipment manufactured to the signaling specifications of Lucent and no other ISDN equipment would work on the 5ESS. The same was also true of the Nortel DMS-100. A higher capacity ISDN is offered through primary rate interface (PRI). PRI provides a 23+D capability. As in the case of BRI, PRI B channels are 64 Kbps. The D channel is also 64 Kbps. ISDN is a dial-up service. The subscriber pays an installation fee in addition to the monthly service charge. When a telephone number is dialed through an ISDN line, usage charges are incurred, either local, toll, or long distance charges.

ISDN is a versatile and robust communications service. It can be used exclusively for voice, data, or video. ISDN telephones are digital telephones with numerous features. They include feature buttons that provide for a one button functionality and digital display sets that can capture caller ID (CLID) from the public network. The caller's telephone number can be displayed to the subscriber or downloaded to a computer database. Information such as the CLID or call setup information are sent down the D channel. ISDN telephones have most recently found a practical application with Centrex. Centrex is a telephone system that is provided via the LEC CO. With Centrex, the LEC CO is partitioned via software so that telephones on a subscriber's trunk lines are enabled with PBX features such as transferring and conferencing. An attractive feature of ISDN is that a PC can be plugged into an ISDN telephone to take advantage of the second B channel. Since this is a completely different channel, the data does not interfere with voice communications. This is very attractive for the concept of small office/home office (SOHO). Many companies find that certain positions are best supported from home. For instance, a field sales force, where most of the job entails travel, is a position that is perfect for SOHO applications. ISDN can provide a feature-rich telephone and enough bandwidth for data connectivity to support most computer applications.

As a dial-up data service, ISDN offers one or both channels for a total aggregate bandwidth of 128 Kbps. The D channel is used for either signaling or packet data transmission. In a business environment, ISDN is practical when online connectivity is not needed 100 percent of the time. In an environment where data connectivity is required eight hours a day, five days a week, a dedicated line is less expensive. Moreover, the task of connecting to the distant end can be complex and time consuming. ISDN is accessed through a circuit card installed in a PC. It is interesting to draw a comparison between the ISDN telephone and the ISDN PC card. Both offer the capability to transmit voice and data. The PC card has a jack for telephone input while the ISDN telephone has a jack for PC input. Both examples will demonstrate how the disciplines of voice and data communications are continually merging.

For data applications, ISDN has also found acceptance as a dial backup service. Since business is now conducted on networks, many companies find it necessary to provide a dial backup capability in the event that the primary data circuit fails (see Figure 2.7). In this example, the primary circuit is a frame relay

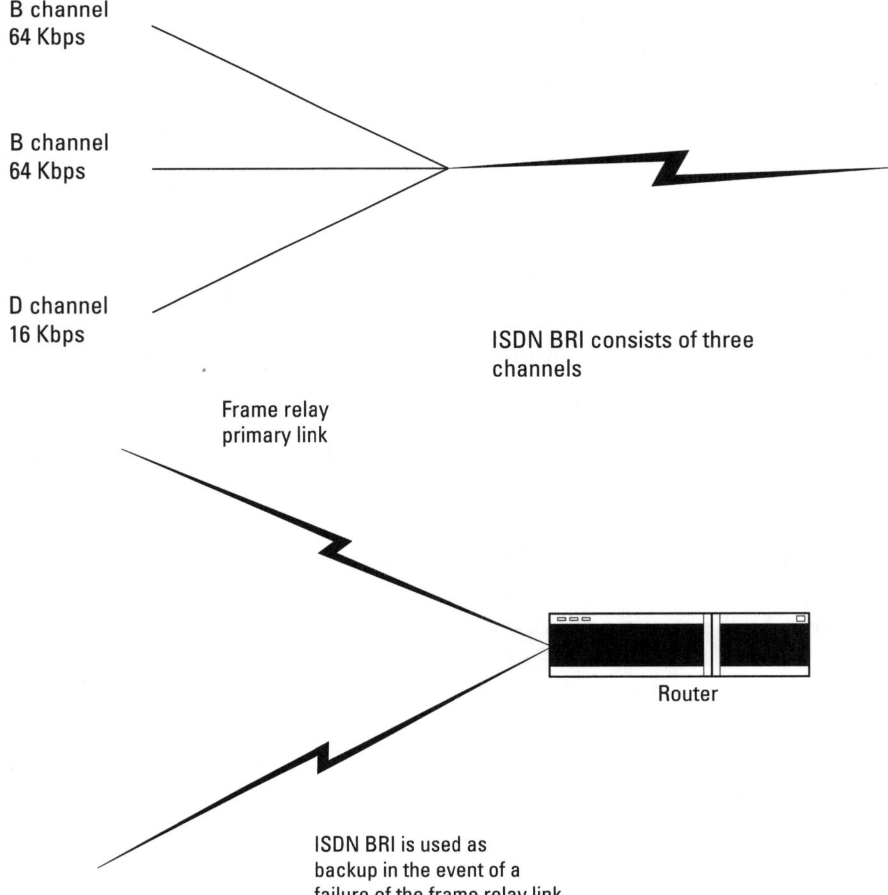

Figure 2.7 ISDN basic rate interface.

link that is connected to a router. When the frame relay link fails, the router is programmed to seek an alternate path, which is an ISDN line. The ISDN line is more expensive to use than the frame relay, but this design ensures continuous service in the event of a frame relay circuit failure. Hopefully, under such circumstances, the primary link will be restored before substantial dial-up costs are incurred.

ISDN has also become popular for Internet access. For as much as the Internet has been heralded, it has also become notorious for its poor performance. The Internet is a prime example of how the concept of client server computing has introduced chaos and uncertainty into the discipline of data communications. PCs are capable of storing gigabits of data and Internet users access large graphic files and video, in addition to basic text. When large numbers of users are accessing the Internet, the transmission of these files deteriorates overall performance of the network. While there is no way to compensate for the overall performance of the Internet, some compensation can be made by increasing the local access speed. Hence, ISDN has been finding popularity with heavy Internet users because the speed is far greater than that offered by analog modems.

ISDN has also become an enabling technology for videoconferencing. When the concept of videoconferencing was first introduced, it was provided via analog transmission lines, and performance was dubious at best. As digital capabilities became available, performance was improved, however, this was over 56 Kbps lines. Video transmissions, of the quality that is offered by major networks, requires 90 to 150 Mbps. In order to transmit video over a 56 Kbps lines, a tremendous amount of data needed to be compressed. Quality video transmission simply required more bandwidth and compression techniques needed to be improved. Enter ISDN equipped with two B channels at 64 Kbps. Video compression advanced to the point where the quality was acceptable at either 2 times 56 Kbps (112 Kbps) or 2 times 64 Kbps (128 Kbps). At 15 frames per second (FPS) this offered acceptable quality for conducting business meetings. Higher quality videoconferencing is available at higher data speeds and 30 FPS. A common method of obtaining higher bandwidth is through the aggregation of multiple ISDN lines. An inverse multiplexer (IMUX) takes the B channels from three ISDN BRI lines (six channels total) to form an aggregate bandwidth of 384 Kbps. At 15 or 30 FPS, this offers acceptable video quality for business meetings and the movement of participants is not distracting.

PRI is used for high bandwidth applications, very often being a trunk interface to PBXs. One application that is becoming more prevalent is as a local access for toll-free services. The D channel accepts signaling information from the public network, and this information is used by the ACD or a computer to route calls or process information about the call.

LECs also offer high capacity transmission capabilities for a variety of applications because they are connected to virtually every building and residence within their area of coverage. One such concept is the metropolitan area network (MAN). Loosely defined, the MAN is a method of interconnecting buildings that are scattered around a general metropolitan area (see Figure 2.8).

An MAN can be used to route voice, video, or data and it can also be supported via private cable installed independently of the LEC. However, MANs provided through the LEC are more common because there is no need on the part of the subscriber to procure right-of-ways to install cable. An MAN can carry voice communications, but most MANs are installed to support high capacity data requirements. An MAN is a concept, not a specific technology. A

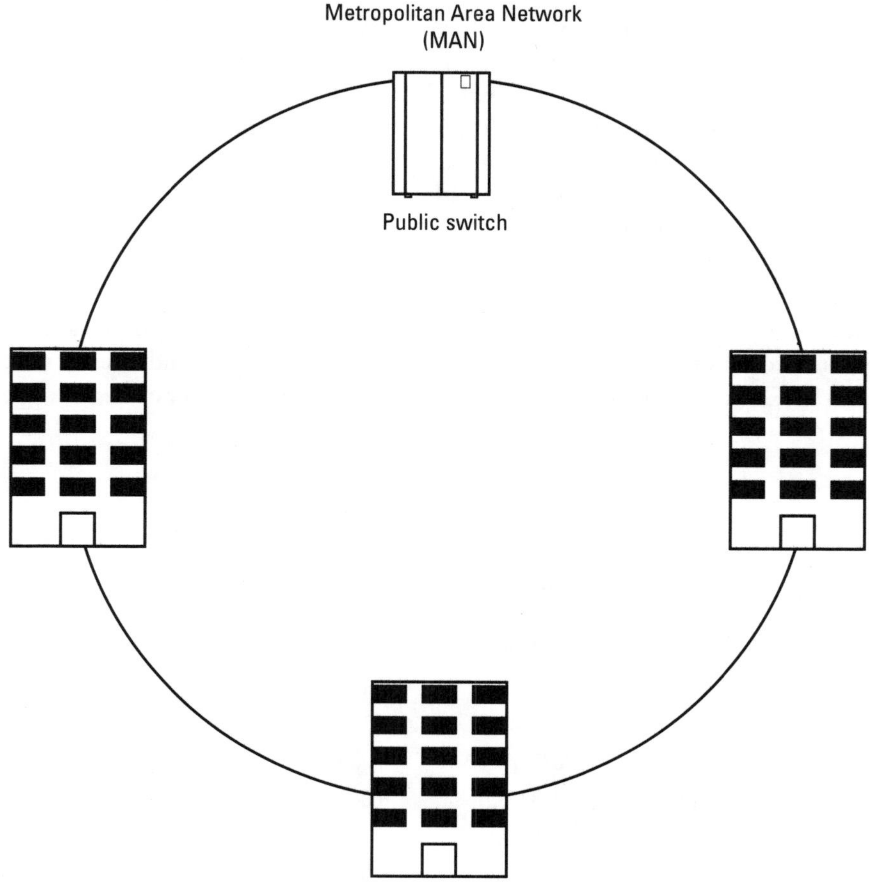

Figure 2.8 A metropolitan area network (MAN).

variety of technologies can be used to connect the sites that constitute an MAN. The interconnecting technology can be digital or analog, T carrier or several high capacity services offered by LECs.

Switched multimegabit data service (SMDS) is a connectionless high-speed data transmission service offered by LECs that was originally developed to run in an MAN environment. SMDS is a connectionless service, meaning that data can be sent without first establishing a connection. An example of this is internet protocol (IP) data traffic, where data is sent to a distant location without establishing a connection. SMDS can send data up to a rate of a T-3 (45 Mbps). It is typically provided by LECs for use within the LATA, however, it can be extended throughout the WAN.

Synchronous optical network (SONET) is a transmission technology specifically designed to be used over fiber optic lines. SONET utilizes a family of transmission rates known as optical carrier (OC). OC capabilities start at an OC-1 rate of 52 Mbps (see Table 2.2) and can go as high as an OC-48 at 2.5 giga (billions) bits per second (Gbps). SONET is used for a variety of applications, but the more common application is as a network backbone capability. This may be as part of an MAN or as a high capacity backbone in a campus environment. SONET is typically sold as a service by LECs, but private networks can be implemented.

As previously cited in this chapter, local loops are still, for the most part, copper installed by the old Bell System. Fiber optics will provide more bandwidth, however, it may be years (if not decades) before fiber optic local loops become common. There is still an immediate need for high bandwidth services from LECs. A method of providing a high-speed capability over existing copper

Table 2.2
Optical Carrier Hierarchy

OC Level	Line Rate	Capacity
OC-1	52 Mbps	28 T-1s or T-3
OC-3	155 Mbps	84 T-1s or 3 T-3s
OC-9	466 Mbps	252 T-1s or 12 T-3s
OC-12	622 Mbps	336 T-1s or 18 T-3s
OC-18	933 Mbps	504 T-1s or 24 T-3s
OC-24	1.2 Gbps	672 T-1s or 24 T-3s
OC-36	1.9 Gbps	1008 T-1s or 24 T-3s
OC-48	2.5 Gbps	1344 T-1s or 48 T-3s

is through digital subscriber line (DSL). DSL is a generic name applied to a family of services, which includes asymmetric digital subscriber line (ADSL), high bit rate digital subscriber line (HDSL), and single pair symmetrical services line (SDSL). Each of these names is a variation on the DSL theme. DSL works in the same way that ISDN does, except that it offers an 8 Mbps transmission rate from the subscriber's premise to the CO and there is a return rate of 1 Mbps. DSL technology is a relatively new offering and little service has been installed. As it becomes more readily available, telecommunications departments may find this service suitable for telecommuting and SOHO applications.

2.3 Interexchange Carriers and Services

As mentioned in the previous section, it is becoming increasingly difficult to define an LEC. The same can also be said of IXCs, who are now making aggressive moves into local markets. Regardless, both IXCs and LECs still maintain core businesses that are definable. In the distant future, there may come a time when the terms LEC and IXC are obsolete and we will only think in terms of carriers. At the present time, there are still clear differences, in spite of recent regulatory changes.

IXCs provide voice and data services beyond the LATA. The largest and most recognizable IXC is AT&T. Other major carriers are MCI Worldcom, Sprint, and LCI. IXCs are carriers that build, support, and maintain their own networks. There are other players in the long distance market. Resellers purchase bulk discount long distance services and resell them to other businesses. They add a percentage cost to each call, thereby realizing a profit. Aggregators are companies that resell the service of a single IXC. For instance, a company might sign a large contract with AT&T, having negotiated a low corporate rate. The aggregator will then re-market AT&T service to other businesses.

The most fundamental service provided by IXCs is basic long distance service, also known as direct distance dialing (DDD). The long distance market is highly competitive. After divestiture, rates declined steadily for a number of years. It was only a few years ago that this trend bottomed out and rates slowly began to increase. Still, the market remains fiercely competitive and marketing campaigns often offer specials that virtually give the service away to new customers. Because the market is so competitive, telecommunications departments have much leverage in negotiating favorable rates for their companies. It is also very easy to change long distance carriers; an order is issued to change the PIC on local lines and long-distance carrier is changed immediately. Often, LECs will make the change based on a verbal order. IXCs also offer access to foreign locations in the form of international direct distance dialing (IDDD).

Because the long distance business is so competitive, and because it so easy to change services, long distance carriers have tried to counter this by offering attractive rates in exchange for long-term commitments. For a long period of time, AT&T rates were regulated and more expensive than other IXCs. AT&T countered this situation by introducing special tariffs that provided customized pricing for individual customers. A tariff is a public document that is filed by a telephone company with either a PUC or the FCC, specifying services and pricing. The document is not legally binding and is only accepted by regulatory bodies, not enforced by them. Tariffs can be struck down in a court of law, but only when an interested party initiates a lawsuit. AT&T offers Tariffs 12 and 15 as their customer specific agreements. Other IXCs followed suit and it is now common (at least with medium to larger-sized businesses) to negotiate low rates via long-term contracts.

Long distance service is billed in a variety of ways, but there are fundamental elements that are common to all IXCs. There are normally fixed costs in the form of a monthly service charge, access charges, or both. This is normally the smallest portion of the bill, the largest being usage.

Access to long distance services has a direct relationship to rates. The most basic form of long distance service is known as switched access, where the LEC enters the PIC at the request of the subscriber for local telephone lines. Changing the PIC is an easy undertaking, and most LECs will make the change through a verbal order. This has created a problem for telecommunications departments because unscrupulous aggregators and resellers will telemarket their services to various departments and branch locations of a company. Very often they will call the LEC and order a PIC change, based on a conversation, but not necessarily a sale. The process is known as slamming and legislation is currently being considered to make this practice illegal, but the practice is still prevalent. The PIC can designate basic DDD service or a special tariffed service.

Companies that use high volumes of long distance service can achieve lower rates if they subscribe to dedicated access products. A T-1 is used to directly connect the subscriber's telephone system to the IXC POP. In the case of switched access, each CO trunk line typically will carry a nominal service charge of $5. In the case of dedicated access, the T-1 will be a large expense. The lower rates usually offset the cost of the dedicated circuit. IXCs prefer the dedicated access for their high volume customers because it is more difficult to change services.

The rate structure of long distance is based on time and distance. In traditional DDD rates, the longer the distance of the call, the more expensive it will be. Calls made during the day will be more expensive than evening rates, and night/weekend rates are usually the least expensive of all. The telecommunications department also needs to be concerned with the billing increment. A company may propose lower rates, but the billing increment may be 60 seconds, which means that every call is rounded to next highest minute. For example, a call that is three minutes and 10 seconds long would be billed at four minutes. If this concept is applied to thousands of business calls, it could negate much of the savings. In addition to the billing increment, the type of rate needs to be considered. Rates can be either distance sensitive or postal. Distance sensitive rates are less expensive for short distance calls and more expensive for long distance calls. Postal rates are a flat rate, regardless of the distance of the call.

Long-term contracts can offer attractive rates for companies but there are a few caveats. First, the negotiated rate is usually based on current market conditions. Five-year contracts are very common, but five years is a lifetime in the long distance business. The rates that look attractive when the contract is negotiated could very well become expensive halfway through the contract, based on market conditions. A second problem is that long-term contracts often carry minimums. The telecommunications department needs to negotiate a safe margin, so that call volumes do not fall under the minimum. If this happens, a company will pay for calls that were not made. A third problem is that long-term contracts carry a termination liability. The subscriber is usually liable for the remaining number of months on the contract, should it be prematurely canceled. This is an amount equal to the monthly minimum.

In addition to offering various rate plans, IXCs also offer virtual private networks (VPNs), which offer a myriad of features in addition to discounted rates. For instance, a company may employ a proprietary dialing plan, allowing their employees to dial seven digits from one company site to another in lieu of a standard 10-digit long distance number. Features such as speed dialing are also offered in addition to authorization codes to prevent abuse. Account codes offer the capability to identify usage per department or individual. In essence, a virtual network allows a customer to use the IXC network as if it were their own private network. As described in Chapter 1, larger companies used to build private networks based on point-to-point leased lines. T-1 and T-3 lines were also employed. If properly designed and managed, these networks provided a low cost per minute. VPNs offered the same functionality through dedicated or switched access, and the point-to-point leased lines were eliminated. The IXCs were also able to duplicate the features that were inherent to private networks.

VPNs offer more attractive rates than other long distance services. There are two basic types of calls: on-net and off-net. An on-net call is an intracompany communication, an off-net call is to a location that is not on the VPN. Dedicated access to the VPN (e.g., T-1 access) offers lower rates than switched access. Both offer full functionality; however, users must first dial "700" from the switched locations in order to access the VPN. Each IXC markets their VPN under a different name: AT&T Software Defined Network (SDN), MCI Vnet, and Sprint VPN Premiere.

In addition to basic switched and dedicated long distance service and VPNs, many companies may still be using an older service known as wide area telecommunications service (WATS). Basically, WATS is a bulk discount service. Prices are based on monthly service charges, installation charges, access line fees, and usage charges. WATS can be outbound or inbound (toll-free). AT&T has maintained a substantial presence in the inbound market because they held a monopoly on toll-free services until only a few years ago. Portability is now possible for toll-free services and AT&T is facing increased competition.

Telecommunications departments are also often asked to provide calling cards for their companies. This can range from ordering individual cards for various employees to maintaining a large database of cards that are part of a large corporate agreement. Calling cards can also be part of a VPN plan and offer many of the same features and functionality. Supporting calling cards in a large corporate environment can be a large administrative task. Cards need to be issued, ordered, canceled, and tracked. There are also security and abuse issues that need to be addressed.

WATS used to be marketed as banded services. That is, there were specific geographical areas of coverage offered at different price ranges. Billing for banded WATS lines was complex and required careful analysis. For example, Band One provided coverage of the states closest to the state of origin and offered a low cost per minute. Band Five covered the entire country, but at a more expensive rate. Depending on the carrier offering the WATS product, rates could also be usage sensitive. In other words, increased discounts were factored for increasing levels of usage. Telecommunications professionals needed to analyze company calling patterns and determine what mix of banded WATS would yield the best cost per minute. WATS prices were also affected by the access method, which could be dedicated analog circuit, T-1 access, or the more recent switched version (a PIC on local CO lines). The banding concept eventually came out of vogue and was replaced by the virtual concept, which meant that no calls were blocked and pricing was distance sensitive. WATS are no longer actively marketed by the major IXCs, but telecommunications staff members need to be aware of the concept and the terminology in order to support legacy systems and marketing programs offered by smaller carriers.

Toll-free services have become a common and essential business tool in the later decades of the twentieth century. They are used for order entry, service, and information gathering or dissemination. Toll-free numbers generate, maintain, and protect revenue and there are businesses (e.g., catalog sales) that rely on them as their primary link with their customers. As the economy continues to become more service based, toll-free numbers are an essential tool for many businesses. Consequently, the concept of the call center has evolved; that is, a network service, business site, and telephone system that are specifically designed and managed to answer a large number of calls. Businesses that utilize complicated call center designs require telecommunications professionals with specific expertise. Consequently, many telecommunications departments have developed the position of call center specialist. These people maintain a high level of knowledge that is exclusive to toll-free numbers, ACDs, ACD reporting systems, and issues such as staffing and call center business practices. A quick glance at toll-free usage will indicate the growing importance of these services. National toll-free usage increases by a factor of one billion calls per year.

At a basic level, toll-free services are specific types of numbers that are offered to the general public (either 800, 888, or 877). The concept is unique from other long distance services in that the call is charged to the subscriber of the toll-free service, not the caller. As in the case of outbound long distance, there are various rates structures that are associated with toll-free services. Toll-free services are often "bundled" with other long distance services and rates are gravitating toward flat rates with time-of-day pricing (i.e., day, evening, night, or weekend). When services are bundled, all types of long distance traffic are written into a master agreement with the IXC. With the deregulation of local markets, local services can also be included in the mix.

Toll-free numbers are portable. It is now possible to change carriers without changing the long distance number. Up until 1993, toll-free numbers were exclusive to carriers. If a subscriber wanted to change carriers, there was no choice but to change toll-free numbers. Because toll-free numbers generate, maintain, and protect revenue, many companies were reluctant to do this. Today, a subscriber can obtain toll-free service from any responsible organization (RespOrg). The RespOrg assigns toll-free numbers to a customer or client, and this assignment is often registered with national directories. A RespOrg can actually be any type of organization. They are registered with the services management system 800 (SMS/800) and pay a fee for the right to access the national database. Typically, a RespOrg will be an IXC.

Access to toll-free services is either switched access or dedicated. Switched access means that the toll-free number is directed to either a local telephone or ISDN line. Dedicated access is via a point-to-point leased line, or T-1 access. Switched access rates are typically more expensive than dedicated access. Information is often sent through the public network to the access links so that specific callers, or types of calls, can be recognized and processed by the ACD.

There are a myriad of features associated with toll-free numbers. In fact, in companies with multiple call centers, where toll-free numbers service a mission-critical function, toll-free services become a separate network with dedicated personnel and support criteria. The following features will demonstrate the plethora of capabilities that modern toll-free services provide. A casual glance at these features will give an indication of how complex a toll-free network can become.

The single number concept is simple: One number services all locations (see Figure 2.9). This is accomplished by the use of routing features; area code routing, exchange routing, and 10-digit routing. Consider that a call center may be located in Connecticut and services the New England states. The area codes that service New England are programmed by the IXC to route to the Connecticut call center when the toll-free number is dialed. Continuing with the concept, when a caller dials the same toll-free number from a state in the mid-West, the call is routed to a call center located in Chicago. By employing this type of routing technology, the company has provided a one number

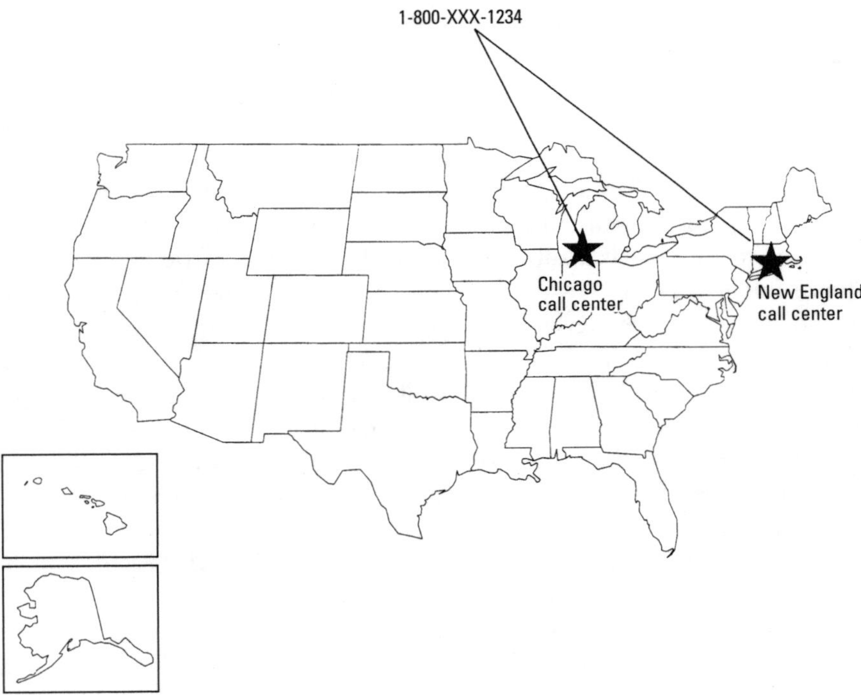

Figure 2.9 A single-number toll-free network.

service. This has many benefits for a company. For example, advertising campaigns only require one number. This is also easier for the customer to remember. Because sales and service territories do not always dovetail over area code coverage, more precise routing can be obtained through exchange, or 10-digit, routing. In the case of these features, an area code can be split and routed to two different call centers.

Continuing with the example shown in Figure 2.9, companies can increase hours of coverage by employing time-of-day routing. When the Connecticut call center closes at 5 p.m. eastern time, calls from the New England area can be routed to the Chicago call center. Even more efficiencies can be gained by employing day-of-week routing. For example, Saturday may be a day when the company receives relatively few calls. However, there is still business conducted. All calls in the United States can be routed to a single call center on Saturday. The company still operates, but at considerably less expense. Day-of-year routing offers the same efficiencies for specific days of the year; for instance, the day after Thanksgiving.

Calls can also be routed from a call center under emergency situations. Once again, consider the example shown in Figure 2.9. Assume there are three call centers listed in this example. If one of the call centers becomes overburdened with calls, the subscriber can either completely reroute traffic to another call center or employ percentage routing. In this case, a percentage of calls can be distributed to other call centers from a disabled call center. In addition, all-trunks-busy routing can automatically route calls if the trunk access is busy. Ring/no answer routing can provide an alternate route if a call is not answered.

Toll-free services can also be configured to accept international calls. Exactly what number is called is dependent upon the country where service is being offered. The NANP is a dialing plan that is exclusive to North America (i.e., Canada and the United States). Other countries employ a variety of dialing plans. Subscribers need to work with their IXC to determine if international coverage is available from a given country and what the dialing plan will be. The International Telecommunications Union (ITU), an international organization that helps to develop global telecommunications standards, has recently developed an international toll-free numbering plan. The plan would call for a 3-digit NPA, a 4-digit exchange, and a 4-digit subscriber code. In theory, the number would work from any country and would provide inter- and intra-country routing. At the time of this writing, this service is in the developmental stage.

Telephone system features have found their way into toll-free networks. For instance, a company can subscribe to a service whereby a call can be transferred to another location or a conference call can be initiated. These features are executed by the subscriber dialing a feature code. IXCs can also offer features such as queuing and en-route announcement. In the case of queuing, a message is delivered to callers when all trunks are busy. In this case, the toll-free network offers a feature that is usually provided through an ACD. In the case of en-route announcement, a message may be delivered to the caller before the call is delivered to the subscriber's site. Messaging (e.g., night service) and voice mail are also available.

As stated in Chapter 1, the lines that distinguish voice and data communications are blurring. One very prominent example of this blurring is transaction processing. The concept is to have customers enter their own orders via telephone. Also known generically as interactive voice response (IVR), this application interfaces a toll-free number directly to a computer. The customers are prompted to enter digits via the touch tone pad on their telephone. This application can also be used to obtain general information or account status. For example, the banking industry implemented this technology on a wide scale basis so that customers could check account status without speaking with a live person. Voice recognition technology can also be provided by the IXCs to service callers who still have rotary telephones. Service bureau functions can be provided by IXCs through toll-free services. For example, a polling service can be used to advertise a number and, through the feature of transaction processing and/or IVR, the IXC will log the responses from the callers. Political organizations have been large users of this function.

Toll-free services also offer the capability to identify either the caller or a type of call. Dialed number identification service (DNIS) is used to identify different types of calls. For example, two toll-free numbers might be published, one for English speaking customers and one for Spanish speaking customers. Both toll-free numbers share the same trunk group (see Figure 2.10). The IXC forwards specific digits to the ACD so that the dialed number can be recognized and the call can be routed to the call center agents who can best service them. Automatic number identification service (ANI) is used to identify the caller. The CLID is forwarded to the ACD and the caller's telephone number is used to route the call to a specific agent (or group of agents). The CLID can also be downloaded to a computer database and the call and customer profile can be simultaneously offered to the agent (known as screen popping). The fusion of the ACD and computer is known as computer telephony integration (CTI). This concept also offers the capability to transfer a computer screen and telephone call at the same time, in addition to simultaneous voice and data conferencing.

In the data arena, IXCs provide both digital and analog services. Dedicated analog services are offered via point-to-point leased lines or multidrop lines. Point-to-point lines are conditioned to offer more error free rates. They are typically connected with analog point-to-point modems. A multipoint line

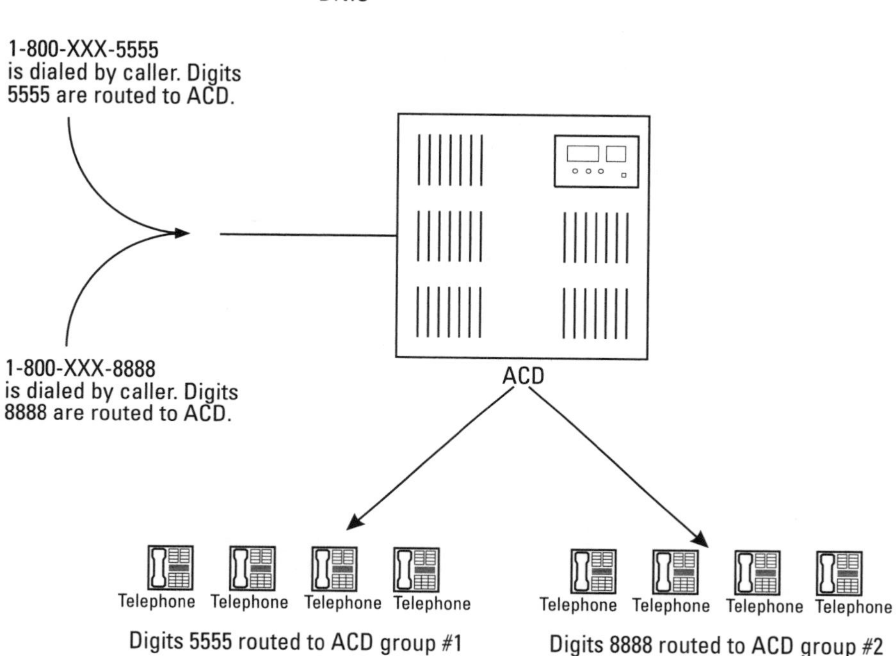

Figure 2.10 Dialed number identification service (DNIS)

circuit (see Figure 2.11) can be thought of as a tree with branches. Usually less expensive than a standard point-to-point line, this circuit services many sites (or drops) from the same circuit. IXCs offer T carrier services, the same as LECs. Since these services are billed according to mileage, T-1 circuits are far more common than T-3, which tend to be expensive. Also, few companies require T-3 bandwidth for their WAN applications.

Packet switching networks have been offered by IXCs for a number of years. Before the popularity of TCP/IP, telecommunications managers struggled with connecting disparate computer systems. An additional problem was connecting small offices or single terminals from foreign locations. X.25 is a set of standards set forth by the ITU to connect various types of computing devices through a packet switched network. Packet switching breaks data into blocks and assigns addresses to the blocks so they may be routed through the network. The interface to the network is known as a packet assembler/dissasembler (PAD). When data is sent through the X.25 network, it may take a number of routes, but is reassembled before being delivered to the final destination. Packet switching networks also offer protocol conversion for disparate systems and are

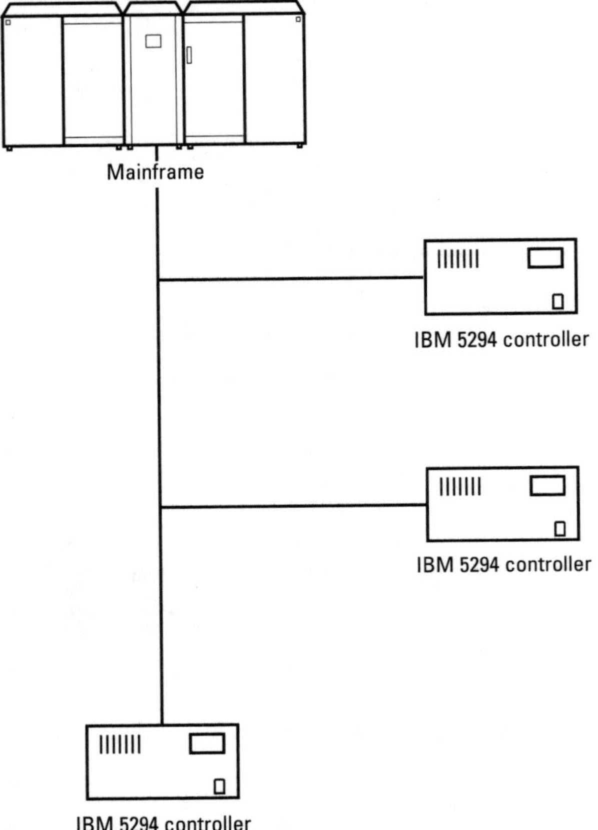

Figure 2.11 Multi-drop circuit.

available in many foreign countries. Packet switching networks are often referred to as value added networks (VAN).

A service that has gained quick acceptance in the business world for high-speed data communications is frame relay. This service is also offered by LECs, but frame relay has found its most practical application in the wide area network (WAN), the domain of the IXCs. This is due to the high transmission capacity of frame relay (as opposed to analog leased lines) and its cost saving potential over its analog counterparts. A basic concept of frame relay is to provide a digital access port (56 or 64 Kbps) that is provided by an LEC or CAP to the IXC. The port speed is the maximum potential speed of a frame relay circuit. The subscriber then chooses a committed information rate (CIR). This is an essential component of frame relay pricing. Subscribers can select a CIR from zero up to the port speed. The CIR is the guaranteed rate, but many companies will opt for a lower CIR which is less expensive. The reason for this is that frame relay will "burst" up to the port speed when the IXC network is not congested. If the network becomes congested, then the IXC will reduce the speed of the frame relay circuit to the CIR. In order to do this, they often discard packets of data. There is an inherent degree of risk with choosing low CIRs. For example, a zero CIR would be the most cost effective alternative. However, a zero CIR guarantees nothing. The subscriber is gambling that the network will never be congested, and the frame relay circuit will always burst to the port speed. Many IXCs, in fact, do over-engineer their networks so that congestion is indeed rare. Unfortunately, no IXC will guarantee that a low CIR will never discard packets. Conversely, because of the bursting capability of frame relay, many companies obtain high capacity data circuits at a reduced price. The design of frame relay networks involves a balance of price vs. performance, but there is also a degree of calculated risk.

It is important to understand that a frame relay circuit is not a dedicated circuit. The data is segmented into frames. This is also referred to as a "best effort" transmission because a series of packets sent through a frame relay network may actually take different routes during the course of the day before arriving at the same destination. There is an address at the front of the frame that guides the packet through the network. In this case, the network is thought of as a cloud (see Figure 2.12), because there are no dedicated channels and the circuits are all virtual. The data link connection identifier (DLCI) is a circuit number of a frame relay link. Each site on the frame relay network has a DLCI and network design and operation requires the inclusion of the DLCI into the routing tables in data communications devices. An advantage of frame relay is that sites can be added or deleted to network routing software. This offers a service that is less expensive and more versatile for the telecommunications staff.

As in the case of T carrier services that require specific multiplexing equipment, frame relay requires a specific physical interface. Stand alone units are called frame relay access devices (FRADs). Other devices such as a router can have the frame relay access built in.

Frame relay has become a popular WAN service with many companies. When first introduced, many companies were hesitant to place all of their traffic on this service because of the "best effort" concept and the risk of losing packets of data. In addition, legacy types of data traffic such as IBM's Systems Network Architecture (SNA) required continuous communications between the remote site and the host site. Under SNA, if the host site does not receive continuous confirmation data, there is a session time out. Consequently, many companies developed two networks; one comprised of dedicated point-to-point lines for legacy systems, and one comprised of frame relay for IP traffic. Router

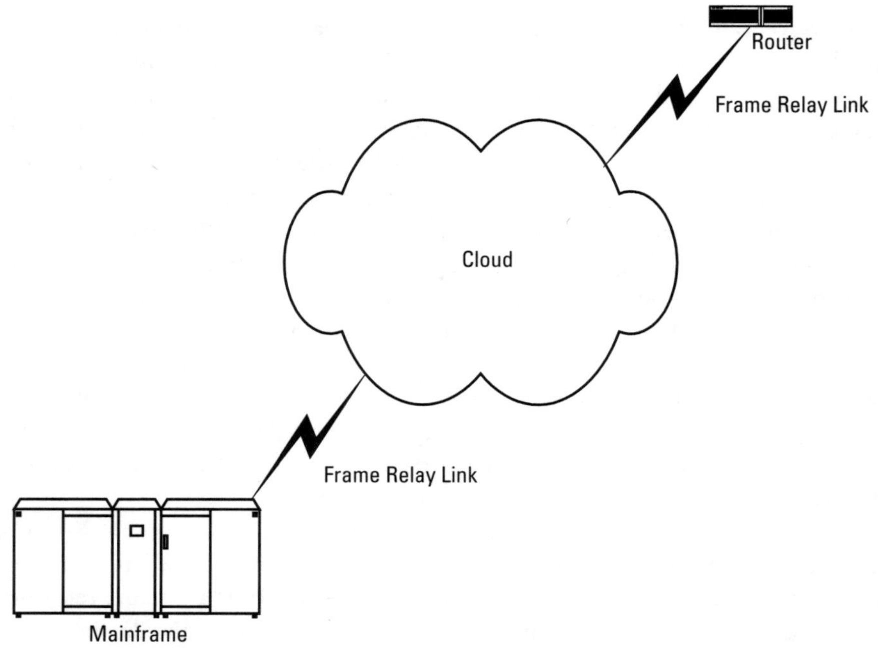

Figure 2.12 Frame relay.

vendors addressed this issue by developing a software feature called datalink switching (DLSw), which allowed for the continuous communications that is required for SNA.

Frame relay offers many advantages, but the reason it has become so popular is probably more related to economics than anything else. Over the past few years, many telecommunications managers have saved considerable amounts of money by changing from leased lines to frame relay. Frame relay pricing can actually be very simple. There is an installation charge, a port charge, and a permanent virtual circuit (PVC) charge. The higher the port speed or CIR assigned to the PVC, the higher the cost of the frame relay service.

An added advantage of frame relay is that it also can carry voice communications. Once again, this demonstrates how voice and data communications are continually merging. The choice to use voice over frame is almost always an economical decision, for voice communications is highly reliable and there is seldom a need to change long distance carriers because of poor transmission quality (although international calling may be different). By inter- leaving voice and data over a PVC that is already paid for, it is possible to obtain free long distance service. However, there are trade-offs. Because packets may be discarded, this could have a devastating effect on voice communications. In addition, performance can be affected when large amounts of data are transmitted through a frame relay network.

Frame relay is categorized as a fast packet technology. Included in the same category is a network service that has gained much notoriety, but has not been as widely deployed: asynchronous transfer mode (ATM). ATM is a transmission technology that was developed to carry all types of traffic: voice, data, and video. Frame relay can carry various types of traffic, but this is not why it was developed. As previously mentioned, there are caveats that must be considered when various types of communications traffic are blended over frame relay. ATM, on the other hand, is a robust transmission technology that offers capabilities interleaving various types of traffic.

ATM is offered as a WAN service by the larger IXCs. At the time of this writing, the service is offered in only a limited number of major cities and is expensive. Data communications devices, such as routers, require an ATM interface in order to connect to this service. Only companies with high bandwidth requirements find use for this service. At the present time, most companies find frame relay to be sufficient.

2.4 Telephone Systems and Providers

The telecommunications department is responsible for providing telephone systems which encompass a variety of requirements, ranging from very basic systems that provide minimal functionality to complex ACDs that require constant and complex analysis. Telephone systems can be classified into four basic categories: key systems, PBXs, Centrex, and ACDs. Adjunct equipment can provide additional functionality in the forms of IVR or voice mail.

Telephone systems are purchased from three possible types of providers: manufacturers direct, distributors, or LECs. A manufacturer direct is a product marketed directly by the manufacturer and this type of vendor offers a number of advantages. First, the entire company is structured to produce and support the same type of product. Also, a high degree of expertise resides within the entire organization and the overall experience level is strong. The largest examples of manufacturers direct for PBX systems are Lucent Technologies and Siemens. When technical or support issues surface, the entire organization is better able to address the issues, in lieu of a distributor that may not have the experience, direct connections with R&D and technicians, or the same levels of expertise. The downside to using a manufacturer direct is that they may be constrained by their own product line. For instance, a company may have committed to manufacturing and marketing large PBX systems, but offers no products for smaller applications. They may also be lacking in certain applications. A distributor could offer a full range of products (albeit from different manufacturers) in order to cover all possible applications. When one manufacturer begins to lag behind in either technology or support, the distributor can seek other products. Conversely, when manufacturers realize deficiencies in their product lines they often have two choices; to either upgrade the current product line (which can be very expensive) or have another company manufacture product as an other equipment manufacturer (OEM). When equipment is offered as an OEM, the manufacturer direct takes on the role of a distributor, even though the product is stamped with their company name.

A distributor markets the products of other manufacturers. Distributors can vary greatly in quality. There are quality, stable organizations with years of experience and expertise, and there are companies that can only be described as "fly-by-night." However, there are many manufacturers of quality systems that rely completely on distributors to market their products. There are pluses and minuses to both manufacturer direct and distributors. When a product and vendor are selected, the telecommunications department must weigh the quality of the product against the quality of the vendor. This process will be covered in more detail in a later chapter.

Key telephone systems (KTS) are normally marketed for smaller and simpler business applications. Key systems are not as feature rich as PBXs, are less expensive, and are easier to maintain. Technological and business trends, however, are changing this situation. Because smaller businesses are demanding more, and because of technological advancements, smaller systems are now approaching PBXs in features and functionality. More advanced key systems are also commonly known as hybrid key systems. While this nomenclature is common in the field of telecommunications, key systems have become very powerful in recent years and many approach full PBX functionality.

Historically, key telephone systems were generally installed for applications calling for 50 stations or less. In addition, a distinguishing characteristic was the visual functionality. For example, a user might have a series of buttons on his telephone set, known as line appearances. In order to access an outside line, the user would simply depress a button, seize dial tone from the CO, and dial the number. In a PBX, the user would pick up the telephone and dial an access code (typically a 9), and the PBX would be programmed to select the next available CO trunk line. Modern hybrid key systems, however, offer such features and so definitions have blurred considerably.

The basic components of a key telephone system are the key service unit (KSU), telephones, a main distribution frame (MDF), and CO trunk lines. The KSU is the intelligence of the system. It is the central processing unit (CPU) that is a minicomputer. As is the case with a PBX, the system is computer-stored and program controlled. The physical appearance of the KSU is of a cabinet that holds electronic circuit cards. Each of these cards serves as a specific interface to the key telephone system. For example, CO trunk lines interface to CO line cards and telephones interface to station cards. The power circuit also is installed in the rack, and many systems offer redundant power supplies as an option in the event of a primary power-pack failure. Cards have a specific capacity, normally expressed as a ratio. For instance, a station card that is "eight on one" would offer eight ports for telephone connection on a single card. Trunk circuit cards can be either analog or digital. Smaller systems are now offering ISDN interfaces, either BRI or PRI. Typical CO trunk lines are analog. Station cards are also offered as digital or analog.

The circuits cards are installed on a shelf that resides in a cabinet. Key systems have a maximum capacity that is expressed in stations and trunk lines. If the system capacity is expressed as 12 by 24, it would mean a maximum capacity of 24 stations and 12 trunk lines. Once the maximum capacity of a system has been reached, it often means that the system will have to be replaced. Understanding the capacity of a telephone system is a critical responsibility of the telecommunications department. Otherwise, systems require premature replacement because they were not sized properly.

Software controls the features of the key telephone system such as station features (line appearances and feature buttons), and system features such as speed dial lists, call pickup groups, hunt groups, and long distance routing. As in the case of any modern telephone system, a customer has the option of being trained and making his own adds, moves, or changes, or paying the vendor a fee to perform the changes on a case by case basis. Smaller installations often require few changes, and it is often less expensive to pay the vendor for infrequent changes than to pay for a class plus travel expenses. In addition, when changes are infrequent, the telephone system administrator often forgets how to use the system.

Key systems can also be installed as an interface to Centrex. As explained earlier, Centrex is a partitioning of the LEC CO. The subscriber is offered basic telephone features, such as transfer and conferencing. Unfortunately, Centrex has traditionally offered only analog telephone sets, which requires the end-user to use a switch hook and dial feature codes in order to execute system features. Many end-users prefer electronic or digital telephones that offer feature buttons. By installing a key system behind Centrex, the telecommunications department can provide feature rich telephone sets.

Telephone sets on any system can be divided into two basic categories: analog and digital. An analog set is a basic set, very seldom offering feature button functionality. The most basic business set is the basic analog 2500 set.

Because we speak in analog and the signal from a microphone is analog, there needs to be a conversion from analog to digital in order for the telephone system to be able to process the information. This is accomplished via the use of a coder/decoder (CODEC), which samples an analog signal at 8,000 times per second. The very same technology is used for digital PBXs. In modern digital telephone systems, there is an analog to digital conversion from an analog telephone to the CPU and digital to analog conversion when the signal is sent over an analog trunk line.

Digital telephones offer features such as digital display, multiple line appearances, and feature buttons. Analog telephones are much less expensive than their digital counterparts, but most businesses prefer digital because they provide a more user-friendly environment for the end-users. Basic features that are commonly offered on digital telephone sets via one-button functionality are:

- Call transfer to another party;
- Call hold;
- Three-way conference calling;
- Consultation hold (one call is placed on hold while another call is initiated);
- Station speed dialing;
- Speakerphone capability.

Some of the major manufacturers of key telephone systems are Nortel, Lucent Technologies, NEC, Inter-Tel, and Telrad, Siemens, and Tie Systems. Key systems typically cost approximately $500 per port. A port can be defined as either a station port or a trunk port. This base figure can be higher if advanced features or peripheral equipment such as voice mail is employed.

PBXs are used for larger applications than the key system and offer higher levels of sophistication and more capacity. PBX is the generic name applied to large telephone systems. Some manufacturers have marketed systems under the guise of other names such as private automatic branch exchange (PABX), computerized branch exchange (CBX), and digital branch exchange (DBX). Essentially, these names all mean the same thing and are vendor specific. The most commonly used term is PBX and will be used throughout this book.

The basic components of the PBX are similar to those of the key system. There is a CPU, telephone sets, cabinets with shelves that hold circuit, logic, and power cards, an MDF, trunk lines, and an operator console. In the case of the key system, the operator console is not always used. In the case of a PBX (because it supports much larger applications), the operator console is almost always employed.

A modern PBX can be thought of as a computer. When the system is initially installed, it is programmed and the configuration is stored on a hard drive, in much the same fashion as a PC program. Modern PBXs are digital and changes are often made via software. For PBX installations, there is often a person at the company site who is designated to be the system administrator. This also often entails overseeing cable technicians who may need to install new cable to support new telephone or trunk line installation. Software changes are then made to facilitate new telephones, features, or routings.

As in the case of a key system, various models of PBXs have size limitations. Maximum capacities are expressed in terms of the maximum amount of stations and trunk lines. When maximum capacity is reached, the system might need to be upgraded to a larger version (known as a forklift), or a complete system replacement might be in order.

Because the PBX supports large and more complex applications there is a wider variety of trunk line interfaces that are used. CO trunk lines are normally set up as two-way circuits (although separate groups of inbound and inbound trunk groups are sometimes employed). DID is also used to offer direct dial capability to individual employees and take much of the call volume away from the operator console. In large installations, this can be an overwhelming task, if not impossible for the operator to answer calls for every employee. Other trunk line interfaces may be in the form of analog tie lines or T-1 access to other PBXs, or as an interface to an IXC POP for long distance service. PRI may be employed as an access for toll-free services.

The telecommunications department is responsible for designing the PBX configuration and the corporate network. Consequently, the PBX software plays a primary role in controlling long distance and local costs. In a PBX environment, users do not control where, or how, a call will be routed. Routing tables in the PBX software are programmed to route the call over the least expensive route. If a T-1 interfaces to the IXC, offering a very low cost per minute, the PBX should be programmed so that all long distance calls attempt the T-1 route first. If the T-1 is busy, an alternate route may be offered, or a busy signal may be given back to the caller.

As in the case of the digital key system, PBXs also offer analog and digital sets. Analog is, once again, less expensive, but digital sets are feature rich and easier to use. Most companies opt for digital sets because they enhance productivity and minimize user error.

The major players in the PBX market are Lucent Technologies, Nortel, and Siemens. Typically a PBX is priced at $1,000 per port, but because the market is so competitive, the per-port price will usually come in at much less.

Centrex can be thought of as more of a telephone service rather than a system. Provided by LECs, the CO is partitioned via software to offer basic telephone system functionality. There is an installation charge, a monthly service charge (also called the common equipment charge) and a per-line subscription charge. The subscriber is responsible for purchasing his own station equipment, which can be basic 2500 sets. Because Centrex is the use of a CO switch, one of the drawbacks has been the availability of a digital multibutton telephone set. Recently ISDN BRI has addressed this limitation.

Centrex is often thought to be expensive, ranging from $20 to $25 per line. Combined with the purchase of station equipment and the common equipment charge, Centrex costs can indeed be more expensive than a key system or a PBX if the operating and capital costs are calculated over five years. However, the picture is not that simple. First, many LECs are now offering long-term contracts with rates as favorable as $10 to $12 per line. Second, there are some inherent advantages to Centrex. Consider that the CO has an advanced and well-designed disaster recovery plan. In the case of a PBX, the telecommunications department must often purchase an uninterruptable power supply (UPS) in order to maintain communications during a power outage. This is an inherent feature of Centrex at no extra cost to the subscriber. Software upgrades to the CO are done by the LEC at no charge to subscriber. Also, in the case of a campus environment, where many buildings are serviced by the same CO, Centrex can save on network service charges and cabling costs.

Peripheral devices are also sold to augment PBX functionality. One of the most commonly used devices today is voice mail. Voice mail is essentially a large digital recording device that interfaces to the PBX or key system via a tie line interface. As in the case of the PBX, voice mail systems are computers that are programmed and administered by the end-user. In the modern business world, voice mail has found acceptance as a productivity tool. It is estimated that nearly 50 percent of all business calls miss their intended target on the first attempt. Voice mail is a method whereby an employee can capture messages without the intervention of a secretary. It interfaces to the PBX so that a message waiting light is lit when a message is left, and the caller can also exercise options such as "zeroing out" of the greeting to be forwarded to a secretary.

Voice mail systems are provided according to their capacity and feature set. Capacity is measured in disk space and number of trunk lines. The PBX administrator also programs the voice mail, which is often provided by the PBX manufacturer. Most of the major PBX manufacturers offer a voice mail system.

Automated attendant service is a concept whereby a device front-ends the PBX. The caller hears a menu, inviting them to dial the intended party's extension or to use a company menu. The caller dials the first few letters of a name on their touch tone pad and normally the person's name will be spoken back to the caller. The system will ask for confirmation and then direct the call to the intended party.

IVR has also gained in popularity in recent years. In essence this is a processing device that connects to a PBX and also to a computer. While automated attendant service ultimately is used to connect people to people, IVR is used to provide computer-based information to people. Common applications are people accessing a bank to check on account balances.

The PBX is used to support basic administrative business needs. When large volumes of calls are directed to a specific group of people, the PBX becomes a call center. As stated earlier in this book, the concept of the call center has become an essential tool in the service-oriented economy. A critical component of this concept is the ACD (automatic call distribution). ACD can be offered either as a stand-alone unit or as an integrated software package for a PBX. ACD capability requires more processing power than a basic PBX. The concept is also fundamentally different from a basic PBX extension.

In an ACD application, groups of ACD extensions are programmed into functional groups. These are also known as gates or splits. When calls arrive at the ACD, the calls are routed to one of these groups which is comprised of agents. The ACD is programmed to calculate what agents have had the most calls and what agents have had the least amount of calls. Calls are theoretically always routed to the least busy agent.

The difference between an ACD station and a basic PBX station is that when an ACD agent arrives for his shift, the first thing he does is log onto the system. By logging onto the ACD, he is telling the system, "I am here to accept calls." In theory, a call should not be offered to an unattended station. A PBX station receives a call whether the station is attended or not. If the call is offered to an unattended station, the call is subject to PBX call coverage parameters (ring/no answer or busy—forward to another station or voice mail).

If no agents are available to answer a call in an ACD environment (because they are on the phone or unavailable), the call is queued. Under these circumstances, the ACD delivers messages indicating that all agents are busy. The caller then hears music on hold (MOH) and possible subsequent messages. If other ACD groups are available, a queued call can be overflowed to another group.

ACD activity is recorded through the ACD management information system (MIS), which includes a monitor that displays real-time activity within the call center and records daily, weekly, and monthly reports. The reports record activity, which includes number of calls, service levels and abandoned calls for trunk groups, ACD groups, and individual agents.

Stand-alone ACDs are powerful systems that have nearly all processing power dedicated to ACD capabilities. The largest suppliers are Rockwell and Aspect. These systems, while more powerful and versatile than ACD provided via a PBX, are also more expensive. Price per port can range from $3,000 to $5,000.

Call center applications, which include ACDs, are usually the most difficult voice communications applications to support for the telecommunications department. Complex toll-free networks and ACDs require constant analysis and tweaking. This is also one of the most highly visible applications because it is normally a direct conduit to the customer base and a concept that generates, maintains, and protects revenue.

2.5 Local Area Networks

When PCs were introduced to the business world, it was inevitable that they would need to be connected, within the building, the campus, or throughout the corporate network. On a local basis, the local area network (LAN) became a concept that was eventually adopted by virtually every business. LANs introduced more complexity into corporate data networks and a myriad of new duties to the telecommunications department. New data communications protocols needed to be understood, in addition to new network devices, and structured cable systems. Unlike the days of slave-host data communications, the networks were more complex, more unpredictable, and more expensive.

There are various types of LANs available on the market today. The two most popular are Ethernet (originally developed by Xerox) and Token Ring (developed by IBM). Defining a LAN can be difficult, but there are five key elements that are common to all LANs:

- High communications speed;
- Very low error rate;
- Geographically bounded;
- A single cable system or medium for multiple attached devices;
- A sharing of resources, such as printers, modems, files, disks, and applications.

High communications speed has traditionally been defined as millions of bits per second. Although LAN speed is defined in 10-16 Mbps, emerging Fast LAN technologies are offering speeds of 100 Mbps and higher. Lower speed LANs are very common today. Ethernet typically runs at 10 Mbps and the newer versions of Token Ring run at 16 Mbps. Higher speed LANs such as Fast Ethernet or 100 VG AnyLAN are now being offered in the form of 100 Mbps capabilities. Because LANs transmit over private cable, they are not constrained by the limitations of the public network, which was designed primarily for voice transmissions. In addition, there is a very low error rate because the transmissions are localized and therefore subject to less interference. LANs are geographically bounded, meaning that they usually do not reach beyond the walls of a building. There are also distance limitations for transmissions. If devices are installed beyond the prescribed limitations, typically the signal will attenuate and fail. In these instances, the signal needs to be regenerated through the use of a line modem.

LANs are associated with specific types of cable and connecting devices. For example, a common form of Ethernet is also known as 10BASE-T. The data rate is a baseband (meaning digital as opposed to broadband, which means analog) of 10 Mbps over twisted pair wire, hence the "T." There are other forms of Ethernet, including 10BASE2 which operates over a thin wire form of coaxial cable. The physical connecting device for 10BASE-T is a network interface card (NIC) that is installed in a PC equipped with an RJ-45 connector (basically a telephone plug with eight wires in lieu of four). The Token Ring cable is a special, unique adapter known as the Token Ring adapter (although an RJ-45 Token Ring interface is available).

LANs offer the capability to share devices such as printers, modems, or larger, high powered PCs known as servers. This is done through special software called a network operating system (NOS). The NOS is installed on a server, a high capacity PC that serves as a control center for the LAN via the NOS. NOS software is also stored on each PC connected to the LAN. The NOS offers security capabilities (users need a password to log onto the LAN), messaging (e.g., confirmation of print jobs being processed), e-mail, and file sharing. LANs are used to connect desktop PCs to other desktop PCs, to file servers, to mini or mainframe computers, to other LANs, and to WAN-based services. Novell, Banyan, and Microsoft are some of the larger companies that produce NOSs for LANs.

LANs are designed to have both a physical and a logical topology. The common physical topologies are bus, ring, star, or a combination of these topologies (see Figure 2.13). A logical topology is how the signal travels on the wires. Logical topologies are also bus, ring, or star.

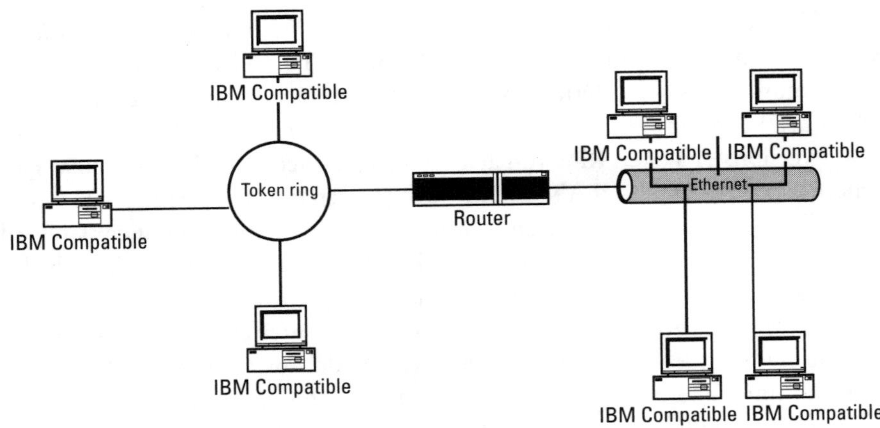

Figure 2.13 Token ring and ethernet connected via a router.

The physical medium that supports LANs is a structured cable system. Because LANs have become larger and more complex, it became necessary to design cable systems that would support existing devices, future expansion, and physical moves. This entailed a project for the telecommunications department that encompassed the design of the cable system, the installation, and the ongoing management. Management of the cable system included physical moves and documentation (if the size warranted it). Because every desktop is usually serviced by a telephone and a PC, many organizations now install a universal structured cable system that supports both voice and data.

There are five categories of cable that can be used for LAN applications: unshielded twisted pair (UTP), shielded twisted pair (STP), coaxial cable (50 Ohm), coaxial cable (75 Ohm), and fiber optic cable. UTP is the least expensive and is provided in various grades which also provides different levels of performance. STP offers higher performance levels than UTP because each cable pair is wrapped in a metallic sheathing. It is also more expensive than UTP but performance is enhanced greatly. The 50 Ohm coaxial cable is also known as thin-wire and is used on two forms of Ethernet: 10BASE-5 and 10BASE-2. Both of these types of Ethernet are continually being replaced by 10BASE-T. The 75 Ohm coaxial cable is the same type used for old IBM 3270 terminals and cable television. This thick, black cable is used almost exclusively for broadband networks. Fiber optic cable has not found wide acceptance for most LANs, however, it is used considerably for interconnecting LANs as the physical layer for network backbones (see Figure 2.14).

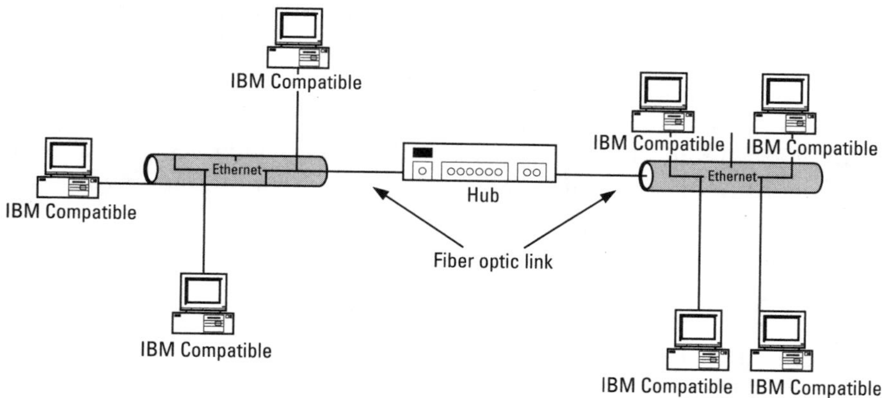

Figure 2.14 Hub connecting two ethernet LANs via a fiber optic backbone.

The Electronics Industries Association/Telecommunications Industry Association (EIA/TIA) has defined UTP cable standards for use in LANs. Known as the 568 standards, the standard defines five levels of UTP and one type of STP. Organized into levels or categories, Categories 1 and 2 are specified for voice transmissions, while Categories 3, 4, and 5 are specified for LANs. Most modern LAN installations use UTP as specified by the EIA/TIA standards.

LAN cable, in most modern installations, terminates on a hub. On a very basic level, a hub can be thought of as a wiring concentrator. One of the very first hubs was marketed by IBM for Token Ring called the multistation access unit (MAU). The MAU acted as a wiring concentrator and manager, but hubs were soon developed into intelligent devices that held electronics and software. The reason for this was that as a simple wiring device, a failure of one PC connected to the LAN could bring down the entire LAN. With an intelligent hub, a failed link could be eliminated so that the rest of the LAN could remain functional.

Modern hubs reflect a myriad of features and capabilities. They can act as repeaters or provide media conversion capabilities to connect different LANs (e.g., Ethernet to Token Ring). The most advanced hubs are called intelligent hubs or LAN switches. These switches offer the capability to connect a large number of LAN segments. This will be covered in detail in the next section.

In baseband LANs, all users access the same channel. Therefore, techniques must be employed that allow for a sharing of the channel. This is accomplished via LAN standards called media access control (MAC). LAN standards are defined by the Institute of Electrical and Electronics Engineers (IEEE) in the 802 series (see Table 2.3).

LANs, like all data communications technologies, have specific data protocols. A data protocol can be thought of as a specific set of rules, procedures, or conventions relating to timing and format of data transmissions between two devices. Two data devices must have the same protocol in order to communicate. If the devices do not use the same protocol, then a protocol conversion must be performed.

The protocols of LANs occur in Layers 1 and 2 of the open systems interconnection (OSI) model (see Table 2.4). Layer 1 is divided into the physical protocol (PHY) and the physical medium dependent (PMD). This is the physical interface level that defines such factors as transmit and receive power levels. Layer 2 is also divided into additional layers, the logical link control (LLC) and the MAC. The LLC layer defines how the LAN frame is formatted and addressed. In data communications networking, data is divided into functional units called packets, frames, or cells. At a most basic level, data is made of bits that are formed into data words known as bytes. For transmission purposes, the bytes of data are formatted into a functional units for transmission. In the LAN environment, a frame is used to transmit data. Each frame has a header and trailer that is used to find the intended destination point. Continuing with the two most popular LANs, Ethernet utilizes a standard transmission method known as carrier sense multiple access w/collision detection (CSMA/CD) as defined in IEEE standard 802.3. Token Ring uses Token Passing (IEEE 802.5).

Table 2.3
IEEE 802 LAN Standards

Standard	Subject
802.1	Describes various LAN standards and their relationship to the OSI model.
802.2	Defines LLC and MAC standards.
802.3	Defines bus standard for CSMA/CD.
802.4	Defines Token Ring bus standard.
802.5	Defines ring configuration for token passing.
802.6	Defines MAN standards.

Table 2.4
OSI Reference Model

7. Application
6. Presentation
5. Session
4. Transport
3. Network
2. Link
1. Physical

In the case of CSMA/CD, all users share a single channel on a contention basis. Dividing the concept into two parts, CSMA can be explained as each device "listening" to the channel. If the channel is idle, the device will send frames. Included in the frame is a 48 bit address, which directs the data to its intended destination. When a frame is transmitted, it is broadcast to all stations on the LAN. The address determines which station should receive the data. The station will store the data in a receive buffer (a buffer is a method of storing data), and all other stations will ignore frames that are not addressed to them. It is also possible to use a broadcast address that sends a frame to all stations.

The CD portion is a method of controlling LAN traffic. If two stations transmit at the same time, it is known as a collision. When a collision occurs, all stations cease transmission immediately and begin a restart sequence. A random number is then generated by each station that is used as a timer. When the timer expires, each station begins to "listen" for an idle channel. Because the idle state is generated by a unique random number for each station, the stations attached to the LAN do not all start at the same time. CSMA/CD also offers a feature known as cyclic redundancy check (CRC), which allows for error detection. When an Ethernet LAN becomes congested, it is typically subdivided into two LANs and interconnected with either a bridge or a router. The telecommunications department may, or may not, be responsible for the administration of LANs. Very often, this is a function of a separate department that oversees PCs, LANs, servers, and printers. Regardless, the segmentation of LANs requires interconnection that is the responsibility of the telecommunications department.

Token passing is the protocol used for Token Ring. Under this system, a station attached to the LAN sends a "token," which is a series of bits specially formatted for transmission through the LAN. The token is sent to each station on the LAN. Each station has the opportunity to send data with the token. If not, the token is simply passed on to the next station. One station on the LAN is designated as the control station, also known as the active monitor. If a token is corrupted, the active monitor will resend it. Data cannot be sent from a Token Ring station until it receives a token.

There are a number of other LAN technologies available, however, Ethernet and Token Ring have become the most popular. While telecommunications departments may, or may not, support LANs directly, they must understand how they work and how they are interconnected. This includes knowledge of legacy systems and emerging systems.

Ethernet (10BASE-T) and Token Ring are considered to be low-speed LANs in today's business environment. New fast LAN technologies offer bandwidth in the 100 Mbps range. These are deliverable technologies; however, LANs in the gigabit range are being tested and will soon be available. Fast LANs are used either as a backbone technology or for high-speed station requirements. Backbones can be thought of as major arteries for corporate networks. They are the links that carry the heaviest traffic loads. Network backbones are more permanent parts of the network. LANs may be added or removed, but the backbone usually remains a permanent fixture.

There are also workstations that are far more powerful than the average PC, such as the SunSparc workstation manufactured by Sun. Workstations are used for applications such as computer-aided design/computer-aided manufacturing (CAD/CAM). These types of computers are designed to work with large graphics files. For example, an automobile manufacturer may have engineers generate a three dimensional design of an engine. This large graphic file, comprised of millions of bits of data, may be sent through a LAN to another engineer for review. On a low speed LAN that is shared with other users, the transmission of such a file would probably bring the LAN to a standstill. When such files are transmitted, the only alternative is to meet the demand with bandwidth.

There are three basic technologies that have found acceptance in the fast LAN category: Fiber distributed data interface (FDDI), Fast Ethernet, and 100 VG/AnyLAN. To classify FDDI as a fast LAN might be misleading. FDDI has actually found more acceptance as a backbone technology to interconnect LANs. For example, many organizations that have campuses use FDDI as the backbone for interconnecting all buildings. ATM may also be classified in the fast LAN category, because it also has high bandwidth capability, in addition to being able to carry multiple types of traffic, such as voice, video, or data. Of the aforementioned fast LAN technologies, FDDI enjoys a large installed base. Fast Ethernet is rapidly being deployed by many companies because it is cost effective, in addition to being familiar technology. This makes it very attractive to many telecommunications managers. Because 100 VG/AnyLAN has limited installations, this does not necessarily reflect upon the quality of the technology. Many promising technologies do not gain wide acceptance for a number of reasons. For a variety of reasons, 100 VG/AnyLAN has not gained the widespread acceptance of FDDI and Fast Ethernet. While ATM may be classified as a fast LAN technology, it may be better to define it as a versatile transmission technology that provides a number of capabilities to the telecommunications industry.

FDDI can be thought of as a 100 Mbps Token Ring designed to operate on fiber optic cable. Copper distributed data interface (CDDI) was developed to provide the same capability on a copper medium, however, the majority of installations are fiber based. With FDDI, a token is passed along the fiber to devices that are connected on the ring. It should be noted that any device connected to an FDDI ring must have a specific FDDI interface. Usually these devices come in the form of bridges, adapters, or concentrators. FDDI bridges are commonly used to connect various LANs (up to 80). Prices range from a modest $3,000 to in excess of $40,000. There are actually two types of stations that can be attached to a FDDI ring: class A/dual attached stations (DAS), and class B/single attached stations (SAS). DAS provides for dual connectivity for a single station. SAS offers a single connection.

Redundancy is inherent in the design of FDDI. There are two counter-rotating rings. While it is possible to attach a device to both rings, typically the second ring is used as a spare, in the event of a primary ring failure. FDDI is designed to automatically reconfigure itself when a failure occurs in the primary ring. Transmission automatically switches to the spare ring.

FDDI has a number of strengths and weaknesses. Regardless of the weaknesses it presents, it is still a deliverable and established technology. For many telecommunications professionals, this says a lot. A great danger in choosing technology is that it may become obsolete quickly. When this happens, the product is no longer manufactured and technicians are no longer trained. Support of the obsolete technology then becomes a nightmare.

Fast Ethernet is also known as 100-BASE-T. Because it is built on the 10BASE-T standard, it has become a logical migration path for many companies that have standardized on 10BASE-T. Fast Ethernet utilizes the same CSMA/CD technology and also runs on UTP. In comparing Fast Ethernet to FDDI, one obvious advantage to Fast Ethernet is the ability to use existing house cable rather than installing expensive fiber optic cable.

The product 100 VG/AnyLAN offers a 100 Mbps capability and will run on fiber, UTP, or STP. The service utilizes a different method of access from the aforementioned technologies called demand priority. In a 100

VG/AnyLAN environment, stations are connected to intelligent hubs. A station on the LAN sends a request to transmit, either high or normal priority. A round- robin system is then used to decide which station has the right to transmit next. High priority requests will take precedence over normal requests.

A single LAN becomes a node on a corporate network. Interconnecting LANs becomes a primary responsibility of the telecommunications department. LANs are often administered locally, just as a PBX might be, although telecommunications departments might carry this responsibility. Interconnection of LANs to each other and to WANs is the responsibility of the telecommunications department. There are two fundamental components required to build a LAN: NIC cards and cable. The cost to connect a PC to a LAN, considering software, hardware, and cable, is generally between $100 and $700.

2.6 Data Communications Equipment: The Connecting Devices

LANs provide the network connectivity on a local basis. However, once there is a need to connect to something outside the building, or from LAN to LAN, the signal must traverse some form of data communications device. For LAN interconnection there are four basic devices that are used: repeaters, bridges, routers, and gateways. The roles of these various devices are more clearly illustrated by examining the OSI Reference Model. The International Organization of Standardization (OSI) developed a model in the 1970s for computer network compatibility. The model (see Table 2.4) divided computer networking into sub-tasks, or logical groupings. The OSI reference model defines seven logical tasks. In order for two computer devices to communicate, all seven tasks must be compatible.

Notice at the physical layer, the interconnecting device between LANs are repeaters, media converters, or range extenders. As illustrated in the previous section, when a device reaches specific distances from the LAN, the signal will attenuate and fail. At the physical layer, the signal can be regenerated through these devices so that the LAN may be extended.

Bridges operate at the second layer, known as the link layer. A bridge links similar LANs, such as Ethernet to Ethernet or Token Ring to Token Ring, either within the same facility or over the WAN. Translating bridges, however, can link dissimilar LANs (e.g., Ethernet to Token Ring). Bridges filter and forward frames between two or more LANs. This is done on a store and forward basis via the Layer 2 addresses found in LAN frames. In the LAN environment, every station connected to the LAN sees every frame. If the address attached to the frame is not intended for the station, the station will ignore it and pass it on. When a bridge connects LANs, it looks at every station on every LAN to which it is connected.

Bridging is a relatively simple and inexpensive method of networking. A bridge connects as a station to a LAN and can range in size from two to 30 ports. When a bridge is connected it must "learn" what stations are on the LAN. The source address of each NIC card is monitored and then stored in a forwarding table. Once the bridge has learned all stations that it is interconnected to, it will check the destination address of a frame in order to determine what LAN the station is on. The LAN to which the bridge is a station is ignored, a process known as filtering. When a frame is intended for a connected LAN, the frame is buffered and then forwarded.

Bridges are often used to segment LANs. When a LAN becomes congested, it is a common practice to divide it into two segments so that there is not as much traffic traversing each segment. Two segmented LANs at 10 Mbps each offers better performance than a single large LAN at 10 Mbps. Beyond segmenting, bridges are also used as a basic LAN interconnecting technology. Local bridging is when LANs are connected within the same facility. Remote bridging is when the bridge is connected to a network service, such as point-to-point private line or some other WAN-based service.

Bridging becomes problematic when, for instance, a data frame is not defined in the address table. The bridge will attempt to locate the station by transmitting the frame to every port on the network. Known as flooding, if the station in question is found, the address will be added to the routing table and data will subsequently be transmitted. Flooding can become a major network problem when a destination address cannot be located. In such instances, large bursts of data flood every possible link in the network, rendering the network inoperable. Known as a broadcast storm, this is one of the major caveats of networking with bridges.

The basic premise of bridging is to connect two networks. Beyond two connections, errors and broadcast storms become common. These problems are addressed via the use of two possible software features: spanning tree algorithm (STA) and source routing. STA defines a root path and standby path. Source routing employs the use of an explorer packet which learns the best route. Unfortunately, source routing does not alter routing during transmissions, even if the network becomes congested.

There are a number of problems that telecommunications departments encounter with bridges. There are the challenges of interconnecting dissimilar LANs, varying frame sizes, network delays, and several data protocols that cannot be routed via bridging. The most common examples are Digital Equipment Corporation's LAT and IBM's NetBIOS.

Because of the limitations inherent to bridges, the preferred method of interconnecting LANs (at least on the local level) has become intelligent hubs and LAN switches. Hubs and LAN switches, just like bridges, work at Layer 2 of the OSI model. A large advantage of hubs over bridges is that intelligent hubs and LAN switches are designed to be high capacity network nodes, capable of transmitting millions of bits of data per second. More recent versions have back-planes in the gigabit range. Because bridges are most often basic PCs fitted with specific bridge cards and software, they offer potential bottlenecks on high capacity networks.

As illustrated in the previous section, hubs began as dumb devices (e.g., the IBM Token Ring MAU) that were more of a wiring concentrator and manager than an intelligent device. A distinguishing feature between "dumb" hubs and an intelligent hub or LAN switch is that dumb hubs work at Layer 1 of the OSI model. Another fundamental difference is that intelligent hubs offer a management port. That is, a port that interfaces to specific LAN management software that will provide traffic statistics and error rates. Monitoring this LAN traffic becomes a key task of the telecommunications department, making recommendations for segmentations, or the implementations of fast LAN technologies as traffic patterns dictate. LAN switches are more sophisticated technology than intelligent hubs. They are distinguished from intelligent hubs by two basic capabilities: store and forward and cut-through capability and the capability to support multiple MAC addresses on each port.

Store and forward capability is simply a method of storing data and forwarding the transmission if the destination is on a different LAN than the LAN switch. If the destination is on the same LAN, the data is filtered. The LAN switch has a table of MAC addresses stored, which is used to determine what addresses should be forwarded. LAN switches can also translate LAN address formats (e.g., Ether to Token Ring). Cut-through capability is a real-time transmission as opposed to storing the data prior to transmission.

Intelligent hubs and LAN switches offer the capability of creating a virtual LAN (VLAN). A virtual LAN is defined in software, and theoretically, any LAN station can become part of a virtual LAN, regardless of type (Token Ring vs. Ethernet) or location. Some of the major vendors of intelligent hubs and LAN switches are Bay Networks, Cabletron, Cisco, and 3Com.

Routers have come to be one of the most common networking devices being used in modern corporate networks. Routers work at Layer 3, the network layer of the OSI model. They are more sophisticated devices than bridges and are also more expensive. Unlike bridges, all routers have the capability to interconnect dissimilar LANs. Whereas bridges will send all data traffic, routers have the intelligence to differentiate what traffic is sent to various links in the network.

Routers have the capability to route multiple types of protocols. The most common type in use today is the Internet protocol (IP), the same protocol used on the Internet. Up until this point, data traffic has been described in functional units such as frames and cells. IP traffic utilizes a unit called a datagram. Routers will extract the IP datagram from a LAN frame and wrap the frame in a WAN protocol. As data travels through a router-based network, it can be translated many times as it traverses a multitude of LANs and various WAN-based services.

In order to manage routers, it is necessary to understand various types of protocols. Transport control protocol/internet protocol (TCP/IP) is the protocol of the Internet. It operates at Layers 3 and 4 of the OSI reference model. TCP/IP is actually a robust series of protocols that offers the capability to interconnect many dissimilar types of computer networks. Understanding different data communications protocols and how to program routers to connect dissimilar computer systems via LAN or WAN services is a fundamental responsibility of the telecommunications department.

When designing and supporting router networks, telecommunications professionals are heavily involved in IP addressing. In a routing network, each device (e.g., workstations and server) will have a unique routing address. An IP address is a 32 bit binary number divided into 4 octets (an 8 bit byte). There are three basic parts to an IP address; the address class indication, the network ID, and the host ID. IP addresses are typically written with decimal points separating each octet. For example 132.45.3.3.

A central group, the Network Information Center (NIC), is responsible for assigning network IDs. There are three classes of IP licenses: A, B, and C. The classes of licenses relate to the size of the network where the addresses will be used. A network ID is assigned to a specific class. The user is then responsible for assigning the host ID within the network ID. Companies can develop their own proprietary brand of addressing, however, they cannot connect their addressing scheme with the Internet.

The network ID designates a specific network to a router, or group of routers. In this case, a network can be a myriad of devices that comprise the entire network. The host ID identifies specific devices, such as workstations or servers that reside on the network. Each device requires a unique host ID. Host IDs are defined by the telecommunications department. A router will only process a host ID if the intended host device is attached to it. The Host ID may also be divided into subnetworks. Defining subnetworks is also a responsibility of the telecommunications department.

Routers contain routing tables, which are programmed by telecommunications department staff members or a chosen vendor. These tables determine what data traffic will be sent through the network as opposed to a bridge that will forward all traffic. In addition, routers can also send traffic to an alternate route. For example, in a bridged network, a failed circuit means that the bridge cannot send data down the failed path. A router can send data to an alternate path, if available, and update its own routing tables. This allows for redundancy in a network design.

As previously stated, bridged networks are often susceptible to broadcast storms. Address resolution protocol (ARP) is a feature of TCP/IP designed to locate LAN addresses for incoming IP datagrams. When a router receives a datagram, it will try to match the IP address with the LAN address in its tables. If it cannot make this match, it will try to discover the LAN address via ARP. The station with the corresponding address will recognize the ARP command, send an acknowledgment to the router, and the new address will be placed in the router tables. Inability of the router to locate a LAN address can result in an ARP storm.

There are a number of routing protocols employed in router use. While they are marketed with various names, they all fall into two basic categories: distance vector protocols and link state protocols. In a distance vector protocol, the router chooses the best route via a distance vector. The router stores a number of hops (routers within the network) and will always pick the route with the least number of hops. Routers are also capable of "learning" how many hops are on a network via distance vector protocols. This process, known as advertising, entails each router sending its table to every router on the network. This is a timed parameter. The most common distance vector protocols in use today are routing information protocol (RIP), internet packet exchange (IPX), and Xerox networking services (XNS). Distance vector protocols are sophisticated but they are not without limitations. Network capacity is consumed during the advertising process, the protocol does not always use the fastest or least expensive route, and network failures can involve long recovery times.

Link state protocols differ from distance vector protocols in a number of ways. First, advertising does not take effect. Rather, when the network is changed, routing table updates are executed. This capability frees the network from the constant transmissions that occur with advertising. While the distance vector protocols use the number of hops in the network to determine the best route, link state protocols use a more complex and sophisticated criteria for determining a route including cost, security, delay, and link speed.

Some examples of major manufacturers of routers are Cisco, Bay, Proteon, Cabletron, and 3Com. Routers are provided in a variety of configurations by various manufacturers, ranging from small, simple devices to ones that are hybrid router/hubs capable of supporting multiple types of LANs. There are varying software levels for routers that allow for prioritization of data traffic along with supporting multiple types of data protocols. Certain types of routers are also used for remote access and LAN connectivity. This allows employees to dial into a corporate network with their PCs. Routers are equipped with various physical interfaces, software levels, and through-put capabilities. Depending on the configuration, prices for routers can range from as low as $500 for a simple branch office router to $80,000 for high capacity backbone routers.

Network management capabilities are available through specific protocols that monitor network activity, errors, or a failed link. The simple network management protocol (SNMP) is part of the TCP/IP suite of protocols. Remote network monitoring specification (RMON) provides a similar capability. Many of the companies that manufacture data communications devices also market network management software, such as Bay Networks Optivity or CiscoWorks. These systems provide a real-time depiction of network activity in addition to historical statistics. For example, when telecommunications staff members see that utilization on a particular frame relay link is getting high, they may opt to increase the CIR.

Routers, intelligent hubs, and LAN switches are provided from manufacturers directly or through distributors. Under the rubric of distributors is a new type of vendor in the field of data communications known as a systems integrator. A system integrator may supply devices from a number of manufacturers. What they are specifically selling is their expertise in terms of network design, installation, and support. Such organizations may carry a whole force of field engineers to support and install the various products. They also usually develop a long-term relationship with a company. Many systems integrators require that a company sign a long-term contract. This situation has pluses and minuses that will be discussed in later chapters. For smaller organizations with little expertise, systems integrators can provide valuable expertise and support. Even in large organizations, they can be valuable for short-term projects that require complex analysis and increased manpower that the telecommunications staff does not have.

Continuing with basic data devices, a gateway is a device that is specifically designed to convert one data protocol to another so that different computer systems can communicate (e.g., IBM to DEC). This conversion takes place at Layers 6 and 7 of the OSI Model. The gateway strips off the header and trailer and translates the information so that it is compatible with another computer system. Gateways are installed as front-end devices to mid-size and mainframe computers.

A final and yet very common data communications device is the modem. Modems come in two basic varieties: dial-up and point-to-point. A dial-up modem is used for data communications requirements that are normally of short duration. A dial-up modem incurs long distance charges and there is always the possibility of connectivity problems. Point-to-point modems connect dedicated point-to-point circuits and are less expensive than dial-up connections. Because they are connected 100 percent of the time, they are also more reliable.

Inter-networking various computer systems and LANs presents a multitude of problems for the modern telecommunications department. When different systems are connected (e.g., Token Ring to Ethernet) a protocol translation must be executed by an interconnecting device. Because the size of the frames differs between various transmission systems, this often entails fragmentation and re-assembly of frames. Network addresses may also need to be translated. In addition to the translations that need to take place, a corporate network is comprised of various types of data communications services that run at varying speeds. A flow control capability must also be used to ensure that buffers do not overflow. When data transmissions encounter varying network speeds, buffering is required to allow for the differences. There must also be capabilities for end-to-end reliability, part of which is error detection and correction.

In addition to connectivity issues, each device must be inventoried. The devices are either purchased or leased, and software upgrades are common in order to provide enhanced capabilities or to fix bugs. Devices also need to be replaced or upgraded as business requirements change across the WAN. Because data communications devices require programming, there are also security issues that need to be considered. All of these issues equate to a myriad of duties on the part of the telecommunications department.

2.7 The Internet and Providers

There is no technology in the later part of the twentieth century that has been more hyped than the Internet. The proliferation of PCs, the advancement of multimedia technology, and the increase in computer literacy amidst the general population have all contributed to the explosive growth and popularity of this technology. The Internet can be defined as a global network of computers connected with standard protocols, the most common being TCP/IP. As described in Chapter 1, the origins of the Internet can be traced to ARPANET, the computer network that was originally commissioned by the Department of Defense (DOD). Today, Internet coverage is global with the exception of a small group of Third World countries.

Data is routed through the Internet via a connectionless adaptive routing system. Packets of data are dynamically routed throughout the network, selecting the best available route. Because routing is dynamic, if a server goes off line, it will be bypassed. Packets on the Internet can take a number of routes and pass through numerous systems before reaching the final destination. The destination address keeps the data on course and the protocols allow for the dynamic routing.

There are two types of vendors that support Internet services: Internet service providers (ISPs) and backbone providers. End-users connect to ISPs via dial-up (e.g., basic POTS or ISDN) or dedicated connections such as T-1. The role of ISPs is to provide service to end-users and to connect end-users to the Internet backbone. Backbone providers provide connectivity between ISPs. There are also regional carriers such as CERFnet in San Francisco, which provide robust networks within a specified region. These regional providers also connect to the backbone.

The most common way for small or individual users to connect to the Internet is via modem through standard analog telephone lines. Larger corporate users opt for high-speed connectivity via T-1 or T-3 access. Prices for Internet access are always in a state of flux due to fierce competition between ISPs. Up until 1996, Internet access was normally usage based. In October of 1996, America Online (AOL) offered unlimited usage for $19.95 a month. Since this initial foray into unlimited access, ISPs have experimented with various plans. On a basic level, there is a start up fee, a monthly service charge, and sometimes additional fees for various enhanced services. ISPs are constantly experimenting with their fee structures, trying to market attractive packages to gain new users, and balance those fees against large users that are hardly ever profitable. At the time of this writing, the norm is a fixed monthly fee that offers unlimited usage.

When an individual or company subscribes to Internet service, they are assigned an Internet name and address. The name is actually a linkage to a numbered address. Names can be, and often are, selected by the end-user. It can be the name of a company, person, or a catchy name used for marketing purposes. The name is followed by a top level domain name that relates to the type of organization that is subscribing to the Internet service. For example, .com is used for companies, .org is used for organizations, and .net is used for networks. This system is known as the domain name system (DNS) and is administered, at the time of this writing, by Network Solutions, Inc. (NSI).

The Internet provides a variety of capabilities to modern business. On a very basic level it is an opportunity to provide information to the general public via computer. Companies provide a web site and publish the address, inviting Internet subscribers to dial into the server. On a basic level, a web site is simply a PC (although it may be a very powerful one) that is connected to an ISP. Companies may have multiple web sites that provide a variety of functions. For example, a company may market a product that is highly technical in nature. Because use of the product requires a high level of technical expertise, the company needs to constantly educate the customer base. One method of providing this information is via a web site. A customer can access the site and choose from a menu of various subjects. When the customer finds the subject in question, the technical paper can be read on line or printed for future reference. Libraries of information can be stored on web sites, allowing people to access vast amounts of information without having to leave their desks.

In addition to providing services to the public, many companies provide web sites to their own companies, known as an intranet. An intranet is a proprietary network, designed to only service members of an exclusive organization, such as employees within a company. In theory, an intranet does not have to use licensed TCP/IP addresses. Companies can develop their own addressing scheme. The problem is that they cannot connect through the Internet. Intranets are often used to provide information to the employees such as human resources. A common feature of web sites is to offer a section of frequently asked questions (FAQ). By providing a FAQ section, the human resources department can be freed from the mundane task of answering these common questions.

The telecommunications department supports Internet access in a number of ways. First, access may be provided in the form of network services and CPE. In a large business environment, this is often via T-1 access. Second, the department may be responsible for negotiating the ISP contract. This contract, however, is not always the exclusive domain of the telecommunications department. Third, the department is often responsible for developing the addressing scheme that will be used to interface to the Internet. This entails developing the naming conventions that will be used for corporate addresses and individual e-mail addresses. Lastly, the telecommunications department may provide a firewall that will reside on the server that interfaces to the Internet. A firewall is a combination of hardware and software that provides management and control capabilities for Internet access.

A firewall can be thought of as a boundary between two networks. Because the Internet is a public network, security becomes an issue when a company connects their corporate network to an Internet server. If no precautions are taken, hackers can easily access sensitive and proprietary information. A firewall controls Internet traffic, both outbound and inbound. It prevents outsiders from accessing proprietary information and it also prevents corporate users from accessing web sites that have no business purpose. One of the most obvious and common uses of a firewall is to prevent access to adult web sites. A firewall also provides management reports, which may expose employees who use the Internet for nonbusiness related activities. Each company has its own Internet policy, and this policy must be clearly conveyed to the telecommunications department so that the proper control and security measures can be implemented.

2.8 Cable

When the Bell system was divested, in-house cable was still the responsibility of the RBOCs. In a very short period of time, however, the Modified Final Judgment (MFJ) made cabling the responsibility of the end-user. Companies found that they had to generally support two types of cabling systems, voice and data. At the time of divestiture, these were generally thought to be two completely different disciplines. Voice communications was fairly simple and straightforward. This consisted of twisted pair cable that was normally included as part of the PBX installation. Data communications was a bit more complex. Because voice communications technologies were developed by the old Bell system, standards had been created for telephone cable systems. Computer systems, on the other hand, developed via the free market system. Therefore, there were different manufacturers, data formats (e.g., IBM's EBCDIC vs. ASCII, which was used by most other manufacturers), communications protocols, and cabling systems. In many companies, it was not unusual to see basic telephone cable installed next to coaxial cable. It was also not unusual to see much of this cable crudely installed. A hole was drilled in the floor, or a baseboard, while computer closets became a rat's nest of tangled wires.

Two factors began to quickly change this situation. First, technological advancements began to introduce innovative new products. For instance, the balun (an acronym for balance/unbalance) was a small, inexpensive device that performed a signal conversion. Data devices could then be installed on standard telephone cable, which was less expensive and cumbersome than thick, bulky coaxial cable. The second development was innovation based on marketing ideas. Cable companies began to spring up that specialized specifically in the field of communications cabling. They wanted innovative systems to market to their customers. End-users also favored cable systems that were uniform (supporting voice and data communications) in addition to being user-friendly. The concept that eventually evolved was the structured cable system.

In 1985, it was apparent that there were no industry defined standards for telecommunications cable. The Computer Communications Industry Association (CCIA) requested that the Electronics Industries Association (EIA) develop standards. The standard was published in 1991 as the EIA/TIA-568. Subsequent modifications to the standard were then published to include UTP, modular jacks, and patch cables. The purpose of the standard was threefold:

1. To establish a generic telecommunications cabling standard for commercial buildings;
2. To provide for a systematic approach to the planning and installation of a structured cable system for commercial buildings;

3. To establish technical and performance criteria for various cabling system configurations.

The EIA/TIA standards brought logical designs to a chaotic part of the corporate telecommunications environment. The standards specified minimum requirements for an office environment, recommended topologies and maximum cable distances, media parameters for performance, and connector and pin assignments to ensure interconnectivity. If the specifications were followed, it was estimated by the EIA/TIA that the useful life of a given cable system would be ten years. The EIA/TIA-568 standard broke a structured cable system into six sub-systems: building entrance cable, equipment room, backbone cabling, telecommunications closet, horizontal cabling, and work area.

Building entrance cable is the cable that enters a building from the outside. This cable can be provided through a local carrier or it can be private cable, installed via a private cable company. The cable might support LEC-or IXC-based services or it might simply be backbone cable that interconnects buildings in a campus or in a metropolitan area (e.g., a MAN application). If the entrance cable is provided by the LEC or IXC, it is usually maintained by those organizations. Private cable is the responsibility of the telecommunications department.

When the cable enters a building, it typically terminates in an equipment room. This is a room that is normally dedicated to telecommunications devices and cabling. It is also a room that should be kept under lock and key so that unauthorized personnel cannot tamper with cable or telecommunications devices. This will be discussed in further detail in a later chapter. The equipment room is usually set up to provide environmental conditions that are favorable to telecommunications equipment. This includes a dust free environment, specific levels of humidity, and exact temperature ranges. In addition to the electronic devices, there is also cable management equipment in the form of racks, patch panels, cross-connect blocks, and patch cords. The design in Figure 2.15 depicts a series of three racks. These racks are seven feet tall and mount standard 19 inch racks. Within the racks are patch panel fields. In this example, there are three patch fields in each rack, each having 48 RJ-45 jacks. Notice that there is a rack-mounted intelligent hub in rack one. A patch cable is used to patch from the hub to one of the jacks in a patch field. This provides an Ethernet capability to a workstation somewhere in the building. Behind the series of patch panels is a main distribution frame (MDF) which is a series of punch down blocks that are used to interconnect local telephone service to the PBX, and patch panels to the backbone cable. This is only one simple example of how an equipment room can be designed. There is no right or wrong way,

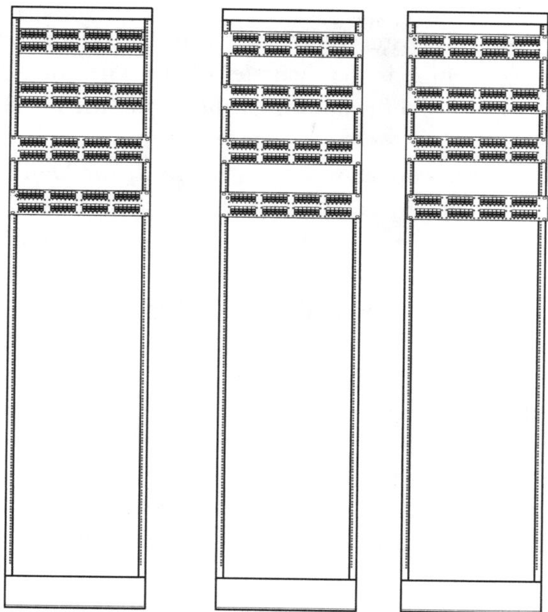

Figure 2.15 Rack-mounted patch panels for a structured cable system.

and in spite of the standards set forth by the EIA/TIA-568, there is much flexibility on the part of the end-user in designing an equipment room.

Backbone cabling connects equipment rooms, telecommunications closets, and buildings. The components of backbone cabling are the actual cables, cross-connect blocks, mechanical terminations, and patch cords or jumpers used for backbone-to-backbone connection. When backbone cabling connects floors within a building, it is known as riser cable. There are four basic categories of cable used for backbone cable:

1. 100 Ohm UTP (24 or 22 AWG) for a maximum distance of 800 meters;
2. 150 Ohm STP for a maximum distance of 90 meters;
3. Multimode fiber optic cable for a maximum distance of 2,000 meters;
4. Single-mode fiber optic cable for a maximum distance of 3,000 meters.

Ohm is a electrical term that is a practical unit of resistance. It is the electrical resistance that allows one ampere of current to pass at the electrical potential of one volt. AWG stands for American Wire Gauge, and the number stands for the gauge, which is the diameter of the wire core. The numbering system for wire gauges is confusing in that the higher the number, the thinner the wire.

In addition to the copper standards set forth for UTP and STP, fiber optic lines are very often used as a backbone physical medium. Multimode fiber is less expensive than single-mode fiber and has a thicker core. Its name is derived from the fact that it allows many modes of light to propagate down the same path. It is used for shorter distance applications and the connecting devices such as couplers are less expensive and easier to install due to the larger connecting surface.

Single-mode fiber allows only a single mode of light to propagate. Whereas multimode uses a light emitting diode (LED) to transmit a signal, single-mode fiber requires the use of a laser. It is also more expensive and difficult to install because it has a thinner diameter than multimode fiber. Single-mode is used for longer distances and higher bandwidth applications than multi-mode.

The telecommunications closet serves as a wiring hub to feed horizontal cable to the workstations. The equipment room feeds the telecommunications closet(s) via backbone cable. An equipment room can also serve as a telecommunications closet, and equipment can be installed in any telecommunications closet. The equipment room is usually where most of the main equipment is installed (e.g., PBX or router), while the various telecommunications closets in a building can house equipment such as intelligent hubs or LAN switches.

The telecommunications closet can actually mirror the layout of the equipment room. There is a distribution frame where punchdown blocks are installed, commonly referred to as the intermediate distribution frame (IDF). There are also racks equipped with patch fields along with patch cables that are used to cross connect from hub to station, or from backbone to station.

Horizontal cable is fed from the telecommunications closet to the work area. The EIA/TIA standards specify a 90 meter maximum length for cable, from the closet to the telecommunications outlet. The patch cord that connects devices, either voice or data, is specified at three meters. Horizontal cabling is defined for four components: the cable itself, the telecommunications outlet, cable terminations, and cross-connections.

Horizontal cable is almost always copper. It is connected from backbone cable to patch fields via cross-connect blocks. There was much publicity given to fiber-to-the-desktop several years ago, but the establishment of UTP categories has eliminated much of the need for fiber-to-the-desktop. Current standards for Category 5 allow for up to 100 Mbps of data to be passed within the specified distances. In addition, laboratory tests have passed data over UTP

Category 5 in the Gigabit range and beyond. Since few business users require bandwidth beyond 100 Mbps, Category 5 has proven to be sufficient for most applications, and is certainly less expensive than fiber.

The telecommunications outlet is a faceplate or utility box equipped with one or more terminations which are also known as jacks (see Figure 2.16). The horizontal cable terminates into this outlet on industry standard jacks. One of the most common jacks in use today is the RJ-45 jack which consists of eight copper wires. The RJ-45 jack can accommodate a telephone or most data connections. For example, Ethernet 10BASE-T uses a standard RJ-45 jack for connectivity. A very common design for the outlet is to provide one Category 3 cable and two Category 5 cables. In a standard business environment, most employees now require a telephone and LAN connection. Category 3 is often a standard used for voice communications while Category 5 is certified for data up to 100 Mbps. When designing a cable system, telecommunications professionals need to consider current requirements and future growth. Future growth includes bandwidth requirements and the number of jacks required to support devices above the requisite telephone and PC.

The final category of the EIA/TIA set of standards is the work area. This is divided into three components: station equipment (e.g., PCs, telephones, printers, etc.), patch cables, and adapters such as baluns. All devices are connected via some form of modular plug making adds, moves, and changes easily executed.

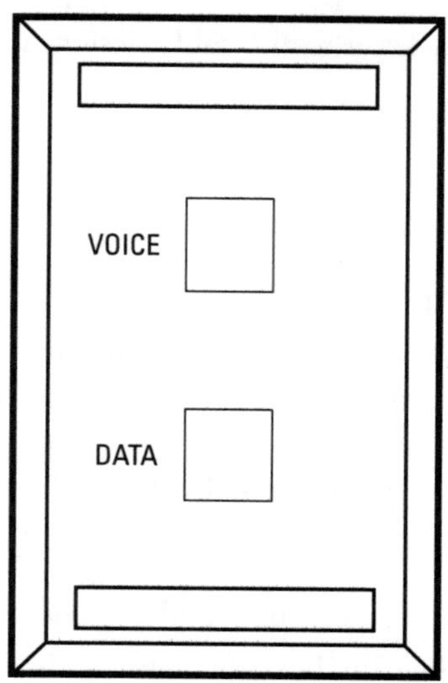

Figure 2.16 Station configuration for a structured cable system.

Structured cabling systems can range from the very simple (encompassing only a few dozen stations) to large campuses that interconnect dozens of buildings with thousands of stations. In the larger installations, the telecommunications department might carry a number of dedicated cable technicians to service the large number of adds, moves, and changes. In smaller applications, the telecommunications department might only make modular changes via patch cables and hire a cable company to perform the infrequent additions.

Cabling systems require procedures for making changes and subsequent documentation. If cable is not documented, and if changes are made on an ad hoc basis, the cable quickly becomes an unmanageable quagmire. The unfortunate result of cable mismanagement is that a major project surfaces to clean up the old cable system.

In recent years, many cable vendors have offered structured cable systems that are classified as "certified" or "warrantied." The premise behind this concept is that each component of the structured system is tested and the result printed. The customer is provided with a printout of all test results and the cable vendor guarantees that the system will support all transmissions as specified in the EIA/TIA 568 specifications. The certification and warranty do not cover damage to the cable system.

Cable vendors can range from a few people working out of their garage to established businesses that carry a staff of experienced and certified design engineers and technicians. Telecommunications professionals should be careful to select quality vendors that have both the experience and expertise to support modern structured cable systems. Experienced vendors bring valuable ideas to the table that can be beneficial to the cable design, implementation, and support process.

2.9 Wireless Communications

As outlined in the examples in Chapter 1, modern workers are becoming more mobile and the business environment is demanding accessibility, regardless of location. As a result, wireless capabilities are rapidly being developed that will support voice and data applications. Though there is still much development that needs to take place, it will one day be possible to provide what many consider to be the ultimate goal of wireless communications. That is, the ability to communicate to anyone or anything, anytime, anywhere. To get to this point, however, there is still much technical and regulatory work that needs to be done. For the present time, the telecommunications department must be cognizant of existing and emerging technologies. Department members must also understand the inherent limitations to the technologies and networks available today.

One of the most obvious drawbacks to present wireless technologies is the lack of ubiquitous coverage. When coverage is provided within a building or campus, it is via a proprietary wireless private branch exchange (WPBX) or wireless LAN. In the case of a WPBX, the building (or campus) is equipped with base stations that cover the facility with transmissions in the 900 MHz range. Proprietary telephones provide mobility throughout the entire facility along with PBX functionality such as transfer and conference features. In a WPBX installation, the wireless telephones are an extension of the telephone system. However, once a user leaves the facility, the signal is lost. A WPBX is a proprietary system, unlike commercially available cellular telephones that transmit through public cellular networks. WPBX telephones also employ technology that secures the transmissions from being tapped. In the case of analog cellular telephones that operate through public networks, a simple, inexpensive scanner can be purchased at electronics stores to intercept conversations.

WPBXs are installed to support mobile employees. For example, hospitals often have employees who are virtually never at their desks. Even if the employees are paged, without a WPBX, they must still find a telephone to answer the page. The WPBX gives them accessibility, regardless of location. The WPBX interfaces to the PBX via trunk lines. Typically, it is not a replacement for the PBX, but rather an adjunct capability because not all employees will require this kind of remote accessibility.

While the business justification for a WPBX is to support a mobile workforce, the justification for a wireless LAN is usually monetary in nature. Because the requisite adds, moves, and changes for cable can be costly, a wireless LAN allows for a station to be moved to any area within the facility without regard to cable. In environments where employees frequently move, a wireless LAN can save a substantial amount of money on cable costs. The process is also faster because there is no need to place an order and schedule technicians. In a wireless LAN environment, PCs are equipped with special wireless modems. Therefore, the end-user simply picks up the PC and moves it to a different location.

As in the case of the WPBX, the wireless LAN uses base stations. There are a number of designs that can be employed for wireless LANs, but all designs require the installation of base stations that send radio frequencies throughout the facility. Because the wireless LAN is engineered and designed specifically for the facility or campus, theoretically, there should be no dead spots.

Wireless LANs have gained slow acceptance in the marketplace because the initial products were expensive and offered relatively low bandwidth. There was also a lack of standards throughout the industry, which equated to a lack of interoperability between vendors. Network management capabilities were limited and the systems were highly complex to install and manage. Recent innovations have addressed this situation to a certain degree, but few industry segments have such a high volume of adds, moves, and changes to warrant such an expensive system.

There are two basic technologies used in wireless LANs: direct sequence spread spectrum (DSSS) and frequency hopping spread spectrum (FHSS). Of the two technologies, many industry experts consider FHSS to be a more viable technology. There are a number of major vendors in the wireless LAN marketplace, including Lucent Technologies, RadioLAN, and Xircom. Recent developments in the public wireless networks are offering data capabilities that may hinder wireless LAN growth. From a financial standpoint, the major difference between a wireless LAN and public network-based service is that air time on the wireless LAN is free once the system has been purchased. The same is true of the WPBX. Public network-based services incur usage charges. For employees who travel and only require limited access to e-mail or other computer applications, a public network offering can be a viable alternative. This may even be an alternative within a building or campus, if the local network offers sufficient coverage. Users who are on the line all day long, however, would incur large usage bills, probably making the network based service cost prohibitive.

WPBX and wireless LANs gave birth to a concept that received a degree of media attention in the early 1990s; the floating work-point. The concept is simple, but reliant upon wireless technologies. Each employee is equipped with a wireless telephone and a PC equipped with a wireless modem. Employees are free to wander throughout the facility, choosing virtually anyplace they wish to conduct business.

Beyond a customer's facility, network-based services begin with basic cellular telephone service. This has become a common service in most urban areas in the United States. It is inexpensive and used by business and residential consumers alike. Older cellular networks are analog and typically provided by ILECs. The transmissions are not secure and quality may be dubious, based on how well the network has been engineered to cover the geographic area. The name "cellular" is derived from the technological concept where communications towers cover a small geographic area known as a cell. As users pass from one cell to another, the signal is transferred. When users travel outside the network where they have subscribed to service, the signal may be picked up by another network. This concept is known as roaming. Roaming charges are more expensive than normal cell phone charges.

Cell phone rates, both local and long distance, are more expensive than rates incurred through the traditional public switched telephone network

(PSTN). Recently, however, competition has been introduced into many areas, including the introduction of personal communications service (PCS). There are varying plans for cell phones that encompass volume discounts and purchase vs. lease plans for the telephones.

PCS is a fairly recent service that has been introduced into many urban areas. It is based on a different frequency spectrum than standard analog cellular, operating in the 1.5 to 1.8 GHz range. The service is also digital, which makes it a more secure service.

Neither traditional analog nor PCS offer continuous nationwide coverage at the time of this writing. In addition, there are no service offerings that allow for a hand-off from WPBX to a public network service, although the technology does exist. Telecommunications departments do not always assume responsibility for cellular telephone service. It is possible to sign large-scale, corporate agreements, but many departments opt to allow end-users to negotiate their own agreements. When telecommunications departments take on such tasks, there is often a high degree of administrative duties. A quantity of cell phones needs to be ordered and inventoried. Individual accounts need to be set up and tracked, and there also needs to be a system for issuing, returning, and restocking phones. Because this is so labor intensive, many departments opt to act on a consultative basis, recommending specific types of phones or services.

Cellular and PCS networks are more ubiquitous than ever before, but they still have their limitations. The situation is improving, and cellular networks are being augmented while coverage is more omnipresent. Unfortunately, the United Sates. is a large country and there are still many rural areas that lack adequate coverage. PCS will provide more choices and the competition will offer reduced prices, but their initial coverage will only be in the major cities. End-users will still have to contend with handoffs to other networks and dead spots when they travel from one city to another in addition to expensive roaming charges.

Since competition has become more prevalent in the wireless phone market, the opportunity is good for negotiating corporate agreements. This equates to a certain amount of free air-time, lower per-minute rates, lower purchase prices for telephones, and master billing agreements that may be paid from one source or billed to each individual user. This also paves the way for a certain degree of standardization, which simplifies support issues.

Cellular networks may free the employee from the desktop and from the facility, but in a sense, they are still "tethered" to the cellular network, which is restricted to a specific geographic area. The satellite telephone is one concept that may free the user of specific networks. The satellite telephone provides direct access to a satellite network, completely bypassing all other land-based LEC and IXC land based networks. The initial foray into this service was the

COMSAT Hand-held Telephone Service, which was made available in 1996. The service was not inexpensive, but it did offer many advantages. The initial product offering was priced at $2,995 for the telephone while rates were priced at $3 per minute ($180 per hour). The service utilizes a spot beam which requires that a satellite be directly overhead. Unfortunately, with a satellite in geosynchronous orbit 22,500 miles above the earth, users will experience a delay in conversations that many users find annoying. Satellite telephone service is, as of this writing, in its infancy and due to the price and technological limitations, it is only appropriate for specific and limited applications.

While the recent satellite telephone offerings may offer certain advantages, they are still not delivering the promise of anyone, anything, anytime, anywhere. Enter the low earth orbiting satellite (LEOS) system that is now currently being constructed. LEOS will be a series of satellites that encompass the earth in lower orbits than the standard geosynchronous satellites. The advantages will be that the user does not have to be directly under the satellite and that the lower orbit reduces the amount of delay.

As of this writing, a number of companies have applied for (or have already received) licenses to provide LEOS services. The most publicized of these is Motorola's Iridium network. A smaller LEOS system (148/137 MHz) will provide for data applications and will require position location. A larger LEOS system (1.6, 2.4, or 4.0 GHz) will provide both voice and data capabilities. To date, nearly a half dozen satellites have been launched and the full network is expected to be in place by September of 1998.

Data services via wireless public networks have not grown at the rate of wireless telephone services, however, they are still growing at a rapid rate. Wireless data has lagged behind voice because mobile employees can only use wireless data under limited circumstances. Telephones can be used while a person is walking down a hallway or performing other tasks: Data usually requires a person's full attention. However, there are jobs that are enhanced by mobile data communications such as car rental employees or utility workers.

Wireless data capabilities begin with three terrestrial capabilities: packet radio, cellular, and narrow-band PCS. Each service requires that the computer be equipped with a wireless modem. The wireless modem and the miniaturization of computing devices has contributed to the proliferation of wireless data. Smaller laptops that can be equipped with PCMCIA cards are a primary example of this.

Examples of packet radio are ARDIS (owned by Motorola) and RAM Mobile Data (a joint venture of RAM Broadcasting and Bell South). Packet radio is simply X.25 using the international standard for radio (AX.25) and provides for speeds up to 19.2 Kbps. Packet radio, like most public wireless networks, is limited to urban areas. Once outside the range of the signal, the signal simply drops.

A second choice is basic analog cellular. As is the case with all voice grade networks, a modem allows the end-user to transmit data. However, analog cellular networks are known to be unreliable for voice communications, which also makes for unreliable data transmissions. Noisy channels and handoffs to cells also often cause a disruption in the data session, making this a poor choice if the data is of a mission critical nature.

With cellular analog being so unreliable, cellular digital packet data (CDPD) has now entered the marketplace, making use of additional capacity that existed within existing cellular networks. As in the case of packet radio, speeds can range up to 19.2 and IP traffic can be carried, making this offering attractive for corporate networks or Internet access. Special CDPD modems are required that range from $500 to $1,300. But once again, there are trade-offs. The end-user gains more reliability, but the network is limited to urban areas. However, end-users may roam from one network to another. CDPD service is expanding into more markets and major players such as the RBOCs are actively participating in standards organizations.

Wireless capabilities for voice and data continue to proliferate at a rapid pace. But the goal of anyone, anything, anytime, anyplace will not be realized in the near future. Telecommunications managers need to choose the proper technology for the given application, and constantly reevaluate the technologies as they progress.

2.10 Video

Video technology has been around for quite some time, but it has only been in recent years that the technology has become practical. The early systems offered proprietary technology that was expensive and provided dubious quality. Videoconferencing is now a viable technology and international standards have provided interoperability between different manufacturers.

Videoconferencing is the concept whereby two or more locations connect via videoconferencing systems in order to conduct a meeting that is enhanced by, or requires, visual interaction between the participants. Modern videoconferencing systems can also be integrated with PCs or electronic whiteboards to enhance the visual aspect of the meeting.

Traditionally, videoconferencing has been implemented to offset travel costs or to enhance employee productivity. For example, a two hour meeting might be conducted between two company sites. Assume that there are six employees involved, three per site. Prior to the implementation of videoconferencing, the meeting would have required three employees to travel for a day. Assume three employees who would have required airline tickets, meals, a rental car, and hotel accommodations. But because the meeting is conducted via videoconferencing, the company is saving thousands of dollars in travel expenses. Moreover, because the employees did not have to travel, the company does not lose expensive and valuable work time to travel.

Videoconferencing systems can be divided into three basic categories: room-based systems, midrange systems, and desktop systems. Room-based systems are also commonly referred to as video boardrooms. These are rooms that have been specifically designed for videoconferencing. A room-based system is not necessarily a specific system sold by one manufacturer. Rather, it is often a design utilizing audio and video equipment from various manufacturers. Very often, the video boardroom is provided by vendors that specialize in audio/video technologies and is a permanent installation.

The room-based system has cameras and microphones placed at strategic positions throughout the room. There is also a control room where a technician can control cameras and audio. Room-based systems are expensive to build and maintain, and prices normally begin at $100,000. They also normally require the support of dedicated technicians. Usually only the largest companies utilize video boardrooms.

Room-based systems and midrange systems are both used to support group videoconferencing. Room-based systems can support larger groups than midrange systems, because they allow for multiple camera angles and multiple microphones to cover more people. Midsize systems are also known as portable or roll-about units. They have one or two large screen video monitors, a single camera, and microphone. Many midrange systems are equipped with powerful document cameras, and additional ports are sometimes provided for additional cameras, microphones, or peripheral equipment. Midsize systems are small, portable units that can be transferred from one room to another. Because of their size and the single camera, they are normally impractical for large meetings. When there are more than six people per location, the use of mid-sized systems becomes cumbersome. Mid-sized systems can be very reasonable. Units can be purchased for as little as $15,000. Top of the line models can be $75,000 or higher. Considering the reasonable prices of these systems, if only a small amount of business trips are eliminated, pay-back periods can be relatively short. Major manufacturers of midsize videoconferencing units are Picture Tel and V-Tel.

Desktop videoconferencing entails the use of a multimedia PC, equipped with a camera and special video software. Desktop videoconferencing is becoming more common, especially through the Internet. It is a useful tool for document sharing or for individuals to participate in bridged videoconferences (three or more locations).

The standards for videoconferencing are set by the ITU and is documented in the H.320 set of video standards. This set of standards encompasses various transmission media, transmission speeds, signaling, compression, and encoding techniques. All new videoconferencing systems are H.320 compliant. Older, legacy systems may require a protocol conversion via a multipoint control unit (MCU), also known as a videoconferencing bridge.

The enabling technology of videoconferencing is the video codec. The codec takes the analog video and audio signals and compresses them into a digital format that can be transmitted across public network offerings. Full video, of the same quality offered on home televisions, would require bandwidths of between 90 and 150 Mbps. Because the transmission of these bandwidths would make videoconferencing cost prohibitive, video compression is necessary so that less bandwidth is required. A sophisticated algorithm is used to only identify and transmit movement from one videoconferencing unit to another. For example, a full image may encompass a number of objects. There may be a desk, chairs, and pictures on the wall. These are stationary items that will not move during the videoconference. The videoconference system will only detect what moves in the picture. This movement may be something subtle, such as the lips of a speaker, or a more dramatic action such as a person pointing to items on a flip chart. Regardless, the system will not retransmit the entire image. The stationary items will be disregarded and only the section of the picture where movement occurred will be transmitted. This does not provide television quality of the same type that residential users experience in their homes. Because videoconferencing is a dial-up application, there has to be tradeoff of picture quality versus transmission costs.

Two basic factors have a dramatic impact on the quality of videoconferencing: bandwidth and frames per second (FPS). Bandwidth is utilized in multiples of either 56 Kbps or 64 Kbps. The most common applications are two or six times the aforementioned base rates, allowing for aggregate rates of 112 Kbps (2 × 56), 128 Kbps (2 × 64), 336 Kbps (6 × 56), or 384 Kbps (6 × 64). Two of the most common ways to provide this bandwidth is either through switched 56 or ISDN. Basic videoconferencing systems have the ability to aggregate two base channels. When more than two channels are used, an inverse multiplexer (IMUX) is required to aggregate the additional channels. When video transmissions are compressed below 112 Kbps, the quality of the video often becomes poor and cumbersome to use, however, the technology continues to be refined and there are some new single channel systems that offer more acceptable quality.

FPS is offered at two speeds, 15 FPS and 30 FPS. Thirty FPS is the same quality used for residential television service and provides full motion video. While the quality of 15 FPS is certainly not as good as 30 FPS, it is acceptable, especially when the meeting does not involve a lot of movement. A typical reaction of end-users, when they experience 15 FPS for the first time, is that the quality is poor and perhaps unusable. Once people become involved in a meeting, it soon becomes apparent that the system is workable. When videoconferencing is introduced into a company, telecommunications staff members need to convey to end-users that they will not experience full motion video.

Videoconferencing systems are priced according to capabilities and system components. Systems that can transmit at the higher rates and at 30 FPS will be more expensive. There are also issues that relate to required equipment such as document cameras, microphones, scanners and electronic whiteboards. System configurations are dictated by the demands of the business environment. It should be noted that high end systems can communicate at lower speeds, however, a low end system cannot communicate with higher bandwidths. This may be critical when a company is conducting videoconferences with customers who demand high bandwidth meetings.

There are two basic types of videoconference meetings: point-to-point and multi-point. A point-to-point meeting is set up by one system dialing another. This is executed by simply dialing telephone numbers that have been assigned to the switched 56 or ISDN lines at the distance end. Long distance charges are incurred for the use of each digital channel. The more channels that are used, the higher the cost of the meeting.

When more than two locations need to be connected, an MCU (or bridge) is required. Companies have the option of purchasing their own MCU or renting bridge time from a service bureau. The basic premise of the MCU is to connect more than two sites, however, it does offer various features that can enhance the meeting. This is accomplished by having each participating site dial into the MCU. MCUs also have dial-out capabilities. When dial-out capability is used, the participating sites need only have their systems turned on. The MCU technician assumes the responsibility of connecting all sites.

Video bridge services are offered by a number of different vendors. The major IXCs, AT&T, MCI, and Sprint, all offer video services in addition to smaller independent companies. The end-user is charged a usage rate for each MCU port used during the meeting. If the end-user opts for dial-out services, the long distance charges are passed along to the end-user, in addition to the port charges. Companies have the option of purchasing their own MCU or using a service bureau on an as-needed basis. Owning an MCU is a logical progression for a company that conducts frequent multi-point conferences.

2.11 Summary

Detailed telecommunications knowledge is essential for all members of the telecommunications department. This entails voice, data, and video. It also encompasses the services and products, and the vendors that supply them. Telecommunications knowledge must constantly be refined and continuous education is essential in order to support the modern corporation in the Information Age and to run a successful telecommunications department.

3

Departmental Structure and Positions

It would indeed be a perfect world if one could develop a standard telecommunications departmental structure that could be applied to all companies and industry segments. It would also be wonderful if every telecommunications manager was allotted the necessary funds to develop the telecommunications department exactly the way they wanted. If this were the case, there would be a multitude of interesting positions available within every telecommunications department. Workloads would be reasonable and overtime minimal. Staff members would come to the department at an entry level position, performing administrative and hands-on duties, but the department would be structured so that upward mobility was available for those who chose to further their education, and displayed an aptitude for higher level functions. As staff members moved up in the departmental hierarchy, they would gain more responsibilities, which would include project management, financial and technical analysis, contract negotiation, supervisory duties, and budgetary duties. Under such ideal circumstances, the telecommunications manager could use the lower levels as a training ground for the higher level positions. Morale would also be high because the structure would give incentive for staff members to improve so that they could gain a better position and make more money. Moreover, there would be a chance for staff members to move from one discipline to another, making their jobs more interesting and challenging. It would be a win-win situation for all members of the telecommunications department.

Unfortunately, this scenario is rather idealistic and has little basis in reality. Most telecommunications departments are home grown, and the structure is one that grew out of necessity, not as a result of any grand plan. Many businesses add positions within the telecommunications department because they are facing severe operational problems, and the existing staff simply cannot meet the present workload. It is normally not because the company wishes to boost morale or provide upward mobility for their employees. In fact, it would certainly be difficult to justify the hire of an expensive telecommunications professional based solely on those criteria. More realistically, when people are finally hired into a department, it still does not provide enough manpower to address the existing workload. This makes it difficult for staff members to branch into new or different technologies. For example, a voice communications specialist may wish to enter the world of data, but his workload will not allow it. Or a data specialist, who works primarily on router installations, wishes to become more involved with network design. Unfortunately, the data specialist is working 10 to 12 hour days, and sometimes has to work on weekends. If these people have no relief from other staff members, they certainly cannot leave their existing duties to train for another job. Such dilemmas are common in the modern telecommunications department. But the problems do not end there.

Because there is often no grand plan and salary budgets are limited, staff members must wear many hats, in a sense becoming generalists. This results in a constant tug-of-war between specialization and generalization for both the department and the individual. One problem is that many technologies have become so complex, they demand a dedicated specialist. Consider a staff member who must maintain the TCP/IP addressing scheme along with DNS naming conventions. This in itself is a full time position within the company, given the size of the network. Unfortunately, because of limited staff, this person must also oversee the implementation of new routers on the network, and upgrades to, or removal of, existing routers. So not only does this person have demanding project management duties, he must also administer a large and complex database.

When the situation demands generalists (or the department is intentionally structured that way), staff members become frustrated. They are frustrated because they cannot apply themselves to the technologies and learn it in the detail they feel is necessary. They also know that their work may be subpar, because they are stretched so thin. Examples of common complaints are as follows:

1. “I sure wish they would hire somebody. I worked hard to further my education and technical skills and I still spend hours every day doing administrative busy work.”
2. “I wish I had more time to work on network design. If I could apply more of my time in that area, we could make the network more

efficient and less expensive. The problem is that I never have the time because I am always implementing a new device, changing something, or putting out fires."

3. "I would sure like to get more involved in the data arena. Right now I am bored with voice communications because my company is not doing much with it. They also are not offering me much opportunity to move over, even though I have made my wishes very clear. If this doesn't change soon, I don't think I will have any choice but to look for another job."

Conversely, telecommunications managers are frustrated with limited budgets and an overwhelming workload. They often realize that they need specialization in certain areas, if not full-time positions for specific disciplines. Because they have budgetary limitations that have been imposed from above, they know that they are working with a dull tool.

A negative by product of such circumstances is that morale begins to suffer on both sides of the fence; management and staff. Staff members begin to think that there is no upward mobility, or a chance to diversify their workload. Ultimately, they feel as though they are caught in a rut. Moreover, if specialists are not developed, generalists are not able to perform at the level that many of the newer technologies demand. This is a frustrating situation for all concerned parties, including the telecommunications manager who feels that he is facing a Herculean challenge. His people want opportunity and relief, and he is somewhat helpless to provide it.

Unfortunately, upper management or even the telecommunications manager does not always recognize this dilemma. Because telecommunications departments evolve on an as-needed basis, many managers learn by trial and error, often reacting to business requirements and conditions as they occur. Because of this situation, there is often very little thought given to the structure of the department. The telecommunications manager is too consumed with running the business to be bothered with things that apparently have no immediate benefit. Telecommunications managers may also feel comfortable with their situation, in spite of operational difficulties or morale problems. In their eyes, the work is still getting done and they may simply view these circumstances as being the nature of the beast, common to all telecommunications departments. Other factors may also prevent the manager from making changes. Restructuring means change that may affect department performance in the short term. If there are operational problems now, can I afford to adversely affect the short-term performance of the group, even if there are long-term benefits?

The problem is also exacerbated by the philosophy of each individual company. Do they view telecommunications as a strategic technology or a utility where cost control is the driving force? Probably more companies view telecommunications as a utility rather than a strategic business tool. However, when the value of telecommunications technologies is realized, more money is available to develop the telecommunications department. Under such circumstances, it is easier to justify department reorganization and expansion. Unfortunately, this rarely happens unless major personnel issues surface, such as excessive turnover or projects that are not getting completed.

Telecommunications managers who ignore the importance of departmental structure should be aware of employment trends in the world of computing and telecommunications technologies. First, there is a shortage of such people that will be acute by the year 2000. Second, the most common reason that telecommunications professionals leave their jobs is not because of money. Rather, most telecommunications professionals change jobs because they want to experience new and different technologies. In addition, they also strive for more responsibility in the form of project management and supervisory duties. An organization that does not offer growth in these areas becomes a source of frustration to the staff. Inevitably, this situation will result in turnover. In geographic areas where telecommunications professionals are in demand, this could have a devastating impact on the department.

Beyond stability and morale, department structure also has a dramatic effect on department performance. Consider the dilemma of having too many generalists in the group. The danger of having all staff members do all things yields two negative effects. First, the too-many-cooks syndrome takes effect. For example, if there is no central authority for TCP/IP addressing, routing problems soon become endemic. As a result of everybody "doing their own thing" a major project soon goes on the books to clean up the mess. This could have been avoided if there was a central authority within the staff guiding the process from the beginning. The second problem is the jack-of-all-trades-master-of-none syndrome. There are departmental functions that will require dedicated expertise because the technology is too complex for a generalist. Both of these problems are directly attributable to departmental structure, which affects many aspects of the telecommunications department success.

3.1 Departmental Structure

If there are logical arguments for a special type of department structure, what specifically should that structure be? The answer is not in a specific, detailed structure that should be rubber stamped for every telecommunications department and corporation, but rather a conceptual model that provides a number of capabilities in terms of employee opportunity and departmental efficiency. Consider two basic concepts: the flat organization versus the pyramidal structure. The flat organization (see Figure 3.1) has a group of staff members reporting to a single manager. Under this design, there still may be specialization, but all decision-making authority rests with the manager. When a department is structured flat, there is no upward mobility for staff members. Consequently, there is a danger that staff will become frustrated because there is no opportunity for advancement. That does not mean that the flat structure is without positive qualities, or that it can even be avoided. Under the flat structure, there is still a possibility that staff members can move between technologies (see Figure 3.1). Notice that there are separate voice and data departments. If the staff members have the opportunity to cross-train and work into new technologies (and more specific disciplines within those technologies), this satisfies the need of some staff members to experience new and different technologies. While the staff members may have limited upward mobility, job diversification is still an option, which satisfies a basic need of many telecommunications professionals. The worst possible situation is when there is no upward mobility and no opportunity to move into new and different technologies. Under these circumstances, many staff members will become bored or frustrated and seek employment elsewhere. It is impossible to prevent turnover, but the turnover should be the result of staff members being offered exceptional

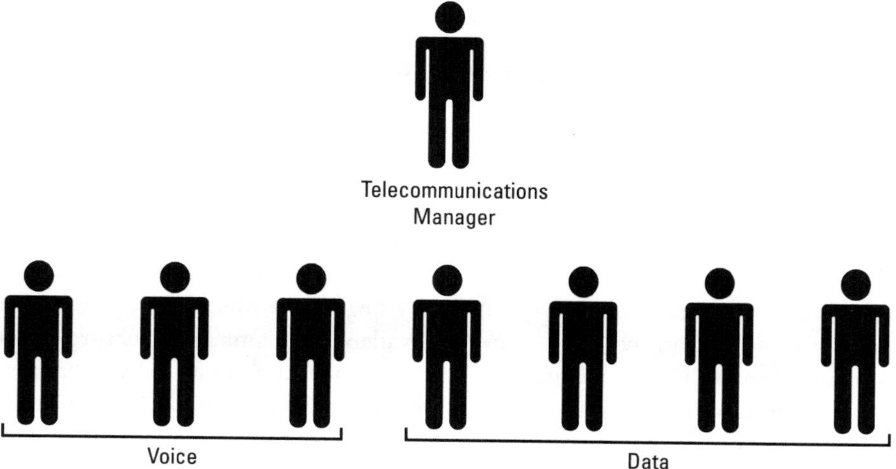

Figure 3.1 A flat department structure.

opportunities, not resignations that are delivered in anger and frustration, because there is no opportunity.

While there are many problems that surface with a flat departmental structure, there are times when there is no other choice. A primary factor that dictates department structure is number of staff. For example, a five person staff will certainly not leave much leeway for positions such as project manager, network designer, or assistant manager. Moreover, if the department does not grow, there will probably never be such opportunities. Under these circumstances, the telecommunications manager must take concrete steps to offer opportunity in the form of job diversification. This would come in the form of new or different project work for the chance of staff members to move into a completely new field.

When the staff is small and structured flat, there should also be a concerted effort to provide an open culture and bring the staff members into the decision-making process. The open culture will be discussed in more depth later in this chapter, but the telecommunications manager should strive to bring cohesiveness to the group. When such efforts are taken, the telecommunications manager provides an environment that meets many of the needs of staff members, in spite of the lack of upward mobility. In fact, staff members may be reluctant to move to a larger organization where they may not have as much opportunity to work with diverse technologies. They may also realize that they will have less say in a larger organization. Larger organizations may offer more opportunity simply because of their sheer size, but smaller organizations can also offer many advantages. These advantages can be aided by the use of the open culture, team work, and employee empowerment.

The open culture brings staff members into many of the operational and managerial processes of the department. When decisions are made, the manager does not make them solely. Rather, they are made through a consensus of the team. While the manager is the ultimate decision-maker, most decisions are made with input from the staff. This promotes a feeling of cohesiveness within the group, a form of esprit de corps. This process also exposes staff members to managerial processes that they might not experience with a larger organization. Unfortunately, this can also be a double-edged sword. When staff members are exposed to managerial processes, they may begin to aspire to a managerial or supervisory position, which may not be available in a smaller, structurally flat organization. Still, if a staff member is frustrated, and chooses to leave because there is no upward mobility, he will probably do it reluctantly. An additional benefit of utilizing the open culture is that staff members are privy to organizational processes. In the event of an unexpected vacancy, with either management or staff, remaining department members can more readily shoulder the burden. I was once a staff member of a department that was structured flat.

The manager made all decisions, and assigned projects and various duties without the input of his staff. He unexpectedly resigned, and the department was temporarily paralyzed, because the staff held very little knowledge of department operations. This was frustrating for the manager who oversaw the department from a higher level. Because staff members were never privy to any decisions, there was nobody in the department who was trained to step into the manager's shoes. The only alternative was to seek a manager from outside the company. Unfortunately, this had a very adverse effect on the staff.

When an organization is larger, there is more opportunity for developing various types of positions. This allows for a pyramidal structure (see Figure 3.2) so that opportunity exists in the forms of job diversification and increased responsibility. There are a number of positions that will be discussed throughout the course of this chapter. Each organization will have different requirements and there is much opportunity for creative design of the departmental structure on the part of management. Regardless of what design or specific positions are used, the telecommunications manager should not lose sight of the two important goals in a pyramidal structure: job diversity and upward mobility. When these two goals are met, morale is higher, turnover is less, and department efficiency is enhanced.

At the top of the pyramid are high level managers. Titles may vary and, depending on the size of the organization, there might be a director, vice president, or a single telecommunications manager. A pyramidal structure is usually only possible in larger organizations. Consequently there will be a number of management positions that address a specific function. For example, there may be assistant managers who will oversee voice or data respectively. It may even be necessary to create further management positions, based on the size and scope of certain technologies, such as Manager of Internet Services. Regardless of how many management positions are defined within the department, the top position manages from a high level, addressing financial, strategic, and operational aspects of the business. For example, the senior ranking manager will forecast, with the aid of his direct reports, the projects and workload for the forthcoming fiscal year. At this level, the senior manager is not concerned with minute details. Rather he is concerned with basic concepts. Are there enough staff members to support the anticipated project and workload? If so, do they have the required experience and knowledge to support the projects? If they do not, should they be offered education? Would it be better to hire consultants, or a combination of both? Does the department structure need modification in order to address the existing workload? Does the head count need to be increased because of operational problems due to inadequate staff? Was the budget met last fiscal year? If not, what were the reasons, and should adjustments be made for the forthcoming year? What is the company business plan

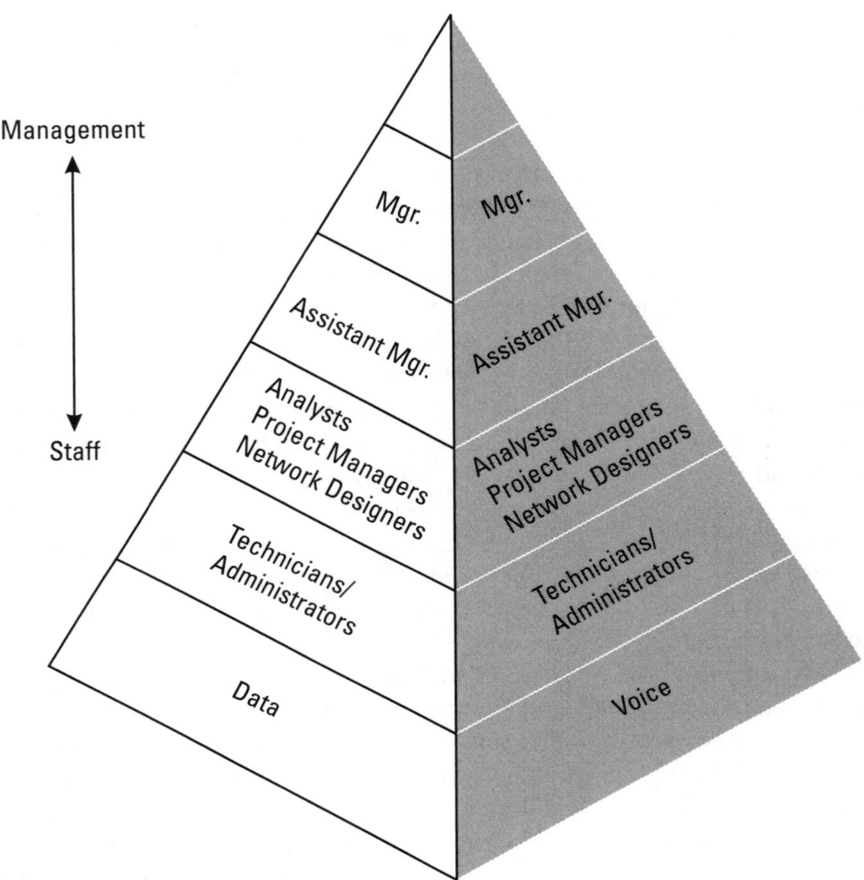

Figure 3.2 A pyramidal department structure.

for the next fiscal year and how will that affect the telecommunications department? Are there any new technologies that should be considered or perhaps refinements of existing ones? What types of financial and technical resources will be required to support these changes?

The highest ranking manager will not be concerned with minute details. For instance, assume that the data group is experiencing installation problems with a new service provider. The vendor does not show up on time, with the proper equipment, or the field engineer (FE) is not properly trained. However, the work is getting done, in spite of the difficulties with the vendor. This is not something that the senior level manager will want (or need) to be involved with, except at a very high level. The assistant manager of data communications would be expected to work with the vendor to correct the operational problems. The senior ranking manager will only want to know that there is a problem, that a course of resolution is being addressed, and when the problem is finally resolved. If the problem lingers to the point where the vendor cannot meet the company's expectations, then there would need to be a high level meeting to determine if the vendor should be replaced.

A further word about the top ranking manager; there is a term that has gained favor in the last decade: the micromanager. Generically, the term refers to managers who spend an inordinate, and unnecessary, amount of time on the details of their staff's workload. There are a number of negative aspects of micromanagement. The micromanager worries about minute details to the point where he often loses sight of the big picture. In addition, because he is spending so much time checking the work of his staff, the work actually is being done twice. This slows department performance in addition to having a negative impact on morale. Staff members feel as though they are being hounded and are perceived as being incapable of performing their jobs without close supervision. Because of the complexity of modern telecommunications products and services, managers cannot be concerned with every detail of department operations. Rather, they must hire and train the staff to work independently, without close supervision. The concept of empowerment is one that has gained popularity in the later decade of the twentieth century. The premise is to empower employees so that they may work independently, making for more efficient operations, and eliminating management as a bottleneck to progress. This is especially apropos for the telecommunications department where complex problems and technical details are a constant. The department needs to be structured so that staff members can solve problems, make decisions, work with, and manage other people. There should be programs that counsel and train staff members to operate in this fashion. This policy then frees management to worry about the larger issues.

Underneath the senior ranking manager may be separate managers who oversee an area of specialty. A very common structure may be to establish voice and data communications managers. Below these layers may be assistant managers or supervisors. These positions complete the top layer of the pyramid that is related to management and supervisory duties.

Management in the upper tier of the pyramid is responsible for department performance. They are a link between the telecommunications staff and the corporation. It is their job to see that the department meets the requirements of the company, which includes executive management and the end-users (various departments and individuals). The upper tier is expected to keep the staff focused on the proper objectives and maintain those objectives within the projected budget.

Below the management layers, the job functions remain at a high level, encompassing project management and analytical duties. Analytical skills include financial analysis, technical analysis, and problem determination, which are often applied to project management. Positions at this level can be project manager or senior analyst. This layer can be viewed as a stepping-stone to management. They are responsible for managing technology and specific projects that relate to the technology. They are not responsible for managing people, although they may have to manage teams of people for specific projects.

Project management is a constant in most telecommunications departments. Only the smallest and most stagnant of organizations do not have a constant and ever changing stream of telecommunications projects on their plate. Project management is so critical to the success of the department, that specific positions have been developed in many companies to oversee and manage the larger and more complex projects. The ability to manage projects successfully requires a unique set of skills. These people must be adept at planning and overseeing the implementation process, while trying to keep the project on time and under budget. This is usually accomplished in an atmosphere that is chaotic and constantly changing. Decisions need to be made quickly, technical and logistical problems must be resolved, bills must be paid, services and equipment ordered and coordinated, and workers managed. Juggling all of these facets is difficult and good project managers soon prove to be invaluable to the success of the department.

All staff members perform project work in one capacity or another. The large scale and highly visible projects should only be trusted to the people who have proven their ability to meet the large-scale challenges. Once they have demonstrated these skills, they can become a driving force in the department and should be recognized as such. This recognition can come in the form of a specific title, a higher pay scale, or project specific bonuses. Project management positions may be specific to voice or data, but may also encompass all telecommunications technologies. The telecommunications manager should strive to develop project management skills throughout the entire department, the lower staff levels perhaps being the training ground for project management positions. Staff members who are at a lower pay scale and position will be motivated to develop and fine tune their project management skills in anticipation of a potential promotion when openings occur.

Management of specific technologies can come in the form of dedicated specialists. Because some technologies have become extremely complex, the role of generalists is diminishing. Also, the scope of the technology may be so ubiquitous in the organization, there is no alternative but to hire and maintain specialists. Examples of technical specialists may be call center specialists, Internet experts, or network design engineers. Because of the sheer scope of work and the complexity of the technologies, these specialists begin to carry on a critical role for the corporation, actually managing a technology, but not necessarily a department or group of people. While technical specialists may not be project managers, they provide critical support to project managers who, like many managers, may not have the detailed knowledge of specific technologies. That is why they are often a critical part of the project planning process.

Financial management is generally the responsibility of managers, however, other staff members may offer support in a number of ways. The overall department budget is the responsibility the highest ranking manager. This entails salary, departmental operating budget, and the cost of various telecommunications products and services. The later may be charged back to departments or individual end-users, however, there are usually some costs that are absorbed by the telecommunications department. Beyond the total cost of operating the telecommunications department are hundreds (if not thousands) of financial issues that need to be managed on a day-to-day basis. There are billing errors, contracts to be negotiated (or renegotiated), internal charge-back systems to be administered, bills to be paid, and financial reports to be written. While every manager is charged with being fiscally responsible, few managers can address the details of every financial issue. The only solution is to disperse many of these duties to staff members. Although it may be inappropriate for staff members to be privy to the entire budget, their contributions are often essential in order for the entire mechanism to function efficiently.

At the base of the pyramid the job functions are administrative and perhaps more labor intensive (such as a PBX technician/administrator, cable technician, or help desk agent). As employees strive to diversify their background and move up in the organization, they must increase their skills in the areas of technical knowledge, project management, supervisory skills, and financial analysis. Staff members who are at the base of the pyramid can be hourly or salaried workers. It is also not uncommon to outsource these positions, meaning that independent contract workers or a telecommunications vendor is used to perform the work. This demonstrates that not all jobs are a steppingstone to higher levels. In fact, history shows that many people who work in these positions prefer to be there. However, it should not be an assumption that staff members in these positions will be satisfied to stay there. The department should always be structured so those staff members who aspire to higher positions have opportunities available.

The pyramidal structure should provide horizontal and vertical movement for staff members (see Figure 3.2). Horizontal movement provides job diversity: a critical objective of many telecommunications professionals. For example, if a voice analyst desires to enter the realm of data communications, he should be able to move into that discipline through educational and apprentice type programs. The telecommunications manager should be open to the wishes of his staff members, helping them to enter new areas and establishing developmental programs to open these doors. This may not always be immediately feasible, based on projects and workload, but the manager should always be striving to create a platform for opportunity.

There are many benefits when a department is structured to provide job diversification. The most obvious is higher morale. Staff members know that their job is not a dead end and that opportunity does exist. At the very minimum, if they do not have opportunity to move up, there is at least potential to move into a new, and more interesting, technology. A second benefit is that staff members are often in a state of cross training. Because the knowledge is spread among the group, if an unexpected resignation occurs, there is a strong possibility that the slot can be filled from within and the turnover will be less problematic. When a position is filled internally, the manager does not need to be concerned with the learning curve that is inherent to new employees. That is, a new employee needs to learn the new company, business applications and practices, and how telecommunications technologies are applied to these applications. This can take from several months to a year (and perhaps longer), depending on the complexity and nature of the business. An internal person does not have a learning curve because he already knows the business. He is more concerned with learning the new technology and applying it to the business that he already understands.

As horizontal movement provides job diversification and department redundancy, vertical movement provides increased managerial responsibility and higher salaries. Remember that the two most common reasons telecommunications professionals change jobs are to experience new and more diverse technologies and for the opportunity for increased responsibility (e.g., project management and supervisory functions). This is where the flat structure falls short and offers little opportunity for advancement. In the flat structure, staff members have to hope that the manager quits or is promoted in order to have an opportunity for advancement. Unfortunately, if the department is structured to be flat, staff members are seldom trained or encouraged to take on more responsibility. Therefore, when a vacancy occurs, it is not uncommon that the position is filled from outside the department. This typically has a negative effect on staff morale.

The pyramidal structure offers upward mobility in graduated steps. As staff members progress from lower to higher levels, the positions carry more responsibility and possibly more specialization. Increased responsibility comes in the form of project management, technology management, financial management, and supervisory duties.

Beyond the high level view of flat versus pyramidal structure, the department should be divided into functional levels that address the specific business needs of the department. Consider the example of Figure 3.2 that delineates voice and data. In this example, there are assistant voice and data managers who report to the telecommunications manager. Below the broad category of assistant manager, specific technologies are addressed. In order to establish positions, titles, and lines of responsibility, specific areas of responsibility need to be established and documented according to technology. Consider the following list for the data communications group.

Network Services

- WAN (private line, frame relay, ATM, T-1);
- Local data (LEC provided services such as dial-up, T-1, SMDS, ISDN, etc.).

Data Equipment

- routers;
- hubs;
- CSUs;
- modems;
- DACs;
- Internet servers;
- LANs.

This simple outline approaches the data communications arena from a high level. In smaller, flat organizations, staff members would probably assume all aspects of administering these services and devices. In larger organizations, it becomes imperative that specialist positions are developed. Consider the first section, network services. In a large, corporate environment there has to be a central authority designated to oversee network design. For example, if frame relay is the standard WAN network service, somebody has to determine port speed and CIR. Also, is the connectivity requirement to a single location or multiple sites? There is also a requirement for backup facilities. This may be in the form of analog dial-up, ISDN, or an alternate DLCI route in the event of a failure at the distant end.

After the design has been determined, there is quite a bit of administrative work that needs to be processed. The frame relay circuit needs to be ordered.

This is through a standardized department procedure, which may assign a control or project number to the order. There will be an installation charge and a monthly recurring charge once the circuit is installed. These moneys must be logged so that they can be charged to the proper departments once installation is complete and then tracked on an ongoing basis.

Once the circuit is installed, there may be the possibility of adds, moves, or changes. Suppose the location began to experience poor response time or there was a change in the computer application. This might necessitate an increase in the CIR, perhaps even an increase in the port speed. Either way, this would require that an order be issued to the vendor. Once again, all aspects of the order and the implementation process would have to be tracked. Beyond installation and administration, there will be ongoing issues such as vendor status (should there be one vendor or multiple vendors), contract status, and service related issues.

Continuing with this concept with data equipment, there are similar issues that need to be addressed. A type of router must be selected, along with a vendor to provide the device. There are also software levels to select and routing tables to be programmed. The router will then be assigned a network address (typically TCP/IP). Once all of these issues have been resolved, there should be a network map and documentation kept in both software and hard copy so that authorized staff members have access to this information when needed.

Considering these various areas of responsibility, it can be seen that specialization is required for large corporate networks. Otherwise chaos will reign in the department. Senior level people need to carry a myriad of high level responsibilities in order to make this system work. They need to set policies, standards, and processes that will provide consistent department performance. Their position dictates that they are responsible for making sure that the processes are executed correctly, on time, and on budget. Because of these responsibilities, they carry specific titles and command a higher wage. Positions below senior management are charged with carrying out the policies so that technology is installed correctly and maintained to provide cost-effective and reliable service.

Consider a company that has just made a major acquisition. There will be two dozen sites added to the existing WAN. In a pyramidal department, the assistant data communications manager will meet with various staff members to establish a project. He will not tell the staff exactly how to execute the project. Rather, he will give guidelines, and allow the senior people to develop a specific plan. He will assign a project manager. This person may carry the title of project manager or may be a staff member who periodically assumes project management duties. It is this person's responsibility to coordinate all aspects of the project: budget, time frames, personnel resources, planning meetings, documentation, ordering, implementation, testing, and billing.

A network engineer, as part of the project team, will inventory the new sites, and evaluate the technical requirements. He will need to know what type of data protocols will need to be supported, along with bandwidth requirements. He may also want to determine an alternate route for the data in the event that the main link fails, or there is network congestion. There may even be an in-house, boiler plate checklist that has been developed to guide him through the process. He may also use a data modeling tool to help him simulate several designs before a final decision is made. It is this person's responsibility to assure that connectivity to the new sites will provide optimum performance for all business applications.

The network engineer will work with several analysts in the group, one who is responsible for maintaining the addressing schemes, another who is responsible for coordinating adds, moves, and changes of network services. They will determine, through a team effort, the proper equipment, reasonable time frames, and specific lines of responsibility. Titles may vary from company to company. These people may be analysts or their titles may be more specific to their functions.

The project manager will conduct team meetings at various times, checking on progress. It is during these meetings that issues are discussed, examined, and resolved through the process of group consensus. Ultimately, the higher ranking staff members, to ensure continuity, will mediate the meetings. The project manager, with the consensus of the assistant manager, will make final decisions when necessary. There should be a method of documenting and tracking progress. This could be in the form of an off-the-shelf software package or perhaps something that was developed in-house. Regardless, there should be a universally accessible software system to track project progress.

When the project is finally completed there may have been a number of issues and problems that surfaced and were solved during the process. Behind the scenes, these were obstacles that the project team had to overcome. Some of these problems were solved through group consensus and some were solved through management decisions. In either case, the structure of the department was instrumental in facilitating the success of the project.

Assume that the project was a success. To the end user, the implementation of the new data equipment and services were transparent with no disruption of service. Even the telecommunications manager was unaware of many of the smaller problems that emerged and were solved during the course of the project. This was because of the way the department is structured, how responsibilities are assigned to staff members, and various departmental policies.

Regardless of the arguments in favor of a pyramidal structure, no telecommunications manager seems to be allotted all of the resources he thinks he should have. Budgets are limited and the workload is demanding. Therefore, it is not always possible to structure the department in the most idealistic manner possible. Where is the answer? It probably lies somewhere in between. Because there are limited budgetary dollars available, telecommunications managers will usually find it necessary to institute a certain level of generalization. But they should also not lose sight of the fact that quality telecommunications professionals are scarce. Therefore, they should always try to strive for a department structure that will provide a certain degree of opportunity and diversification. A step in this direction is to understand various positions, titles, and lines of responsibility that may be incorporated into the pyramid structure.

3.2 Positions, Titles, and Lines of Responsibility

When telecommunications professionals are recruited there is often much confusion regarding a candidate's experience and abilities when compared to his title. I have met people with the title of analyst who virtually ran a telecommunications department themselves, carrying responsibility for both voice and data, in addition to project management, financial, and supervisory duties. Conversely, I have met people who carried the title of telecommunications manager who were nothing more than glorified PBX administrators. Because there are no industry standards, titles can often be deceiving. One company's telecommunications manager is another company's telecommunications analyst, and vice versa. This confusion is caused by a number of factors.

Industry segments are certainly a major factor in affecting the title paradox. For example, health care is an industry that is driven by costs due to a number of variables. The costs of salaries, medical equipment, and drugs escalate at a high inflationary rate. This problem is exacerbated by insurance plans that are constantly changing in an effort to control medical costs. The net result of these problems is that hospitals and medical facilities are always being pressured to reduce operating costs. Consequently, the performance of the telecommunications department is very much driven by cost performance. Under such circumstances, telecommunications managers and staff often find themselves scrambling to find savings, perhaps ignoring larger issues that could potentially benefit the organization. Telecommunications managers under these circumstances might oversee a WAN and PBX that is fundamental in design because their directive has been to cut cost, not drive technology. Perhaps a primary function of the department is to administer a charge-back system so that revenue is realized through patient telephones. Because of the state of this particular industry segment, the telecommunications staff finds that they are spending a major portion of their time on financial matters as opposed to technical issues.

Conversely, consider an organization that is in a more competitive industry segment, such as the mail order catalog business. In this example, telecommunications technologies and services carry a more strategic role in the company's business infrastructure. Consider that a catalog company may have six call centers located around the United States. There is an elaborate toll-free network, ACDs, and a WAN that connects the call centers with the corporate headquarters where the mainframe computer resides. In this application, the telecommunications department is given a directive to keep all telecommunications technologies operational 100 percent of the time. Assume that in this example, it is a highly competitive environment. The company has determined that a call has a dollar value of $250. If this company processes thousands of calls per day, lost calls can quickly add up to high volumes of potential lost revenue. Under these circumstances, the telecommunications department is charged to find technologies that make the business more competitive, refine the existing technologies, and ensure that the telecommunications infrastructure is extremely reliable. Cost is considered only after function and reliability has been defined.

Compare the telecommunications department of the hospital with that of the mail order business and it can be seen that management and staff alike will have different priorities and different skill levels between the two organizations. Assume that the mail order business aggressively pursues technology. This means that staff members must aggressively pursue education and enhance their knowledge base. Their project load is diversified and intensive. While the staff members for the hospital also realize a heavy work schedule, their workload is geared more toward administrative and financial functions. Moreover, they are not driven to learn new technologies because their job performance is dependent upon financial matters, not the ability to implement and support leading edge technology. If the telecommunications manager from the hospital were to change jobs and seek employment with the mail order company, he might find that his technical knowledge requires enhancement for new technologies. He might also find that he has to redirect his priorities because he has spent a number of years directing department activities toward financial matters, and technology was a smaller piece of the pie. In comparing the two positions in this imaginary scenario, the reader can see that two managers—both holding the same title—have very different knowledge bases and skill sets.

The title dilemma is further complicated by the lack of naming conventions for telecommunications titles. As illustrated in the beginning of this section, titles very often can carry very little weight. When companies recruit telecommunications talent, it can never be assumed that the candidate possesses all of the experience, knowledge, and skills required to perform the job, based on what their title suggests. There is actually a plethora of titles within the industry. The following is sampling of titles that may be encountered. Once again, the reader should not be fooled into thinking that a weighty title necessarily reflects quality experience, ability, or knowledge. In addition, seemingly lesser titles might represent stronger candidate capabilities.

- Director of Telecommunications;
- Vice President of Telecommunications;
- Telecommunications Manager;
- Voice Telecommunications Manager;
- Data Telecommunications Manager;
- Assistant Telecommunications Manager (voice or data);
- Manager of Global Communications;
- Telecommunications Consultant;
- Network Engineer;
- Certified Network Engineer;
- Call Center Manager/Specialist;
- Network Designer;
- Telecommunications Administrator;
- Network Administrator;
- LAN Administrator;
- PBX Administrator;
- Telecommunications Specialist;
- Project Leader or Manager;
- Telecommunications Analyst (junior, associate, and senior levels);
- Telecommunications Associate;
- Network Technician;
- Telecommunications Technician;
- Internet Specialist;
- Telecommunications Intern;
- Help Desk Support Technician.

The titles have been listed to reflect the structure of the pyramid. The top seven titles are managerial in nature and there will be relatively few of these in the department. The later positions are those that would be placed at the bottom portion of the pyramid, being more hands-on in nature, but representing a higher number of employees.

At the higher levels, the managerial positions are where the ultimate decision-making responsibility lies. Directors, vice presidents, and managers are more concerned with high level concepts that relate to strategic direction, technological platforms, financial management, human resources issues, project management, department structure, and resource allocation.

3.3 Senior Level Positions

Trying to develop a precise definition of each position listed above would be like trying to hit the proverbial moving target. As illustrated earlier, there are no national standards for telecommunications department structure or support. Consequently, people who carry the same title across companies and industry segments will have varying skills, knowledge, background, and capabilities. For the purposes of this book, and the sake of simplicity, I will approach the role of telecommunications manager as the key managerial figure within a company who oversees and manages telecommunications services for the entire enterprise. Various organizations may have different titles, such as vice president or director of telecommunications, but the senior ranking manager will have three key duties:

1. Supervisory duties over a group of telecommunications professionals;
2. Financial responsibility for department operating costs;
3. Responsibility for evaluating, recommending, implementing, and managing telecommunications technologies to meet company requirements and objectives within reasonable cost parameters.

There may be other managers who report to the senior manager and carry these same three key roles at a more concentrated level. For example, there may be a data or voice manager.

The three basic categories (staff, budget, and technology) can be viewed as a triangle (see Figure 3.3). Each category must be kept in balance with the others. Staff must be sufficient in number and expertise to support the array of technologies. The budget must be sufficient to support the staff and the operating expenses of the department, and the cost of various technologies. Technology must be provided within the constraints of the budget and in a manner that is supportable by the staff. The technology must also meet the current and future business requirements of the company. Balancing these three categories

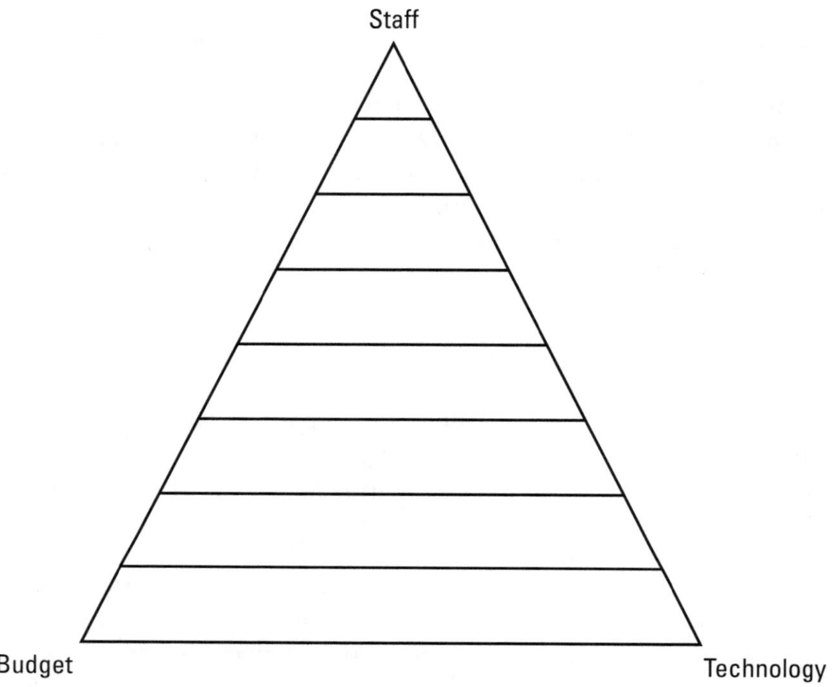

Figure 3.3 Three principles to boost morale and productivity.

requires considerable managerial skills, especially in an environment that is continually changing because of technological advancement. There are operational, logistical, and planning issues that are constant duties for the senior ranking manager. He must also manage his staff in a number of ways. There should be sufficient head count to meet the workload, and the duties should be logically and equitably distributed among the staff. There also has to be a concerted effort to recruit top quality professionals, and an ongoing effort to continually educate the staff on existing and emerging technologies.

The telecommunications manager can be thought of as a conduit between the company and the staff. He is the person who will be given directives by executive management or other departments as to what technologies will be required to support the business environment or what business applications will require support. If the telecommunications manager is told to implement a specific technology, it does not mean that the request is blindly accepted. There may be operational, strategic, or financial problems with the proposed technology. There is then a need for the telecommunications manager to educate his end-users so that they are informed of newer, more appropriate, or more cost effective solutions. This, of course, has to be accomplished in a professional and diplomatic manner. Intracompany and departmental relations and communications will be covered in a later chapter.

The telecommunications manager will also receive directives that are sound technological solutions or strict business requirements. He must work with his staff so that they understand and meet the corporate objectives. Not all projects or objectives will be popular or readily accepted. There will be some projects that will be very demanding and will severely tax a number of staff members. There will also be times when staff members disagree with the direction the company has taken. Telecommunications professionals tend to be extremely knowledgeable and opinionated. If a company chooses a different technological direction, or simply chooses to ignore a popular technology, it may be perceived negatively by a number of staff members. This is when the telecommunications manager, as a conduit between staff and company, must be able to reconcile the differences while still achieving the corporate objectives. This is no easy task and often the manager needs to exercise considerable managerial skills.

While the telecommunications manager must demonstrate considerable ability in managing people, it still does not divorce him from the intricacies of the technology. This introduces a debate that occurs often in the field of telecommunications management. There is a school of thought that telecommunications managers should be managers first and technologists second. This school of thought mandates that the manager only possess a high level view of the technologies. What is the business application? Is it a proven and viable technology? What does it take to support the technology in terms of staff and operating expenses? What are the up front costs? What are the ongoing costs? Conversely, there is another school that mandates that a telecommunications manager possess both managerial and technical skills, with a special emphasis on the technology. Proponents of this school (of which I am one) argue that quality decisions cannot be made without good technical knowledge. If the telecommunications manager does not truly understand the nature of the technology, it is quite possible that he will not even know the correct questions to ask. This scenario holds true for the management of telecommunications products and services, and the management of telecommunications professionals. The telecommunications manager who is not technically astute is vulnerable. He runs the risk of losing the respect of his staff and also of making choices that are uninformed. This may result in inefficient technology and higher costs. Because of the rapid pace of technological advancement, there is little room for telecommunications managers who do not understand the technology.

The problem of management versus technical skills is not an easy one to resolve. Certainly the telecommunications manager needs technical knowledge, but not to the degree of many staff members. In fact, once telecommunications professionals achieve the status of manager, they often find it difficult to remove themselves from the intricacies of the technology and look at the big picture. This they must certainly do or they run the risk of micromanaging the department and slowing department performance. How then can the telecommunications manager judge whether his technical knowledge is adequate enough for him to perform his job properly? And to what degree should he be involved with the details of the technology? There is no precise way to answer this question, but there are guidelines that can be used to gauge whether the knowledge base is sufficient and is being used in the proper manner. These guidelines also gauge the degree of involvement for the telecommunications manager.

Technical knowledge for the telecommunications manager can be gauged in a number of ways. First, is the manager able to effectively communicate with his staff and are they able to effectively communicate with him? Does the telecommunications manager have a clear understanding of the technology in a manner that allows him to deliver clear and precise instructions? Does he have enough understanding of the technology to offer guidance and possible solutions when problems occur? If the staff offers constant corrections or they are often baffled by—or critical of—the instructions, this may be a good indication that he does not understand the technology well. This is not to say that the staff should always agree with the manager, but there is a difference between ideological differences and a basic lack of respect. A manager who does not possess a fundamental technical knowledge will not command the respect of experienced and knowledgeable telecommunications professionals. This will yield a number of negative results.

Low staff morale will certainly be one of the resultant problems. If the telecommunications manager does not have a fundamental technical knowledge, he will not be able to effectively communicate with his staff. This is akin to manager and staff speaking two different languages. This will frustrate the staff who will find it difficult to communicate technical problems and concerns. Moreover, if the manager does not make sound technical decisions because he does not understand what the staff has told him (or he has simply ignored them), they will grow angry and frustrated. I have worked in an environment where my manager was not technically astute and have interviewed colleagues who worked in similar environments. The overwhelming reaction was negative and many of these people expressed a desire to seek employment elsewhere. A telecommunications manager who is not technically astute is not necessarily a bad manager. He may be well educated and carry a plethora of quality business experience. But without telecommunications knowledge he lacks a critical component of the triangle illustrated in Figure 3.3.

A telecommunications manager who has limited technical knowledge is certainly at a disadvantage. But there is also a danger of the telecommunications manager possessing a wealth of knowledge and using it to his disadvantage. For example, I knew of a telecommunications manager who was a registered communications distribution designer (RCDD) and overseeing a large cable installation. During the course of the project he continually occupied himself with checking the work of the cable installers, to the point where he was checking the number of threads on a screw to ensure that the design was adhering to EIA/TIA standards. This is an extreme example (although true), but demonstrates how a manager begins to perform the work of his staff while sacrificing his managerial duties. A manager who indulges in such practices will also demoralize his staff, because they feel that they are not trusted and are being micromanaged.

The telecommunications manager does not need knowledge to the degree that a network engineer does. Nor does he need to know how to program a router or a PBX. He does need knowledge that includes a fundamental understanding of all the basic technologies. This includes a basic understanding of the technology; the cost of the technology; who the major vendors are, what type of staff is required, how many people are needed to support the technology, and how the technology is applied to business applications. The telecommunications manager will not have in-depth knowledge of each technology, but simply keeping up with these high level concepts is daunting enough.

Many telecommunications managers are people who worked their way through the ranks, demonstrating a penchant for the financial and managerial aspects of the job, in addition to their talent for understanding the technology. These people already possess a high degree of technical knowledge. If anything, they must redirect their educational efforts, which will be different than those of their staff members. Rather than being concerned with the intricate details of a given technology, they must be more concerned with managing people who will be touching the technology directly. They must also develop, or refresh, their knowledge of accounting and personnel management.

For the telecommunications manager who has chanced upon the position with little or no telecommunications background, there will be a need to aggressively pursue fundamental telecommunications knowledge. This will not be an easy task, but there are many courses, books, and periodicals available today that will provide a sound foundation. This will be covered in more depth in Chapter 10. The telecommunications staff can also be a wonderful source of knowledge. They can offer advice, technical knowledge, and make recommendations for educational materials. In addition, if the staff sees a concerted effort on the part of the manager to understand their profession, they will tend to be helpful and understanding, rather than critical.

There are many other skills and qualities that a telecommunications manager should possess. Many of these skills are inherent to all managerial positions such as personnel management and basic accounting. But all managers are expected to manage their staff, balance their budget, and accomplish their objectives. What distinguishes the telecommunications manager from other managers is the technology. There are two basic factors that affect the telecommunications manager's role that relate to technology. First, as illustrated in Chapter 1, telecommunications technologies have become an integral part of the business infrastructure. If data links fail, or calls do not reach their destination, it is quite possible that revenue will be lost because business cannot be conducted. Consequently, there is pressure to provide reliable services along with backup plans. The second issue is that the technology continues to evolve at a rapid pace. Not only does the telecommunications manager need to provide a reliable infrastructure, he also needs to provide technologies that will keep the company competitive.

With these two factors in mind, there are a number of traits that the telecommunications manager should demonstrate in order to provide an effective telecommunications department. He should be a visionary, able to extrapolate into the future, and determine what technologies will likely be needed for the future of the business. This is often a difficult and risky undertaking, but essential in the modern business environment. For example, the department may have been given a directive to install a structured cable system to support a corporate campus. The immediate requirements call for bandwidths that will be supported by category 5 cable. The telecommunications manager has seen the company aggressively pursue new computing technology, and it is safe to assume that much higher bandwidths will be required within the next few years. Rather than opt for the least expensive solution, the manager decides to install a fiber optic backbone across the campus, even though it is overkill for the current application. The recommendation is not a blind risk, but rather a projection based on experience and trending data. In this instance, the telecommunications manager is strategically positioning the company for the future.

The telecommunications manager is also the architect of the department structure and working environment. The department will often reflect his personality, his capabilities, and the manner in which he manages people. This will have an impact on department performance in almost every possible way. Has he created the proper department structure to meet the technological needs of the corporation? Does this structure offer opportunity for qualified staff members? This would include creating specialist positions, managers of specific technologies, and project managers, in addition to ensuring that there is enough headcount to support the existing workload. Does he openly and actively promote (and require) continuous education and training so that the staff is knowledgeable and proficient for current and future technologies? Is he a good project manager who can oversee large, complex projects without getting bogged down in the details to the point where it slows progress? Can he accurately project costs, for both projects and the department? Does he demonstrate good problem solving skills without destroying morale or creating animosity with vendors?

Ultimately, the telecommunications manager may be judged by the reputation and performance of the entire department. Are staff members generally perceived as knowledgeable? Or is the general consensus that they are behind the times and often do not know the answers to questions? Do the staff members handle problems well, demonstrating good problem solving skills and calm, even in the face of adversity? Are projects generally on time and on budget? Do the staff members have good interpersonal skills, working well with end-users and helping them with their technological needs? Does the department offer quality technical solutions, or are there often operational and support problems after the installation?

As the reader may ascertain, the telecommunications manager is the catalyst, the guiding force behind the performance and success of the department. He must be able to project a solid professional image and be able to relate to the highest ranking officer within the company. This includes presenting the technology in a user friendly format along with the dollars and resources that will be required to support this technology. In addition, he should be able to convey the strategic advantages of the technology and the benefits it will bring to the company, for the present and the future. Conversely, the telecommunications manager must also be able to relate to technical people who might have no managerial or professional aspirations other than to be a technologist. On the one hand, he must be able to relate to the MBA, on the other hand, he must relate to the engineer. It is not an easy task, and it does take special capabilities.

3.4 Assistant Managers

Assistant managers are more prevalent in larger organizations. In many cases, they tend to be specialists with responsibility for a specific technology. While the senior ranking manager in either the flat organization or the pyramid is usually a generalist who oversees all technologies, the assistant manager takes on a more specialized role, taking exclusive responsibility for a technology segment with dedicated staff. Some of the most obvious classifications are the general categories of voice and data communications. Depending on the size of the organization, there might even be a need for other types of assistant managers. For example, there may be a need for a manager of video or Internet services.

At this point, it should be noted that the title of assistant manager is being presented as a generic classification, the same as the title of telecommunications manager. Just as the senior ranking manager can be a vice president or director, the subsequent layers in the pyramid can carry various titles such as manager of voice communications or senior data communications manager. It is also possible, in the largest of organizations, to subdivide a technology and assign a manager to a specific aspect of managing that technology. For example, there may be a manager of data network design, a person who oversees a staff of designers and network engineers. The opportunity to develop such subdepartments is contingent upon the size of the company and the scope of the technology within that company. A large organization with thousands of locations and a multitude of different communications protocols would require people dedicated to network design.

The assistant manager, being in the upper level of the pyramid, is responsible for the same three categories illustrated in Figure 3.3. The difference is that the responsibility is exclusive to a specific technology or subdepartment within the department. The assistant manager requires many of the skills held by the telecommunications manager. He is responsible for his own budget, a staff, and must manage the technology to meet the corporate business objectives. His knowledge base is more exclusive than the telecommunications manager. He knows what type of people he needs on his staff, what tasks he needs to assign, and has more detailed knowledge than the telecommunications manager. In the pyramidal structure, the telecommunications manager relies heavily on the assistant managers to manage the individual technologies. He also expects them to give him guidance and advice, because he simply does not have time to learn the details and nuances of each technology.

The assistant manager can be thought of as a conduit between staff and the telecommunications manager. He obtains the company objectives from the telecommunications manager and works with his staff to accomplish those objectives. He takes a more hands-on approach to the projects and ongoing administration of his particular field of specialty.

At this point, it bears repeating the effect that the telecommunications manager will have on the department. As mentioned in the previous section, the telecommunications manager is the guiding force for the department, and the department will reflect his managerial style and personality. For example, if the manager is demanding and unreasonable, or perhaps dictatorial in his style, this will reflect upon the assistant managers, and eventually down into the staff. As much as the assistant managers may try to shield the staff from this behavior, they will find it difficult and their frustration will eventually emerge at some point. Or consider that the telecommunications manager does not care to solicit the opinions of staff in terms of strategic or technological direction. At this point, regardless of how frustrated the staff may become, the assistant managers will be relatively helpless to address their frustrations. They have no choice but to follow the direction of their manager, regardless of whether or not they agree with the philosophy. That is why the concepts of open culture, team work, and employee empowerment are very effective tools in the modern telecommunications environment. There will be more discussion of these concepts later in this chapter.

3.5 Project Leaders/Managers

Project management is such an important part of telecommunications department management that many companies have developed specific positions to address this set of unique skills. Termed project managers or leaders, these are people who have demonstrated the ability to oversee and manage projects that are large, expensive, and complex. At a high level, a project encompasses four basic components: a timeline, tasks, resources, and a budget. But this only addresses a portion of what constitutes successful project management. Project managers must be adept at planning, problem solving, personnel management, and documentation. They are people who must be able to vacillate between micro and macro levels at various times during the course of a project. This is necessary in order to ensure continuity so that the proper objectives will be met. Consequently, these are people who must possess strong organizational and managerial skills.

As stated earlier in this chapter, only the most stagnant of organizations do not experience a constant influx of telecommunications projects. Therefore, it is imperative that project management skills be taught to all staff members who may have cause to use them. Most staff members, at one time or another will be assigned some form of project. The projects will be in varying degrees of size and complexity. The largest and most complex projects may be assigned to project managers. Also, they may serve as advisors and consultants to other staff members who may be new to project management.

The position of project manager can be a critical part of the telecommunications department, effectively guiding the department through projects that are highly visible to the rest of the corporation. In fact, the reputation of the department may rest solely on its ability to effectively manage projects. It may not be noticed that telephones are added or moved quickly and accurately. It might also be taken for granted that response time is always good or that there are always sufficient ports for the PCs. These are administrative functions that "are expected to be executed." However, a special project, such as the addition of a new building on a campus, will probably gain much attention, especially if the building supports a critical and lucrative part of the company's core business. If the project is completed successfully there will be accolades for the telecommunications department. If the project is not successful, the department will receive a considerably large black eye. Why is there slow response time? Why weren't PCs and telephones installed for employees who were ready to begin work? Why is the project 20 percent more costly than your original estimates? Project managers are charged with ensuring that such problems do not occur.

Because project managers carry a high level of responsibility, and because their role is so visible, these positions can be effective interim positions between staff and management. The position can carry a higher salary range than most staff positions, making it an attractive target for aspiring staff members. And while project managers do not have supervisory authority over other staff members, they must interact quite often with members of a project team, which is comprised of staff members. A project manager with poor people skills will affect the flow and possibly the success of the project. It is therefore imperative that these people display quality interpersonal skills. This is one reason why project managers may be good candidates for managerial positions. If they can effectively supervise staff members who do not report to them, it is a good indication that they will be effective managers with direct reports.

In the field of telecommunications, project managers require strong technical knowledge, but technical analysis is not necessarily their main duty. Just as managers must maintain a high level of knowledge of the technologies without getting mired in the details, so must the project manager. What is required of the project manager is the ability to understand what resources are needed to address the needs of the project. This includes understanding what technical expertise is required. The expertise may reside within the department, or there may be a need to hire consultants or contract workers. This is part of the planning process when project managers assess the scope of the project. They will assess the situation and make a report to their manager about staff members who are required, the estimated amount of time that they will have to commit, or if the department will have to budget for personnel resources from outside the company.

Project managers must also possess strong financial skills. Very often, they will be asked to assess the feasibility of a project. Part of the assessment will be financial analysis. They must be able to evaluate operating and capital budgets, in addition to preparing reports and making presentations to management. Their preliminary reports are important because they will indicate to management how much money is to be allocated for the project. Once the moneys are approved, the project manager will be expected to complete the project within the budget he predicted.

The planning process can be complex, especially if the department is implementing new technology with which they have no experience. This introduces a certain degree of guesswork, but the project manager is also expected to research information available about the new technology. He may also solicit the opinions of staff members to see if they can offer opinions, have taken a class, or know a colleague who has experience with the technology. This may indicate the viability of the technology and possible challenges or risks.

Challenges and risks are an inherent part of any project. The project manager must be prepared to overcome possible problems or to offer a back-up plan in the event of failure. It certainly does not look good when the department has to back off a major implementation. But it looks twice as bad when there were no contingency plans in place if the implementation fails. Good project managers are able to plan for the worst case scenario. This means that the project may only be delayed in lieu of being perceived as a failure.

Once a project is in progress, there are numerous logistical issues that need to be addressed. In addition, there are many problems that will surface, many of which will be unpleasant surprises. Problems arrive from a number of sources and they are not always the result of poor planning. There are times when vendors do not meet their objectives, when the corporate objective changes, or the project may be something with which the telecommunications department has no experience. Consequently, there would certainly be a learning curve. Regardless of the source of the problems, project managers must be adept at solving problems. They will work with project team members to determine solutions and make reports to management regarding any changes that will be required. There could be an unpleasant surprise about the capabilities of the technology, an unexpected increase in the budget, or something unforeseen that has affected the schedule. The project manager needs to identify the problem and set a course of resolution. Moreover, he must overcome these obstacles while maintaining the cohesiveness of the group. During the course of a high-level telecommunications project, when things go wrong, there is often a great deal of finger pointing. This is between staff members, vendors, end-users, and management. This can be a very destructive force in a project and the project manager must diffuse anger, minimize the emotional element, and keep all concerned parties focused on the project. He must then make his report to management, defining problems, solutions, and changes.

During the course of the project, there will be progress meetings, status reports, documentation, the assignment (and reassignment) of tasks, and brainstorming sessions. Most of this occurs through interaction with end-users, vendors, and staff members. There are times when issues will be difficult and the pressure will be great. Consequently, the project manager will have to exercise good interpersonal skills to keep things on track, even in the face of adversity.

In organizations that support large project loads, good project managers are an invaluable asset. They can assure the success of highly visible and complex projects in addition to mentoring staff members about project management. Also it should be noted that project managers do not always have to be a specific position. A project manager can be assigned for a specific project, but not necessarily carry that title. However, many companies still recognize that the project manager is an important function. While there may not be a specific position assigned to this function, staff members who are assigned difficult and complex projects are often awarded bonuses that are specifically tied to the project. Once again, this makes the position of project manager, even if it is a temporary one, a position that many staff members will aspire to.

3.6 Designers

Large and complex networks often demand that positions be created that are dedicated solely to network design. Corporate networks, whether voice or data, require a central authority in order to address cost, capacity, capability, reliability, standards, and infrastructure. These people assess the business requirements of the corporation, what type of traffic will traverse the network, and then design a system that is logical, efficient, reliable, and cost effective. Once the foundation is laid, these people are responsible for keeping the design current by either fine tuning it or redesigning it as business requirements may dictate, or as technology changes.

The position of network designer is highly technical in nature. Designers are expected to understand, in minute detail, the intricacies of various technologies. This includes legacy systems and emerging technologies, network services (e.g., frame relay, SONET, ATM, etc.), and equipment (e.g., routers, hubs, etc.). They must understand the hardware and software, and the strengths and weaknesses of each service and product. This requires that designers continually read current literature, which includes books, magazines, and vendor-specific literature. This continuing education is also supplemented with various courses and seminars. This is essential, because there is no option of resting on one's laurels. The technology simply changes too quickly.

This in-depth knowledge is essential because the company expects guidance and expertise from the telecommunications department for network performance and reliability. Management will want to know if emerging technologies will provide operational or financial advantages, or if the technology is even mature enough to consider. They will also want to know if the existing network, with its array of technologies, will support the future needs of the company. Network designers are expected to provide such insight. And whether the technology is leading edge or a reliable stand-by, the network designer is expected to provide an optimum performance at a cost-effective price.

The position of network designer is steeped heavily in theory. They must understand how the communications traffic is formatted and how it is affected by various conditions throughout the network. They learn how to blend various types of communications protocols, prioritize the mission critical traffic, or to design alternate routes during congested periods. They also may build redundancy into the network. If a primary WAN link fails, an alternate route may be automatically programmed, or enabled via manual intervention. The designers make these designs based on business requirements. Generally, the more reliable and robust a network is, the more expensive it will be. Therefore, some applications will not be as critical as others.

Many network design positions are related to data, however, there is still a place for voice designers. Consider a number of call centers that are order entry points. A voice network designer will provide a toll-free network, equipped with advanced features so that calls are always routed to a site where they may be serviced. This may entail time-of-day routing, day-of-week routing, day-of-year routing, area code routing, exchange routing, and all-trunks-busy routing, to name a few of the available features. Such a network requires the management of a large database (area codes, exchanges, and 10-digit numbers), the programming of the routing features, and the provision of sufficient trunk lines to carry the offered call volumes. Network traffic must be monitored periodically and features changed based on changes in the toll-free traffic. In a business environment that is highly dynamic, a full-time voice network designer would be warranted. A prime example of this scenario may be the airline industry where large and complex toll-free networks are used to support the reservations side of the industry. In the airline industry, this is the economic lifeblood of the business.

Once a network is operational, network designers may also employ various types of sophisticated tools that provide information related to network performance. By examining this data, they may be able to see problems emerging and take corrective actions before the problem becomes acute. They may also utilize sophisticated software that enables them to simulate network conditions so that they may examine new or alternate designs, based on emerging trends or new business applications. Once again, education is a constant. Network design tools and management packages are sophisticated software that requires training and sometimes a certification process.

Beyond the specific disciplines of voice or data, there are also many companies that design networks to carry all forms of traffic: voice, video, and data.

When these types of traffic are blended, the network has to be carefully designed so that time-sensitive traffic such as voice is given priority over data that may be buffered. Network designers must ensure that there is sufficient bandwidth and traffic is properly prioritized.

Network designers sit somewhere in the middle of the pyramid. Good designers are an invaluable part of the organization. They are well-paid, salaried employees who are looked upon as being "gurus" who offer insight when complex applications and requirements face the telecommunications department. But while these people may be a critical and essential part of the organization, they also represent positions that are often purely technical in nature. This issue represents a quandary that is common in telecommunications department management. That is, the technical specialist who wishes to remain a technical specialist.

Highly technical people may be well-paid, highly educated, and dedicated, but they may not aspire to be anything other than a technical professional. There is certainly nothing wrong with this. Many telecommunications professionals are in the business because they have a passion for the technology. Others arrived in the business for completely different reasons. This does not interfere with the purpose of the pyramidal structure and the capability to offer opportunity. There will be people who will not aspire to supervisory or management positions and will be perfectly content to work in a purely technical position for the duration of their career. The point behind offering opportunity via the pyramidal structure is that staff members have upward mobility should the opportunity arise. It does not mean that everybody will have an opportunity or that everyone will aspire to a management position.

A second issue is that purely technical professionals may approach their duties differently than other staff members. For example, an analyst who aspires to move into a management position may dress in a more formal manner, willing and able to interact with the end-users and management, making presentations to management, and continually developing management skills. A staff member who sits purely on the technical side may not have these capabilities, and may only reluctantly participate in, or develop, these skills. This is not indicative of a bad employee, or a person who is a detriment to the department. Rather, it simply means that this person's role is different. Neither the managers nor other staff members (such as analysts) usually have the degree of technical knowledge held by the network designer. If they did decide to pursue knowledge in such detail, they would find that they had little time for anything else. This demonstrates the dividing lines between technology and management; specialist and generalist. These are important components of the department structure and must be considered in the management philosophy.

3.7 Analysts

One of the most common positions that are prevalent across company and industry segments is the telecommunications analyst. There is seldom a telecommunications department that does not have a quantity of analyst positions. Generally, these people make up a major portion of the telecommunications staff and perform a number of critical functions. Analysts range from junior to senior level and, depending on how the department is structured, this may be a stepping stone to management.

Analyst positions are usually designed to be generalist positions, but describing it this way may be misleading. Analysts can actually have very detailed knowledge and many departments require that their analysts become specialists in one of the many disciplines of telecommunications technologies. Often the specialization will be reflected in their title, for example, senior data communications analyst. Regardless of how specialized or detailed the analyst position may be, the title is indicative of what the analyst does: analyze technology. It is not, however, a purely technical position. There are many duties performed by telecommunications analysts which include financial analysis, project management, and administrative functions.

Financial analysis performed by analysts is always related to specific projects or isolated segments of overall department responsibility. For example, a project may be assigned to an analyst to evaluate new routers for a new section of the WAN. Assume that the company has acquired a new division and the old routers are being replaced. Should the new routers be purchased or leased? If they are leased, should the department exercise an upgrade clause, in the event that the vendor introduces a new product line? What type of maintenance contract should be purchased, a contract that covers Monday through Friday, 8-5 p.m. or a contract that covers 24 hours a day, seven days a week? The latter would be more expensive but would it be needed? The analyst would consider these factors (and other variables) and develop a spreadsheet. He would make a recommendation based on the needs of the end-user, the network design, and the allotted budget. Based on this analysis, he would make his recommendation to the telecommunications manager who would, in turn, make his recommendation to upper management. This example illustrates how a staff member will need to execute work that is high level and managerial in nature. The telecommunications manager very well could perform the work, however, he probably does not have time to operate at this level of detail. Consequently, he should guide staff members (in this case the analyst), teaching them financial analysis and setting department standards for reporting (text and spreadsheets), so that the deliverable items are consistent and the data easily assimilated for decisions.

When reports are provided to the manager, he can quickly assimilate the information, analyze it, and make a decision. As he examines the information, he may also confer with the analyst, asking for opinions and guidance. While the telecommunications manager is the ultimate decision maker, he brings the analyst into the decision making process, thereby promoting the concept of teamwork. This is also an educational process for the analyst, who learns how the management process works.

Beyond specific projects an analyst may be financially responsible for a specific segment of technology on an on-going basis. For instance, the analyst may be given the responsibility to oversee the inventory and billing of the routers that are purchased from the chosen vendor. Although this may be more administrative in nature, staff size may mandate that analysts perform this duty. Also, there may be technical issues involved with the administration of the routers. For example, if the upgrade clause was elected for the maintenance contract, software upgrades or router replacement might require technical analysis in addition to the administrative functions of billing and inventory management.

Project management is very often a primary duty of telecommunications analysts. In departments that do not develop the positions of project managers, this responsibility usually falls on the shoulders of the staff analysts. Even if project managers are part of the department structure, many analysts are still required to perform project work, although perhaps not to the degree that a project manager would. Project management duties for analysts are the same as for a project manager. The difference is that project managers are charged to exclusively oversee projects while analysts will carry other duties. When an analyst finds that his workload is almost entirely dominated by project work, at the expense of technical analysis and education, the telecommunications manager should consider adding a project manager position to the department. As previously stated, project managers do not always need to have the most in-depth technical knowledge on the staff.

The analyst's job is to evaluate technology on a number of levels. This can range from technical analysis for a project to handling trouble calls for existing technologies. The analyst carries a wide range of responsibilities, which includes understanding many technologies in great detail, what they cost, how to implement them, and how to support these technologies after the fact. As in the case of the project manager, this position can be considered a stepping-stone to management, but this is not always the case. Because of the lack of standards in the industry, many companies consider the analyst position to be more hands-on, closer to a technician. Other companies maintain a job description that is concentrated more in the analytical arena, closer to managerial duties.

The position of analyst lies somewhere in the middle of the pyramid. These are highly technical positions, and can be thought of as the backbone of the department. Of all the positions in the department, this is the one that is most common and critical to department performance. Not all departments will have assistant managers, network designers, project managers, technicians, or administrators. They may, however, have a number of analysts who carry out some (or all) of these functions. Once again, the position of analyst can be thought of as one that is a stepping stone to higher positions. The pyramidal structure allows for aspiring and capable analysts to fulfill this objective.

3.8 Technicians

Technicians are not a part of every telecommunications department. They are positioned at the lower portion of the pyramid because they are generally considered to be staff members who work directly with the technology. Technicians may be hourly or salaried employees. Either status is common, however, many technician positions are considered to be a skilled labor job and so it is not uncommon to retain these positions as hourly. Telecommunications managers may want to consider making technician positions salaried, given the long and unusual hours and how much overtime might be paid.

Technicians are usually specialists, assigned to support a specific technology. For example, a medical center may retain a staff technician to administer adds, moves, and changes for a PBX. The technician will also perform basic troubleshooting duties, while major problems will probably be addressed by the vendor. While other members of the staff may have four-year college degrees, technicians often have several years of technical school in addition to on-the-job-training. Vendor-specific training often augments their basic technical training. The technician who administers a PBX attends training provided by the manufacturer of the PBX. The same is true for vendors who support routers or hubs.

Many telecommunications managers choose not to carry technicians on their staff. This decision is based on economical and operational issues. It may be less expensive to outsource the hands-on support of various products rather than carrying expensive personnel. While the hourly rate from the telecommunications vendor may be more expensive than an in-house person, there are other factors that affect the total cost of ownership. Employees are also paid benefits, which can often be an additional 30 percent of the employee's salary. Moreover, there is vacation time that must be allotted, which introduces additional staffing and scheduling issues. The vendor that is supporting the system often has an entire staff of trained technicians. The hourly rate may be higher, but the telecommunications department is not paying benefits, nor do they need to be concerned with scheduling around a staff member's vacation time. The downside to outsourcing is that a complex network often requires a technician who is knowledgeable of the design. Otherwise, he may not know how to accurately evaluate a problem. There is also the factor of employee loyalty and commitment. Your own employee often wants to make sure that work is right or the problem is solved. The dispassionate employee of another company may not attack a problem with the same verve. As is the case with all other positions, the pyramidal structure allows growth for all staff members who wish to enhance their knowledge and business skills.

3.9 Administrators

Administrators perform many of the clerical duties that are a large but important part of departmental functions. For example, there are scores of bills that need to be paid. There are long distance bills, local bills, calling card bills, data circuit bills, installation bills, and maintenance contracts, to name but a few. These bills must be logged, approved for payment, and charged back. In addition, any experienced telecommunications professional will know that telecommunications bills are notoriously prone to errors, in addition to being difficult to read. Just being able to plod through the bureaucracy of a vendor, trying to correct a bill or understand if a mistake was made, can be the ultimate exercise in futility. Even in the information age, the administration of the telecommunications department can be a complex and daunting undertaking that demands large volumes of man-hours to plod through the mountains of computer and paper records. As much as any telecommunications manager tries to streamline his operation or institute policies to enhance efficiencies, the department is still faced with a paper blizzard and large volumes of complex data that must be meticulously managed. Since this work is primarily administrative in nature, with only a minor emphasis on technology, there is a need to establish administrative positions within the department for two reasons. First, the sheer volume almost always demands a full-time position (or perhaps a number of them). Second, if administrators are not hired, other staff members will be bogged down with this work, to the point where they will not be able to perform their primary duties. If this happens, it will have a detrimental effect on the morale of these staff members. There is nothing more frustrating to a telecommunications professional than to be burdened with "busy" work.

Administrators can be full-time employees, but just as often, they can be part-time or even contract workers. Because the position does require a certain degree of technical knowledge, many departments hire retired telephone company employees on a contractual basis. These are people who understand the complexities of vendor billing and therefore require relatively little training. The position may also be developed so that nontechnical employees may enter the department. Perhaps a secretary is looking for a better position within the company, and the position of telecommunications administrator will offer more money. Such a person may gradually learn the technology, perhaps moving up the pyramid, if they have the initiative and talent.

There are many other duties that may be performed by administrators. Long distance calling cards can be a large database that requires constant maintenance. New cards must be issued to new employees, lost or stolen cards replaced, toll abuse monitored, and bills checked. The same responsibilities may hold true for cellular telephones. In addition, the administrator may perform simple software functions on systems such as PBXs. This may appear to the reader to be busy work and, to a large degree, it is. It is, however, unavoidable work that must be executed. A new salesman who does not receive his calling card will become vocal very quickly. If an executive does not receive his cell phone prior to an important business trip, there will certainly be repercussions. Managers, network designers, project managers, or analysts cannot be burdened with these details or they would never be able to provide the high level functions that are essential to department operation.

Administrators reside at the bottom portion of the pyramid. The pyramid is designed to offer opportunity to all staff members, of which administrators are certainly a part. However, just as technicians may never aspire to a higher position, the same may be said of administrators. These positions may be filled by employees who have no aspirations to become a telecommunications professional, and will only learn as much as they need to know in order to perform their job. Or, the positions may be filled via contractor positions, perhaps a retiree who is only looking for supplemental income. Regardless, opportunity is available to all positions via the pyramidal structure.

3.10 Contractors and Consultants

Contractors are not company employees, nor are they a permanent part of the staff. They may appear anywhere in the pyramid below the management positions to serve any of the previously mentioned job functions. Contractors are hired on an as-needed basis. There may be a specific length attached to their contract or they may be retained on an as-needed basis. They may be hired individually or from an employment agency. Consultants and vendors also often offer contractor workers as a part of their array of services.

The role of contractors may be two-fold. First, they may perform work because there is simply not enough manpower within the department. The telecommunications manager may understand that the sudden increase in the workload is only a temporary situation. Therefore, he has determined that it would be less expensive to carry a few contractors on the staff for a short period of time, rather than to hire permanent staff. Once the workload has diminished, the contractor agreement is terminated. A second reason for hiring a contractor is to provide expertise that does not reside in the department, or as an adjunct to existing expertise. Here there is a fuzzy line between the definition of a contractor and that of a consultant. The difference is that consultants are almost always hired for their expertise, whereas contractors are hired to address a staff shortage and to perform actual work.

As illustrated in the previous section, contractors can be retained to perform much of the administrative work in the department. Although these people may not be a permanent part of the group, they may actually work a full-time schedule. This is still advantageous for the budget because there is no contractual commitment, other than the length of the contract. In addition, the contract may not be with the individual, but rather with a contracting firm. When the agreement is with a firm, the telecommunications manager does not have to be concerned with issues such as vacation or absenteeism. It is the firm's responsibility to provide a capable substitute. It is also less expensive to use contractors in lieu of regular employees. While the hourly rate may be high, benefits are not included and the relationship may be terminated when there is no longer a need.

Consultants are retained on an as-needed basis. As previously stated, consultants are hired for their expertise and they may be brought into the department for a number of reasons. Perhaps the department has received a directive to implement a technology of which the department has no expertise. Or the company may have made a major acquisition and the department is charged with merging the existing network with one that is completely foreign. Consultants bring expertise that does not exist within the department and the use of consultants is often a necessity. It should be noted that the process should be educational for the department and a natural method of addressing the workload. However, the telecommunications manager should be careful to stress this with the staff, to allay any paranoia or wariness that the use of consultants may introduce.

3.11 The Help Desk

Telecommunications departments as first level support often provide help desks for problem determination. First level support is defined as the first (and only) point of contact for the end-user. Because telecommunications problems can originate from a number of sources, the telecommunications department needs to define a resource that is consistently available in addition to being capable of solving complex problems. That resource is very often the help desk.

In small departments, any staff member may field calls from end-users. As the department grows, the staff will find that it is inundated with trouble calls, in addition to spending most of their time resolving problems. This takes away from other duties to the extent that it may slow progress for projects or other essential duties. When staff members begin spending an inordinate amount of their time resolving problems from the field, it is time to consider the development of a dedicated help desk.

The help desk provides a number of services to the end-users and the company in general. A basic function is to provide technical advice and support to the field. The most important function, however, is to provide first level support for problems that relate to telecommunications products and services. The trouble reporting and resolution process can entail many levels. At a very fundamental level, the process can be simply walking the end-user through a simple diagnostic procedure in order to resolve the problem over the telephone. However, most problems are not that simple. In many cases, the help desk specialist will have to ask many questions and try to determine the source of the problem. This could be the LEC, an IXC, or one of several equipment vendors. He will then have to generate a trouble ticket with the vendor and track the course of the resolution process. During this process, the help desk specialist will need to gather the necessary resources, and inform all concerned parties within the company. This is certainly not an easy process and there will be times when vendors will point their finger at another vendor. Ultimately, the telecommunications department owns the problem until it is resolved, and the help desk specialist can be thought of as the traffic cop who oversees the process from beginning to end.

Modern telecommunications networks are complex, comprised of a wide array of technologies, vendors, and services. Because of this, help desk specialists must have a high degree of technical knowledge. They must also have a thorough understanding of the network design, in addition to the vendors that support the network and their respective organizations. A fundamental knowledge of company infrastructure and policies is also mandatory. Beyond the knowledge base, the telecommunications manager must provide the necessary resources to provide for an effective help desk. This includes documented procedures, trouble-reporting numbers and contacts, escalation procedures, organization charts, and diagnostic tools that will allow specialists to monitor and diagnose network problems.

Maintaining a help desk for the purposes of trouble reporting offers many challenges to the telecommunications manager. Help desks are often maintained twenty-four hours a day, seven days a week. This makes for long, odd, and stressful hours. Help desk specialists are inundated with complex problems and irate people. They must demonstrate superior interpersonal skills in addition to advanced problem-solving skills. Finding people to fill such slots can be a demanding undertaking. For the most part, people who possess good technical skills have no desire to spend their entire day dealing with end-users that are complaining and often abusive. Consequently, it is difficult to find quality recruits and turnover is common due to burnout. For as challenging as the help desk positions can be, the pyramidal structure can offer them many opportunities. In fact, this may be an aid in recruiting technical talent, citing developmental programs that allow career paths through the pyramidal structure.

3.12 The Open Culture, Teamwork, and Empowering Employees

Although department structure is important for morale and department performance, department and corporate culture also have a critical impact on the success of the department. Supporting a corporate telecommunications environment in the latter part of the twentieth century has become a complicated undertaking. No single person, regardless of knowledge level or degree of experience, can make quality decisions completely by himself. If the telecommunications manager were to assume this duty, how could he keep abreast of all technologies, the financial aspects, how the technology is applied, and what support requirements are necessary? A person who had such a complete knowledge would have to devote his full-time duties to education.

Because the telecommunications manager must also manage a department, his time is precious and he must be selective about how he spends his educational time. Therefore, he must make use of all of his resources. This includes bringing the staff members into managerial processes that were once the sole responsibility of the telecommunications manager. If the staff members are not brought into many departmental processes, it could create many difficulties for the telecommunications manager. The staff members may not readily accept the telecommunications manager's decision because they have an alternate (and possibly more informed) opinion. Because one person makes the decision, it may not be the most informed decision, and could be expensive and ineffective. Lastly, if the decision is not accepted by the staff, they will fight it to the point where the entire project is undermined, making the telecommunications manager's life miserable. The answer to this situation is three-fold: open culture, team concept, and employee empowerment.

The open culture means that there is an open line of communication between management and staff. This includes an open door policy whereby management is readily and easily accessible. It also means that opinions are solicited from both sides of the fence, both management and staff. When this happens, the staff understands that their opinions are valuable. This encourages them to offer ideas, opinions, and to openly participate. Otherwise, valuable opinions may be withheld that could have benefited the department and the company.

The team concept is invaluable to the success of the modern telecommunications department. Teams serve a number of functions that are beneficial. First, they can serve as a committee that continuously addresses the changing demands of specific technologies that are supported by the department. For example, there may be a WAN team that meets periodically to address a number of issues related to the WAN support such as design, performance, monitoring, and alternate technologies. In this instance, the team may be comprised of a network designer, various analysts, and possibly an assistant manager or technician. The team would examine various issues on a periodic basis. There are current issues and problems, which include pending projects, potential new technologies, vendor support issues, and departmental procedures. The WAN team would make reports to the telecommunications manager, always striving for continuous improvement. A committee team meets infrequently, possibly once a quarter, or when issues demand a more frequent meeting.

A second application is to assign a team to a specific project. Project teams are comprised of staff members who have knowledge and expertise in the technologies that are specific to the project. If the department was charged with providing communications to a completely new site, the required expertise might be cable, LANs, PBX, and WAN specialists. The project team would be led by a project manager.

The team concept promotes department unity and serves as a training ground for staff members to move into higher positions. Team leaders must demonstrate good interpersonal skills, often assigning tasks and overseeing projects with people to whom they have no direct authority. They also must demonstrate good organizational and problem-solving skills. This can lay the groundwork for supervisory responsibilities, creating the possibility of future advancement up the pyramid. Once again the open culture is essential to successful teamwork. The team leader cannot dominate in a dictatorial fashion. All team members must be given the opportunity to participate and offer ideas and opinions. If progress falters, then the team leader will be the ultimate decision-maker.

Empowerment is also critical to the success of the department. Staff members are encouraged (and expected) to be independent and able to make their own decisions. This enables the department to become more efficient, and the telecommunications manager does not become a bottleneck. If the telecommunications manager has to approve every aspect of department or project administration, efficiency is seriously impaired. Staff members should be empowered to make such decisions. Lines are then ordered, circuits installed, technicians are dispatched, equipment and software is designed, and meetings are set up and conducted independently. When these things happen, the operation of the department proceeds more efficiently.

A benefit that evolves as a result of the empowerment process is that trust begins to form between management and staff. That is not to say that there will not be difficult times. Employees who are not used to making their own decisions will certainly make their fair share of mistakes. There will even be a percentage of employees who prefer direct and close supervision. Regardless, the empowerment process benefits both management and staff, enhancing the efficiency of the department. Management is relieved to address large-scale issues while staff receives invaluable training that may be the foundation for future advancement.

Because empowerment may not be an easy concept for some staff members to grasp, the implementation of empowerment as a standard policy may be one that is gradual. The telecommunications manager will need to train and council employees, indicating what decisions they can (and should) make, and outlining the issues that should be the exclusive domain of senior management. The boundaries will be different for every company, but managers should always strive to empower their staff as much as possible in order to assure that department efficiency is optimized.

Regardless of the structure (whether flat or pyramidal), these three concepts promote higher morale, greater department efficiency, and more employee responsibility. Staff is better positioned for future advancement, decisions are more informed, and ideas for continuous improvement flow through the organization.

3.13 Summary

There are no national standards for defining telecommunications department positions or department structure. Most companies develop their departments on an ad hoc basis, or as business and technology dictates. There needs to be a logical design to the department structure in order to establish maximum department efficiency and to maintain the highest possible level of morale. The design should allow for upward mobility, offering staff members the opportunity for advancement. Specialist positions should be developed to address complex and widely deployed technologies. When properly structured, and with the correct corporate culture, the telecommunications department becomes more efficient and stable.

4

The Telecommunications Department and the Corporation

The telecommunications department serves many roles in the modern business environment. It can be the sole provider of all telecommunications products and services or serve as a partial provider. There are also times when the role of the department is that of consultant, and end-users have the option of using the telecommunications department, performing the work themselves, or hiring outside help. Regardless of the role that the telecommunications department serves, the telecommunications manager should take concrete steps to ensure that the department provides the optimum value to the company, and that the value is understood and appreciated by all concerned parties.

The telecommunications manager does not always define the role of the telecommunications department. This can be dictated by the overall business plan, policies set forth by executive management, the project load, or the performance and reputation of the telecommunications department. However, whatever role has been defined for the telecommunications department, the telecommunications manager must make sure that the role is served well.

4.1 The Telecommunications Department in the Decentralized Company

Companies are structured in a variety of ways. They can be centralized, decentralized, or some combination thereof. The structure of the company has a definitive impact on the role of the telecommunications department. For example, a decentralized company may be comprised of multiple business units, each one acting independently of each other. Consider a company in the foods business that has six divisions (see Figure 4.1). Each division is its own profit center and is managed independently of the other units. Each division also has developed its own corporate culture and business procedures. Now consider that each division is not responsible for tying into the mainframe computers located at the world headquarters in order to perform daily business functions. They also provide their own PBXs, ACDs, local service, long distance, and toll free service. Basic business functions, such as order processing, inventory, and a number of accounting functions are performed on their own computers. In fact, their only obligation to the corporate headquarters is to batch financial reports to the corporate headquarters on a daily basis. Corporate headquarters will not interfere in the operations of this subsidiary unless their

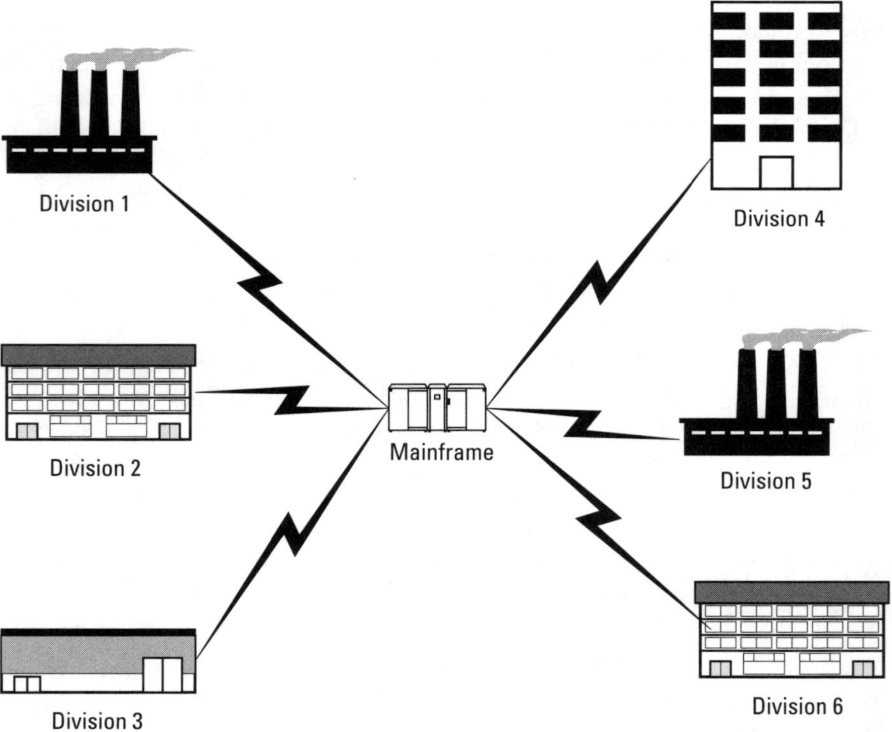

Figure 4.1 Decentralized company structure.

financial performance begins to suffer. Under such circumstances, there may actually be very little need to retain a telecommunications department. In fact, it would not be uncommon to find that nontelecommunications professionals within each subsidiary have assumed all telecommunications duties.

Decentralized companies offer a number of advantages to each business unit. First and foremost is that the business unit is "lean and mean." In other words, division managers can make quick decisions, independent of and unencumbered by the bureaucracy of the corporate parent. There are many more advantages to the decentralized approach but, unfortunately, there are just as many disadvantages. One of the most distinct disadvantages is the lack of central telecommunications planning. Decentralized companies are very common in the business world. In fact there are often trends that argue for centralization versus decentralization and vice versa. When these changes are implemented, they have an enormous effect on the telecommunications department.

In the decentralized environment, it is quite common for the telecommunications department to carry a diminished role. Very often, there is a small department defined for the corporate headquarters with the primary role of supporting the corporate parent. This may include support of a single building or campus. The telecommunications department may also be asked to act in an advisory capacity to the rest of the corporation, or to oversee major projects for branch locations or subsidiaries if they require more advanced expertise. While there may be many arguments for a decentralized structure, the telecommunications infrastructure of the company usually suffers greatly.

Once again, consider the example shown in Figure 4.1. Assume that this company does not have a master long distance contract and each subsidiary has negotiated its own long distance contract. The negotiated rate may or may not be competitive. Whatever rate was negotiated, it is a safe bet that the contract was negotiated by employees who were not telecommunications professionals. Once more, if the aggregate long distance rate of all subsidiaries were to be calculated, they would probably be much more expensive than if a master contract had been negotiated (see Table 4.1).

Continuing with the example shown in Figure 4.1, equipment chosen by the subsidiaries may prove to be inadequate or expensive. Once again, a master contract could be in order, depending on the type of equipment. However, the more glaring problem is that equipment is evaluated and purchased by nontelecommunications professionals. Consequently, many operational problems will probably surface. For example, call centers may have inadequate ACD reporting systems or Internet connections may be via a slow dial-up connection in lieu of a dedicated link. It may also mean that a PBX or key system will run out of capacity, necessitating a "forklift" to a new system before the old system has been fully depreciated.

Table 4.1
Long Distance Charges for a Decentralized Company

	Monthly Bill	Cost/Min.	Master Contract	Cost/ Min.	Savings
Frozen foods division	$1,430.00	$0.11	$780.00	$0.06	$650.00
Diet foods division	$1,527.80	$0.10	$916.68	$0.06	$611.12
Condiments division	$3,285.66	$0.14	$1,408.14	$0.06	$1,877.52
Dairy products division	$2,979.76	$0.07	$2,554.08	$0.06	$425.68
Ethnic foods division	$1,274.68	$0.11	$695.28	$0.06	$579.40
Candies division	$1,576.20	$0.15	$630.48	$0.06	$945.72
Totals	$12,074.10		$6,984.66		$5,089.44

In the decentralized environment the telecommunications manager has two options. First, he can take the path of least resistance. This would mean supporting only the central corporate environment in addition to acting in an ad hoc consultative role to the subsidiaries. To the telecommunications manager, this may certainly be considered "doing your job," but this policy is certainly not in the best interests of the company, nor is it in the best interests of job security. Under such circumstances, the department will probably not grow very much. This is because few people within the company, whether managers or employees, will realize the value or potential of the telecommunications department.

There is nothing the telecommunications manager can do to address the decentralized structure of the company. This is the business plan that has been dictated by executive management. The plan does not mean, however, that all company processes need to be distributed. In fact, closer examination often reveals that a number of functions do remain centralized. This is also true of telecommunications products and services. Consider that a master long distance contract can support the distributed environment completely. Each department or location may receive an individual bill, while a master report may be sent to the corporate headquarters to administer the plan. There are several benefits to the subsidiaries. First, a low competitive rate has been provided, probably lower than the subsidiaries could have negotiated individually. Second, the subsidiaries are able to relinquish a responsibility to the telecommunications department. This is something that they will be glad to do, allowing them to concentrate more on their core business. The situation is a win-win. Each subsidiary has saved money while reducing their workload. The telecommunications department gains credibility and more job security.

Now consider a subsidiary location that needs to replace a PBX. If site personnel are assigned to this duty, it is doubtful that they will have the skills to lead such a project. Moreover, site managers will probably be hesitant to assign valuable personnel to this project. The employees assigned to this project, not being telecommunications professionals, will probably resent the additional duty. They will also be nervous and uncertain of their work. Both management and employees alike will view the project as a nuisance. They will wish that they had somebody helping them, or that they could afford a consultant. The net result is that they have to make a major capital acquisition with little or no expertise. The margin for error will be great, and the site personnel understand this.

A possible solution to this problem is that the telecommunications department arrives at the site with a boilerplate Request for Proposal (RFP). They establish a project team that consists of site personnel and a representative from the telecommunications department. They educate the site personnel on a limited basis, overseeing the various steps of a PBX acquisition. The boilerplate RFP is modified to reflect the specific business applications of the site. It is issued to a number of PBX vendors and the telecommunications department oversees the evaluation of the proposals and the technological platforms, and performs a cost analysis. To the telecommunications department, this is a common and fairly easy exercise. To the end-user, this would be a complex and exasperating experience. Once again, this can be a win-win situation. The subsidiary has gained a new PBX with a minimal commitment from their valuable and limited staff. The telecommunications department has enhanced its reputation, saved the company money, and provided the correct technological solution for the business application.

As these examples illustrate, the distributed structure can and should be supported by a centralized telecommunications department. However, this is often easier said than done. A number of issues may prevent the department from fulfilling these duties. First and foremost is a limited staff. If the corporate department cannot keep up with the workload at corporate headquarters, how can it possibly support the needs of the field? Second, what if executive management has dictated that the department will remain small in order to control operating expenses? Moreover, what if his superiors tell the telecommunications manager that his only duty is to support the corporate environment? Under these circumstances it would appear that the telecommunications manager would have very few options. But that is not necessarily the case. The telecommunications manager should consider a marketing and public relations campaign to address this situation.

4.2 Marketing the Services of the Department

The telecommunications manager should not be content to accept such a minor role for his department. The more diminished the role of the telecommunications department, the more vulnerable it is to outsourcing or staff reductions or both. A first step in countering this situation is to begin preparing business cases that argue for increased responsibilities and staff. Defining operational and financial problems that exist in the decentralized environment does this. The telecommunications manager will want to prepare reports that demonstrate that he is a strategic thinker and that his department has an important role in the overall corporate infrastructure. Offering well-defined solutions will do this along with a specific timeline and quantifiable data.

Operational issues can be many. The telecommunications manager will want to first define categories of telecommunications products, services, and issues. At a high level this is simply dividing the categories into voice and data. Under each major category, services and products are listed that are not under the control of the telecommunications department. The initial analysis should only be to define the high level areas. For instance, consider the examples shown in Table 4.2.

This simple matrix demonstrates a number of issues that can be cause for alarm to executive management. Consider that there are response time issues in certain locations. If these WAN links support a call center that is an order entry point, customers may be queuing on the ACD because the ACD agents cannot service the customers quickly enough. This means that customers may be hanging up in anger, which could equate to lost revenue. Also alarming is the lack of a disaster recovery plan. Once again, if this is an order entry point and

Table 4.2
Charges for a Decentralized Company

Service Product	Issue
WAN Services	1. No master contract. 2. Unnecessarily expensive services (T-1 vs. frame relay) because of inaccurate design.
Routers	3. No master contract. 4. Maintenance contracts are not consistent resulting in excessive charges for parts and labor.
Disaster Recovery Capabilities	5. Lack of plan may result in lost revenue.

transactions occur over the WAN, a failed WAN link can equate to lost revenue. The same line of logic can be applied to all other categories and issues cited in this matrix. A similar matrix is then prepared for the voice products and services. In addition to the operational issues, the telecommunications manager will also want to add financial information. Once again, the high level issues need to be defined first (see Table 4.2)

Once the high level issues have been defined, it is time to define more specific information. The high level analysis, by itself, does not justify increased responsibilities or resources. An astute manager would say, "Okay, you've defined a number of issues that are cause for alarm. But what exactly are our liabilities? Also, if you're going to offer a possible solution to these issues, how will you do it and what will it cost?" Such questions would be well taken. The actual financial or operational liability may be very little compared to the costs and effort it would take to fix them. Therefore, the telecommunications manager must perform solid analysis and offer tangible results.

The telecommunications manager must always consider that the language of executive management is money. What will gain the most attention and yield the greatest results will be reports that are financially oriented. In addition, the telecommunications manager may want to propose a gradual process for change. For example, an analysis of long distance charges for each subsidiary may yield cost savings of $100,000 for one year. In an environment that is not centrally managed, this may not be an outrageous amount of savings. This may also be accomplished without committing many resources from the existing department. The largest effort will be to obtain the billing information from each location. Once bills are collected, the telecommunications manager may want to provide copies of the bills to a number of vendors and write a simple RFP. The resultant proposals will provide information that will define cost savings per location for the entire corporation. The telecommunications manager will pick one proposal and present it to executive management. Key selling points of the proposal are as follows:

1. Reduced operating costs for most sites;
2. Minimal participation by sites in order to implement savings;
3. Minimal resource requirement from telecommunications staff.

Critical to the implementation plan is the fact that the bidding IXC will be working with each site to change over to the new long distance service. This has been negotiated as part of the bidding process that the IXC will act as project manager for the implementation of the contract. In addition, there are no changes in the general operation of each division. Each division has been

Table 4.3
Data Communications – Operational Issues

Service/Product	Issue
WAN services	1. Professionals do not manage services.
	2. No planning.
	3. Response time issues.
	4. Access is either too much or too little.
Routers	5. Designed and managed by non-telecommunications professionals who are not certified.
	6. Disparate router vendors between divisions/subsidiaries.
Disaster recovery capabilities	7. No alternate routing.
	8. Limited or no dial back-up facilities.
Hubs	9. No standard hub.
	10. Capacity problems often surface.
	11.Bandwidth problems rampant.
Security issues	12. No Internet policy in two subsidiaries.
	13. Nonstandard e-mail addresses across corporation.
	14. Lack of firewall, which results in employee abuse.

receiving and paying their own long distance bills. This will not change with the new contract. Changing to the new long distance service will be transparent.

Of course, there is no such thing as a free lunch. The tradeoff is that the telecommunications department will assume responsibility for managing the master contract. This will increase the duties of the department, which will mean that additional resources may be required. This may come in the form of an administrator or perhaps an additional analyst. Regardless, budgetary issues will factor into the decision. Because the telecommunications manager has not justified the savings for his own budget, he may not be allotted an additional person. Critical to this issue is for the telecommunications manager to exercise good public relations skills. If the effort is well publicized, it will be easier to petition for additional resources. This is especially true when most, or all, executive management knows of the efforts. An additional argument for added personnel is that the plan cannot be administered unless additional personnel are allocated.

In the decentralized environment, the telecommunications manager will want to continually pursue operational improvements and cost saving proposals. This will result in a slow but continual migration of telecommunications responsibilities to the corporate telecommunications department.

4.3 The Telecommunications Department as the Sole Provider

The telecommunications department commonly carries responsibility for all telecommunications products and services on a corporate-wide basis. This can be on a national or international basis. Under these circumstances, the value of telecommunications services are more clearly understood by executive management, thus the commitment to a larger department and centralized control. Telecommunications departments in the centralized environment face different challenges than in the decentralized environment. It is not necessary to justify the existence or duties of the department. Rather, the problem is that the workload is often too overwhelming for the limited resources of the department. This presents a number of challenges to the telecommunications manager. First, how to build a department large enough and competent enough to meet the demands of the company's business objectives. Second, how to meet the goals and objectives of the corporation when staff resources are limited and there is no budget to hire additional personnel. An additional issue is the expectations of executive management and the corporation in general. Because there is a centralized department, staffed with well-paid professionals, the telecommunications department is expected to provide effective solutions that are economical. If this does not happen, it casts an unfavorable light on both the telecommunications manager and the department in general.

In the centralized environment, one of the primary duties of the telecommunications manager is to first understand the workload, and then to control it. Understanding the workload means being able to assess the importance of the business requirements and requests that arrive from the field. A telecommunications manager must put procedures into place to analyze and prioritize each request that is made of the department. If all requests were accepted or acted upon immediately, the department would soon be overwhelmed, morale would deteriorate, and turnover would occur. Establishing a procedure for project requests is explained in Chapter 6.

Managing the workload is a delicate business that is often more art than science. Most telecommunications departments are stretched to their limits, and there seems to be no end to the infinite number of requests that bombard the department. The requests range from simple inquiries to proposals for major projects. Regardless of the nature of the requests, few departments have the luxury of accepting all requests and acting on them immediately. Inquiries can be addressed immediately but project work needs to be examined more carefully. The telecommunications manager, with the help of his staff, needs to determine what projects can be addressed immediately, what projects can be postponed for future consideration, and what, if any, projects can (or should) be refused. For any inquiry that is not immediately addressed, the telecommunications manager will need to exercise considerable public relations skills. This will be covered in more detail later in this chapter.

Economics can play a major role in helping the telecommunications manager to control the workflow. Many centralized telecommunications departments are charged with the entire telecommunications budget. Although this is a good philosophy for controlling overall costs, the downside is that end-users do not understand the costs of telecommunications services. Consequently, their requests do not necessarily encompass all of the factors that a quality business decision requires. Included in this is certainly the cost of the requested service. When an end-user understands the cost of a service, and that the cost will be charged to his department, he or she will weigh the importance of the request. If the end-user does not have to pay for the service, his or her request will be based solely on their perception of their business needs, without having to examine his or her own budget.

A charge-back system puts into effect a system of checks and balances that benefits both the telecommunications department and the end-user. When the end-user must pay for the products, services, and possibly the time of the staff members, he or she will make a calculated decision whether they need the service or not. A charge-back system can entail having the vendor bill the end-user directly, or have one master bill that the telecommunications department performs as an internal charge-back. In a typical charge-back environment, the telecommunications department will charge-back all products and services. This will include both operating and capital expenses. What is not nearly as typical is charging back the time of the telecommunications staff. In this scenario, the telecommunications department takes on the role of an in-house consultant. This situation becomes awkward if the hourly rate is perceived to be expensive and the work is deemed to be inferior.

If the telecommunications manager has instituted workflow controls, it still does not mean that the workload will be manageable. In fact, even with controls in place, it is more common to have an unbearable workload that seems overwhelming to all staff members. Still, the work needs to be done, and the excuse that "I don't have enough people" often rings hollow with executive management and end-users when mission critical business applications that need to be supported. There are a number of options that the telecommunications manager can implement to address this dilemma.

A first step toward managing the workload is to strive for standard designs, vendors, and processes. When standards are established, the learning curve is lessened for each project and efficiencies are gained. This is covered in more detail in Chapter 8. For example, when a site is planned, the telecommunications department should have defined processes for all aspects of supporting the telecommunications requirements. There should be boilerplate RFPs, and project management templates stored in software, with a checklist of all high level issues that need to be addressed. When such standards are established staff members can readily attack the new project, rather than having to deal with a long and difficult learning curve. A second option is to use outside sources for project work. This may entail using vendors, consultants, or contract workers. A third option is to outsource the function completely.

Beyond managing the workload, the centralized department will also face the expectations of the corporation. If there is a centralized department, and it is comprised of well-paid professionals, the expectations will be that the department is a source of quality expertise and that quality results are the norm. In the decentralized environment the criteria are normally not nearly as high. That is why telecommunications managers must be cognizant of the quality of their department's work and how the rest of the corporation generally perceives the department.

4.4 The Public Relations Role of the Department

Modern businesses rely heavily on telecommunications technologies. As illustrated in Chapter 1, almost every modern employee is affected in one way or another by telecommunications technologies provided through the telecommunications department. Unfortunately, there are a number of problems that the telecommunications department must face as the sole provider of telecommunications services.

A fundamental problem is a lack of understanding on the part of employees. They may not have been properly trained on the product or do not understand its function. Regardless, these employees may be very vocal about their frustrations, often creating perceptions about the technologies that are inaccurate. The complaints generated by these employees may affect the reputation of the department. If enough complaints are generated, executive management may begin to perceive that the capabilities of the telecommunications department are suspect.

Additional problems are the limitations inherent to some technologies. For example, employees may complain of the poor transmission of their cell phones and the limited coverage. What they do not realize is that all suppliers of wireless technologies have similar limitations. There is nothing that the telecommunications department can do about the limitations of a particular technology, but the end-user may still complain, stating that the telecommunications department provides inferior products. This may seem completely unfair, but these situations do exist.

A third problem is the volatility of certain technologies. For example, the Internet is still in a state of development and has performance problems. Consider that telecommunications software, especially new releases by start-up companies, is often prone to bugs. This can cause operational difficulties that can range from slow response time to a complete failure of a device or service. If a particular department or site has experienced a number of these failures, the blame may be placed on the telecommunications department. While the telecommunications department has no control over software bugs or public network problems, it still does not absolve them from blame in the eyes of the end-user, however unfair it may be. The mind-set that is the telecommunications department recommends these products and they are accountable for what they recommend.

An additional issue may be resource limitations of the telecommunications department itself. This may be a lack of expertise for a specific technology or a simple lack of labor. Consequently, an end-user may have received poor advice or may not have been serviced in timely manner. Regardless of the reason the work did not get done, or was inaccurate, the end-user may be unforgiving. Once again, unhappy end-users tend to be vocal, especially if the problem affected their daily operations or cost them money.

Lastly, there is always the problem of end-users that are "experts" in the field of telecommunications. Because telecommunications technologies are so ubiquitous, and because there is so much literature available in popular publications, the average person has access to tremendous amounts of information related to computing and communications. Because many people have become PC novices, they also read extensively about the technologies that connect them. Consequently, the telecommunications professional may find that his ideas and recommendations are often challenged by the end-user. For example, when a staff member recommends a particular PBX, the end-user may loudly state that he or she feels the recommendation is subpar. Opinions may also be offered for other technologies such as hubs or ISPs. The situation becomes even more frustrating when the recommendation of the telecommunications department does not perform as advertised. Then the perception is that end-users have more knowledge than the telecommunications staff members.

There will be many times when the telecommunications manager and his or her staff will be helpless to defend themselves when complaints are leveled against them. Perhaps there will be times when the criticisms are justified, but there will also be many more times when the criticisms will be unjust. Unfortunately this is the nature of the business, and all telecommunications professionals face unpleasant circumstances at various times in their careers. There is no panacea for these issues, but there are steps that can be taken by the department to reduce the problems.

End-user education and good public relations skills should be an ongoing goal and objective of every telecommunications department. This begins with the simple act of openly explaining technologies to end-users, in the most user-friendly manner possible, when general inquiries are made, or projects are requested. The telecommunications manager should strive to council his staff members about proper techniques for communicating with the field. Although this may seem trite to some people, a little bit of diplomacy can go a long way in the telecommunications business. If end-users are treated fairly and respectfully, they will be more forgiving when problems occur, and less likely to escalate complaints.

One of the first things a telecommunications manager can do is to school his or her staff on the basics of corporate diplomacy. On a very basic level, this means treating the end-users with respect in addition to educating them about the technologies they use as part of their everyday existence. What the telecommunications manager will want to avoid is having staff members who are impatient with end-users, perhaps to the point where a staff member is perceived as arrogant. Consider that an end-user is having problems with response time. A staff member is trying to determine over the telephone what the source of the problem may be, but is frustrated because the end-user is so naïve about telecommunications technologies. What may happen is that both telecommunications staff and end-user become frustrated, the net result of which is a public relations problem that the telecommunications manager must address. Staff members should be coached to be patient with end-users and to offer fundamental, user-friendly explanations. Just as a surgeon does not use esoteric medical terms to communicate with his or her patients, the telecommunications professional must also be aware of his or her audience. Every so often, the telecommunications manager may want to address this issue in a staff meeting. It never hurts to go back to basics and council staff members about their roles. The suggestions can be very basic and should continually be reinforced. For example, do not assume that the end-user understands the concept. Do not hesitate to offer a simple tutorial, and do not be afraid to draw pictures to illustrate the concept. The end-user does not like to be baffled with acronyms or confused with techno-speak. Moreover, end-users may be insulted, thinking that a staff member talked down to them. The preferred situation is for the telecommunications manager to hear that his or her people are friendly and helpful, and offering explanations that the end-user can understand.

There is more to end-user education than simple diplomacy. One of the most common problems is when a new technology is installed. A new telephone system, voice mail, cell phone, or remote access LAN capability may be installed and the end-user may have not been able to attend the classes that were offered. Even if the class was attended, the end-user may not have grasped the technology completely. Such problems are common in the field of telecommunications and contribute greatly to perceptions of the technology and the reputation of the telecommunications department. Regardless of the reasons why the end-user did not attend training, or perhaps did not grasp the concept, the telecommunications department is still responsible for end-user training. Denying this responsibility only leads to acrimonious relations between end-users and the telecommunications department.

Whenever a product or service is installed, training is usually offered to the end-user. For example, a new PBX will probably require that end-users learn about the system capabilities and the feature codes that activate them. Included in the purchase price of the PBX will be a number of classes to teach the end-users. But as any seasoned telecommunications professional will attest, the classes are certainly not a ubiquitous solution. Classes are often not well attended, and people may not grasp the subject the first time around. The telecommunications department needs to plan for follow-up training and on-going training. For example, once a PBX has been installed for a week or two, the project manager may want to schedule several make-up classes for end-users as a refresher course. This may address an immediate need, but what about new employees who will be hired over the life of the PBX? Certainly it would not be economical to hire a professional trainer from the PBX vendor for one or two people every few months. The answer may be twofold. One is to make sure that the local PBX administrator takes on the duties of trainer for new hires. Included in the original PBX package may be a train-the-trainer class. Second, many PBX manufacturers offer video courses that may be purchased for ongoing end-user training. When a new employee is hired, part of his orientation is to watch a videotape of telephone system training. A reference manual and quick reference material may accompany this. Critical to end-user training programs is the philosophy of "walking in the shoes of the end-user." How would any of us feel if we were handed an unfamiliar technology and told to make it work without any guidance? Telecommunications department policy should be strict in that no technology should be released to an end-user without adequate training.

The department's responsibility does not end with providing classes and supplemental materials. As previously stated, the department is still obligated to answer questions and even train users individually, if necessary. There is nothing more frustrating for an end-user than to have a new technology thrust upon them that they do not understand. This situation is further exacerbated by an inquiry to the telecommunications department that results in an arrogant rebuff. The end-user might very well say, "That's some department! They jam a complex technology down your throat and don't even show you how to use it. Then they treat you like you're stupid when you ask an innocent question!" Comments like this tend to escalate to executive management. The telecommunications manager should understand that he or she has more important matters to attend to than public relations issues that are unnecessary.

The telecommunications manager should always assume that an educated end-user is an ally. There was once a time when telecommunications professionals sat in ivory towers, their decisions never to be questioned by the rest of the employees. With advent of PCs, this has changed. The modern employee has more knowledge than ever before and telecommunications professionals may also find that their knowledge and ideas are just as often challenged as accepted. Once again, diplomacy is the rule of thumb. Staff members should not be insulted or acerbic when an end-user possesses knowledge. Rather, they should be prepared to discuss the business applications and proposed technologies as if they were talking with a peer. Chances are, the end-user will have textbook knowledge but no practical experience. This is where the telecommunications staff member must diplomatically explain why a proposed solution will not work or why it was considered and dismissed by the department. There may even be a time when the end-user may teach the staff member something. Regardless, this is a common situation in the modern business world and telecommunications professionals must be prepared to deal with it.

Beyond the personal contact that staff members encounter, there is still the general problem of educating the corporation about telecommunications products and services. There are two educational categories that the telecommunications manager should consider pursuing on an ongoing basis. First, there is general telecommunications knowledge. Second, there is the public relations initiative of advertising major accomplishments. Both functions can be served via a newsletter or through a company intranet, if available.

A company newsletter does not have to be elaborate, nor does it take much effort. It need only be two or four pages long, can be distributed via intercompany mail, and issued on a quarterly basis. Included in the newsletter should be a column that describes basic technologies. For example, What is IP, ACD and Our Customers, the Internet and You, etc. There is no need to write elaborate, technically detailed articles that would appear in a major telecommunications magazine. End-users have no interest in the technology other than what it is, how it works, how much it costs, and how it relates to their portion of the business. An employee who supports the telephone of a help desk may be very interested to know why he or she has to log onto an ACD every morning instead of simply using a basic telephone. The average employee may not care that TCP/IP can interconnect various data communications protocols, but does want to know how and why he can e-mail somebody outside of the company. Bringing this knowledge to the end-user reinforces the concept that the telecommunications department is a support organization, dedicated to helping the end-user and the company. It also helps to break down barriers between end-user and the telecommunications staff. Immediately, the end-user gets the sense that the telecommunications department is there to help him, not there to sit in an ivory tower and spew information that nobody can understand. In addition to the basic technology primers, the telecommunications department will also want to publish the progress and accomplishments of major projects. If the department has just standardized a remote access capability for LANs, then the newsletter may be the perfect vehicle to advertise the accomplishment in addition to educating the end-user on various aspects of the technology. After all is said and done, newsletters can prove to be an invaluable public relations tool to improve relations with end-users and executive management.

4.5 Vendor Relations

Much has been said in this chapter regarding the relationship between the telecommunications department and the end-user. Of equal importance are the relationships that are developed with vendors, which play a critical role in the telecommunication department's performance and reputation. The telecommunications department will want to select vendors that are reputable and have established longevity in the business world. Vendors must demonstrate technical competence, quality business practices, and the ability to provide solid business solutions. Many vendors will develop long-term relationships with the telecommunications department. This will be a result of standards (e.g., the same router or PBX located in each location) or a master contract (all long distance under one contract).

The most important policy that a telecommunications manager can establish with vendors is that the relationship between the vendor and the telecommunications department is strictly based upon business. This must be understood by the vendor but, more importantly, executive management and the end-users must know this. One of the worst situations that can occur is when a technology goes awry, and the end-user has seen that the telecommunications department has been wined and dined during the selection phase of the project. Inevitably, end-users will wonder if the telecommunications staff was working in their best interests. There may even be whispers that the telecommunications department indulges in unethical business practices.

As cold as this may sound, vendors should be viewed as tools. They serve a requirement of the business and, when the requirement changes, they are to be replaced if they cannot meet the most current business objectives. Forging strong personal relationships with vendors, at the expense of the business objectives, is not in the best interests of the telecommunications department. That is not to say that staff and vendor personnel cannot be friends or fraternize in off-hours. In fact, strong relationships are often forged between telecommunications staff and vendors that last for years and transcend the business environment. However, all concerned parties should understand that the relationship between the telecommunications department and the vendor are based on sound and clearly defined business reasons. This means that carefully prepared documentation and decision matrixes will support any business decision, such as the purchase of a PBX or a standard router for the WAN.

In larger companies, there are strong arguments for choosing a single vendor as the standard. For example, long distance costs can be significantly reduced if a master contract is signed. This means that all long distance services, for all locations, are serviced from one vendor. Under these circumstances, the vendor spends a great deal of time with the telecommunications staff. At first the time is spent to help convert all the locations over to the new plan. Then there is an ongoing effort to maintain billing accuracy and to help bring new locations into the plan, or delete sites that have been closed. There is also an ongoing effort to develop new applications that may further benefit the corporation. For example, an elaborate toll-free network may be developed to service a number of call centers. A series of features are eventually added to the network to enhance service, such as time-of-day routing. In a large corporate environment, this can involve tremendous efforts between the telecommunications staff and the vendor. The concept is known as "partnering" and it is an effective way that both vendor and company can mutually benefit. The vendor begins to intimately understand the intricacies of the customer's business, and the telecommunications staff does not need to worry about constantly teaching new vendors about their business. In many ways this is a win-win situation.

In addition to large corporate contracts, there may also be one-time projects that will involve large expenditures. Although this may not establish a corporate-wide standard vendor, the deal still involves a great deal of money.

Even if this deal only applies to a single location, there is still a need to partner with the vendor. After all, if a PBX is purchased, the relationship with the vendor has been established for at least five years.

When there is so much money at stake, the telecommunications manager needs to be cautious and professional about the decision making process. There must be a systematic approach that provides an audit trail so that the decision can be proved to be sound on both a business and ethical basis. That is why all decisions regarding the vendor's products and services should be made with meticulous and well-defined processes. For example, no new equipment should be selected without the use of an RFP, RFI, or RFQ. Certainly the most expensive decisions should entail an RFP. Less expensive decisions can be made via an RFI or RFQ. Regardless, there should be a decision matrix developed for each product, which includes technical, financial, and vendor analysis. The end-user should also be included in this process so that they can clearly see the department processes at work. The end-user cannot then readily challenge the decision that was made.

Consider that a new PBX must be purchased for an existing site. There is no corporate standard for PBXs and so the telecommunications department recommends that an RFP be written. The assigned project manager will meet with the end-user to determine the business needs. Based on these conversations, an RFP will be developed, clearly defining the business requirements, and issued to a number of potential vendors. The project manager will help conduct meetings with the bidding vendors, guiding the selection process, and educating the end-user about the technology he will be using. The project manager will also develop decision matrixes (see Table 4.3) that will help guide the process. The end-user will see that a logical and impartial process is being executed that will guide a sound business decision. The end-user may not want to make the ultimate decision. But whatever decision is made, he will feel comfortable that he has worked as a team with the telecommunications department and that the process has been fair.

Not all purchasing decisions will offer the opportunity to work so closely with the end-user. For example, the purchase of routers or hubs would not require such close interaction with the end-user. This decision may be made independently of most end-users, and this introduces new challenges to the telecommunications manager.

As purchasing decisions are made, there will be ample opportunities for a number of perks. This will be for both telecommunications manager and staff. The telecommunications manager should give serious thought to his or her policy regarding perks. It is one thing to have lunch with a vendor, and certainly nobody would accuse an employee of unethical behavior over a single lunch. It is yet another thing to take time off from work to golf with a vendor, or to be provided dinner in an expensive restaurant when there is no other reason for the meeting. Perks also come in the form of tickets to sporting events, expensive getaway items, and even getaway excursions that are masked as business meetings.

The first thing a telecommunications manager must address is a policy for all members within the department. The person who will be offered the perk first will be the telecommunications manager. If the manager readily accepts these perks, without regard to the staff, the end result will be low morale, if not blatant resentment on the part of the telecommunications staff. In this instance, there may be rumblings from both telecommunications staff and end-user alike that the purchasing decisions are based on unethical reasons. When the telecommunications manager accepts perks independent of the staff, there is a wall that is created between them. This will be especially evident if decisions are made during these meetings, and projects are then assigned to the staff.

There are a number of directions the telecommunications manager can take regarding perks from vendors. There is a strict policy stating that no gratuities may be accepted under any circumstances. Many companies, in fact, have this policy as a corporate-wide policy, setting the limit for gifts at a nominal amount (e.g., a twenty-five dollar limit). Other companies have no established policies, letting department managers establish their own policies. The telecommunications manager should consider implementing a number of policies regarding vendor relations that will establish ethical guidelines and fair practices for all members of the department.

First, if the telecommunications department is not doing business with a vendor, perks should not be accepted in any form. There is nothing more unethical than accepting lunches, dinners, golf, or tickets when no business relationship has been established. This is the surest way for the telecommunications department to develop a bad reputation; especially if the vendor in question is eventually awarded business. Second, if a business relationship has been established, then there should be limits set for perks. It is certainly harmless enough for a staff member to go to lunch with a vendor. If the company is spending hundreds of thousands of dollars with a vendor, a hamburger will have little impact on any business decision. It is another matter if the same staff member is constantly being taken to golf outings, expensive dinners, and sporting events. This invites speculation that the business relationship continues because of the perennial perks. The telecommunications manager is better served when perks are evenly distributed among the group. For example, the telecommunications manager may be offered tickets to a sporting event. Even if the manager wants desperately to see the game, it would be better if he or she offered the tickets to the group via a lottery system. Or consider a vendor that holds a customer appreciation day, which includes golf and dinner. The golf outing is not offered exclusively to the telecommunications department, but to all customers in the area. Here the telecommunications manager may allow any staff member to attend, with the stipulation that a vacation day must be taken. What the telecommunications manager should forbid are things such as expensive dinners that have no business purpose, getaway weekends, or expensive gifts. These are the perks that cause jealousy, resentment, and speculation that the business practices are unethical.

The telecommunications manager must establish this policy with the vendor in the earliest stages of the relationship. He or she must stipulate that there will be limits, and that all perks will be distributed, as equitably as possible, among the staff. He or she must also convey this policy to all staff members when they are first hired, and when new vendors are under consideration. The telecommunications manager does not have to prevent the staff from receiving perks. Rather, he or she must institute processes to ensure that decisions are based on quantifiable data. Once business relationships have been established, there needs to be a series of checks and balances put in place that limits a single person from benefiting. Also, limits must be put into place so that perks are not construed to be excessive.

4.6 Summary

The value of the telecommunications department is not always realized because of a number of factors. The corporate environment may be centralized or decentralized. Regardless of the corporate structure, the telecommunications manager must take steps so that the value of the department is realized. This ensures stability and offers the opportunity for expanding the department. In addition, the telecommunications manager should take steps to communicate telecommunications knowledge and the accomplishments of his department. He must also institute policies so that end-users understand and respect his processes and ethical practices are instituted for the department.

5

Budgeting

Regardless of how well the telecommunications department performs its role, it has to be accomplished within the budgetary boundaries assigned to the department. Any telecommunications manager or department can provide a superior telecommunications infrastructure and support if given an unlimited budget. Unfortunately, this is not the way the business world works, and telecommunications managers often find that they must compromise their efforts in order to meet the monetary constraints dictated to every department within the company. Moreover, they also find that they must constantly petition executive management for more funds as the workload increases, new projects are assigned, or new technologies are adopted.

There are two fundamental aspects of telecommunications budgeting: managing the operating costs of the department itself, and managing the costs of telecommunications products and services. The former places limitations on personnel resources, while the later places restrictions on products and services. Ultimately the telecommunications manager needs to provide optimum performance for both of these categories, while still staying within the constraints of the budget.

The telecommunications manager cannot always perform all budgetary functions. This is because there may be too much work for one person to address. Consequently there needs to be consideration given to who should be involved with the financial aspects of the department. The department budget itself should ultimately be controlled by the telecommunications manager and, to a certain degree, his or her direct reports that carry out supervisory duties. The budget that relates to products and services may be delegated further down the pyramid to analysts, project managers, and even technicians. Many managers maintain that only a central authority should control the entire budget, and that authority should be the telecommunications manager. While this does offer the opportunity for stricter controls, it may also place a greater, if not impossible, burden on the telecommunications manager. A small department and/or company will allow for complete centralized control of the budget. It will be nearly impossible with a larger company. The telecommunications manager needs to ask the question, what is the most effective way to manage the budget? If it is discovered that he or she is spending most of the time working on the budget, there will not be time to address other important issues, such as technology planning or personnel issues.

There needs to be a balance between all aspects of department management, however, budgetary duties cannot be addressed in the same manner as other responsibilities. The staff member who is assigned a specific budgetary duty must have financial skills and be responsible to the budget. Not all staff members will be adept at financial analysis, nor be interested in doing them. Many telecommunications professionals apply their efforts almost exclusively to understanding technology. Their only concern is understanding the cost of the technology, not necessarily the overall budget. Consequently, they may lack the advanced financial skills required to control a budget. These are also the people who may choose not to move into management.

5.1 The Department Budget

The department budget is a summation of all charges required to run the department. This includes salaries and benefits, travel expenses, general operating expenses, capital expenditures, and educational costs. The department budget is not related to telecommunications products or services, although the department may be responsible for these items. In this book, for demonstrative purposes, it will be assumed that the department budget is separate from the technology budget.

The telecommunications manager usually controls the department budget exclusively. This is usually the most efficient way for a number of reasons. First, salaries are proprietary. It does not matter that everybody in the company knows the cost per minute of long distance, but it would be a problem if staff members all knew what each other made. Second, the telecommunications manager must assess the total budget and apply the monies where they can be put to the best use. Budget dollars are precious and, if one section of the department finds that they will come under budget, the telecommunications manager will need to determine the best way to apply the excess funds in other areas. This is necessary because excess funds will probably be taken away from the budget in the next fiscal year. That is why it is always better to spend all available funds. As convoluted as this logic may seem, it may even be better to go over budget. Barring the fact that the telecommunications manager may have mismanaged the budget, this can provide ammunition to justify additional funds for the future. A budget that is always balanced, or yields an overage, will give senior management the impression that no changes are necessary, or that the budget should be reduced.

Managing the salary budget requires the balancing of three factors: present staff and salaries, projected promotions and raises, and inflationary factors. If the department is reasonably stable, one of the contributing factors will be reasonable and fair compensation. If there is a great deal of turnover, and the cause is inadequate compensation, the telecommunications manager will need to lobby for additional funds in order to bring the pay scales in line with fair market value. If the compensation is reasonable and fair, then the most critical issue will be salary increases and optional compensation such as bonuses.

The percentage of salary increase, or size of performance bonus, will also have an affect on department stability and morale. Here is where the telecommunications manager may be caught between the corporate budgetary philosophy and the market value of telecommunications professionals. If the corporate budget mandates average raises of 3.5 percent, but the industry average for telecommunications professionals is between six and seven percent, it would not be long before the staff would fall behind the industry average for compensation. In a market where telecommunications professionals are in demand, turnover would begin to become a problem. It would be difficult to keep people and to recruit new talent.

If the telecommunications manager needs to increase the monies allocated for raises, there needs to be a justification that includes market analysis and historical data related to staff stability. Market analysis includes average salary and raises for various positions in the general geographical area (see Table 5.1). This analysis should be divided into industry segment and job descriptions. In addition, if there has been turnover because of salary or raises, this information will give more credibility to the market analysis. The telecommunications manager may also want to point out specific high level projects that would be in jeopardy if key individuals were to leave the staff.

Regardless of how well documented the analysis is, it may not be recognized that telecommunications salaries and merit increases should be any higher than other employees in the corporation. The telecommunications manager must then exercise a degree of creativity in order to avoid turnover based on compensation issues. There are, of course, a number of ways that these issues can be addressed. One method is to not replace staff members when they resign or retire. The extra monies can then be applied to raises for existing staff

Table 5.1
Average Salaries for Telecommunications Professionals

Position	Average Industry Salary*	Average Industry Raise*	Current Salary Levels	Estimated Average Raises
PBX Administrator	$40,000	7%	$42,000	4%
Voice Analyst	$55,000	7%	$56,000	5%
Data Analyst	$60,000	8%	$63,000	6%
Senior Voice Analyst	$65,000	7%	$63,000	5%
Senior Data Analyst	$72,000	8%	$75,000	6%

*Based on averages in the Northeastern U.S.

members. Of course, this also introduces several complications. First, if a staff position is not filled, will it mean that the existing workload will become unmanageable with the remaining staff? Second, the human resources department will have clearly defined job descriptions and grades that are associated with specific salary levels. The grade level and associated salary ranges will need to be rewritten to support the proposed increases. Last, the corporate guidelines may not allow the telecommunications manager to redefine the salary levels for the existing grades. Therefore, it may be necessary to use a bonus plan in lieu of a larger salary or increase.

Another method is to work creatively with job descriptions and titles. If a senior analyst resigns, the telecommunications manager may be able to promote an analyst into the senior position at the lower end of the salary range. Consider that the senior analyst's position may be $70,000. An analyst is promoted to the vacant position for $63,000. The analyst position then opens. Assume that the position is filled for the same salary. The telecommunications manager has saved $7,000 from the budget while still maintaining the existing staffing level. In addition, the analyst who was promoted received a raise because of the promotion.

Inflationary factors that affect the salary budget can come from a number of sources. However, the primary source in the field of telecommunications will be the local, and to a certain degree the national, market for telecommunications professionals. In the previously mentioned example, it was assumed that the analyst position could be filled for the same salary. Unfortunately, this will not always be the case. When a vacancy unexpectedly occurs, it is common to hire new staff at a higher range, simply because candidates may have a choice of positions from other companies. In addition, a candidate may receive a counter offer from their existing employer. If the telecommunications manager truly wants an exceptional candidate, he or she may become engulfed in a bidding war with the existing employer, or perhaps another perspective employer. Conversely, the same can happen for existing staff. A staff member may one day walk into the telecommunications manager's office and offer a letter of resignation. Suppose that the staff member is essential to supporting a mission critical application. Or assume that he or she is the project manager for a key project that is not complete. The telecommunications manager may then have to go to executive management to solicit additional funds. Consider that an analyst was making $65,000 per year and has an offer for 10 percent more, bringing the new salary to $71,500. The problem may not stop there. Consider that the staff member may be offered a signing bonus of $5,000. If the telecommunications manager wishes to retain a valuable staff member, he or she will have to find a $5,000 bonus and an additional $6,500 from the salary budget. Moreover, the staff member is still due a raise for his or her annual review. This will incrementally increase the raise because the salary is higher.

When such occurrences happen, the telecommunications manager needs to carefully examine the issues and weigh them for the budgetary forecast. For example, if the past two years has shown a 15 percent turnaround, and a new hire typically costs 10 percent more to fill the vacancy, this should be factored into the new budget. In addition, if the manager has had to make counter-offers, there should be monies available (perhaps in the form of bonuses) to compensate for this. The telecommunications manager can attempt to maintain staff stability, but he or she cannot work miracles. In spite of a matching offer, a staff member may be offered better future opportunities, better benefits, or be lured by a more interesting work environment. Any of these factors could render the manager helpless to retain his valuable staff. Consequently, he or she must have as much ammunition in the arsenal as possible.

The operating budget will also include standard departmental operating expenses such as utilities, office supplies, etc. Budgeting these items for the telecommunications department is no different than any other department. What is different for the telecommunications department are capital and operating expenses for office automation and technology that is required to support the telecommunications infrastructure. For example, there is a difference in a PC that an average employee might require and a PC that a telecommunications professional would require. A telecommunications professional may have to use graphics software that requires a large screen with high resolution. The software may also require a higher memory and processing power than the average employee might need. These types of software packages are used for network diagrams that utilize high-resolution graphics. Such a PC may cost two to three times as much as the basic PC used by other employees. If the telecommunications manager has a staff of 20, his PC budget may be $5-6,000 per machine as opposed to $2,000 per machine for other employees. In fact, the PCs required by the telecommunications department may fall outside of the company's approved list of PCs. The telecommunications manager will have to write a justification for the larger PCs.

It should be noted that if the high-end PCs are not provided to the staff, it might have negative repercussions. I once worked for a telecommunications manager who would not provide large monitors because of the price. The staff members were required to use a graphics package to perform network design work. Because some of the diagrams were so complex and small, many of staff grumbled that they were getting headaches or that their eyes were constantly burning. Although this example may seem trivial to some readers, this may be one factor that contributes to staff turnover. A staff member may interview at another company, and be immediately struck by employees working at state-of-the art machines that offer large screens and high processing capability.

There are many other items that may contribute to employee comfort and productivity. For example, staff members may require laptop PCs equipped with modems in order to dial into network software remotely. Staff may also require pagers and cell phones so that they are always accessible to the company. Special software may also be required to document, monitor, and troubleshoot the enterprise network. Regardless of the type of software, the telecommunications manager needs to understand what types of software are required to support the environment, who within the staff will need licenses, and what computer hardware will be required to support the software. In any given fiscal year, this could culminate in a rather large budgetary increase. However, it is better to submit such costs in the budget forecast than to have to purchase unbudgeted items because the staff cannot provide the proper level of support. During the course of the present fiscal year, the telecommunications manager needs to solicit the opinions of his staff in order to determine what office automation will be required to support the enterprise network in the forthcoming fiscal year.

There are other issues that may affect the budget that may fall outside the "standard" definition of required office equipment. One of the best ideas I have ever encountered was when a manager I reported to provided white boards in the cubes of every staff member. Whenever there was a discussion about network design or the connectivity of devices, staff members would migrate to the boards and begin brainstorming, rather than having to seek a conference room with a white board. Productivity was increased considerably for a relatively minor investment.

The budget will also have educational items. This will be divided into training, books, videotapes, software, and subscriptions. This can actually be some of the best money that a telecommunications manager can spend because the staff will need to constantly reinforce their knowledge base. Because the industry changes so quickly, there needs to be a constant influx of educational materials coming into the department. A well-stocked corporate telecommunications library can be worth its weight in gold, offering valuable information to all members of the department.

Finally, most telecommunications departments require a travel/business expense budget for either end-user support or education. The telecommunications manager will need to assess the projected project load for the forthcoming fiscal year, along with travel that may be required to support the standard operating environment. There are several issues that should be discussed regarding travel expenses. First, if the company policy will allow it, travel expenses should be reasonable, but not policed in excruciating detail. For example, I once worked in a department where there was a tremendous amount of project work that required travel. Many of the staff spent a great deal of time away from their families. A new manager was brought on board and he began an austerity program to prove his worth. He would demand that staff members buy super saver plane tickets that required weekend stays, connection flights in lieu of direct flights, and would not authorize rental cars. Although he was able to save a small amount of money, he was not successful in gaining any productivity, and actually had incurred the wrath of most staff members. Because they were spending so much time away from their families, they did not want to spend the weekend required for a super saver ticket purchase. The connection flights made their trips longer and more wearisome, and the restriction on rental cars meant that they had to spend long hours waiting for courtesy vans or taxis.

It is usually better to give staff general guidelines when budgeting for travel. For example, offer the staff members the option of super saver, but do not make it mandatory. Although there may be some staff members who will want to take the extra few days, many frequent travelers will not. The choice should be up the staff member, not the result of an enforced policy. Connections should not really be an issue because, if given an option, a direct flight will almost always be selected. A rental car will enable staff members to move about more efficiently. They can come and go as they please and also drive to a different restaurant should they choose to have a change of scene.

Budgeting for travel should be simple and straightforward. It should be a series of averages that will provide staff members with comfortable travel accommodations (see Table 5.2). In this example, the telecommunications manager has budgeted, based on the projected project load, six business and

Table 5.2
Budgeting for Travel

Business Trips*	**Food**	**Airfare**	**Car**	**Hotel**	**Park**	**Entertain-ment**	**Misc.**	**Totals**
48	$7,200	$36,000	$8,640	$18,000	$1,440	$2,400	$720	$74,448
Education†								
16	$2,400	$12,000	$2,880	$6,000	$480	$800	$240	$24,816
Total	$9,600	$48,000	$11,520	$24,000	$1,920	$3,200	$960	$99,264

* Average 3-day trip
† Average 5-day trip

two training trips per staff member. The average business trip is three days long and the average class is estimated to be five days long. There are averages weighed for each aspect of travel: food, airfare, car rental, hotel, parking, and miscellaneous. Food is averaged at $50 per day: $10 for breakfast, $15 for lunch, and $25 for dinner. Airfare is rounded to $750. It will be less expensive for shorter trips or more expensive for longer ones. Over the years, the average price for a plane ticket has proven to be $750 and so this is what is used for budgeting purposes. Of course, if all trips are from coast to coast, or there is international travel involved, the average price could be higher. Rental cars are weighed in at $60 per day and hotels are averaged at $125. Parking is budgeted at $10 per day and there is an average miscellaneous charge of $5 per day. Entertainment is factored in at $50 per trip. In most cases, it will be difficult for the staff to exceed these prices and travel should be less of a burden which will help to maintain morale.

5.2 The Corporate Communications Budget

Beyond the department budget is the cost of telecommunications products and services. One of the first charges that any telecommunications manager should have is to understand what the company is paying for all telecommunications products and services. For the new manager this would be a new project to understand and identify all costs. To the incumbent manager, it is a priority to ensure that all costs are identified, that all costs are managed, and to make sure that the monies are spent wisely and allocated properly.

Telecommunications costs can be divided into three basic categories: capital costs, one-time costs, and recurring costs. Capital is the outright purchase of equipment, such as PBXs, voice mail systems, hubs, or routers. One-time charges are costs such as installation charges or labor for a specific task. Recurring charges are costs related to services, such as long distance, trunks, lines, circuits, or maintenance charges.

Capital costs are usually high-ticket items that need to be budgeted for the forthcoming fiscal year; usually in the form of a major acquisition or upgrade. The telecommunications manager will need to extrapolate this from various sources of information. For example, he will need to maintain communications with various corporate disciplines in order to determine what projects are anticipated for the pending fiscal year. It is not enough to simply monitor past growth patterns. The telecommunications manager must understand new business ventures or applications, and growth or changes in existing applications. This will allow him to estimate what new equipment, or additions to existing equipment, will need to be budgeted for the forthcoming year. There is no single source within the company that will provide this information. The telecommunications manager will need to establish communications with various disciplines, such as Human Resources, Facilities Management, Operations, and Finance. All of these departments will be able to provide insight into what changes are scheduled in the forthcoming fiscal year. In addition to new systems, or systems that require additions or upgrades, there will be systems that are obsolete or not meeting the needs of the business. It is usually not necessary to seek this information because the problem will be brought to the telecommunications department.

The telecommunications manager may be directly responsible for the capital costs, or the costs may be charged to the end-user. Either way, the telecommunications manager will be responsible for forecasting the costs. For example, a new division that will be housed in a single building may have the following costs projected. The PBX with voice mail is estimated to be $500 per port. For a 200-port system, this equals $100,000. The router will be leased and does not fit into the capital plan (as opposed to the operating budget). Cable is estimated at $200 per station and hubs have been budgeted at $5,000 each. It should be noted that estimating is certainly not an exact science. These are round figures that are based on the projections provided by other company disciplines. Because these people may not have an accurate idea of what the final design will be, or the final plan may change at the last minute, it behooves the telecommunications manager to buffer his estimate. At the very minimum, managers should project 10 percent. If history has proven that end-user estimates tend to be extremely conservative, managers may want to apply an even larger buffer.

There really should be relatively little justification to provide capital costs for a new business application. After all, the new business will require some form of communications. It is simply one aspect of the cost of doing business. The telecommunications manager's concern is to provide the right system for the application at the most cost-effective price. This will be accomplished via the RFP process. The RFP, however, will only refine the decision making process in terms of cost. There will have to be an estimate made prior to the RFP process. If the estimate falls far short of the actual cost, it will reflect badly on the telecommunications department.

Although estimating is often thought of as a black art, there are processes that can be used to help refine the exercise and make it more accurate. First, the telecommunications manager needs to use the resources and expertise within the department. Because no single person can know everything, the best approach is to formulate an estimate based upon the collective experience within the department. The telecommunications manager can estimate the capital costs based on his or her knowledge, but must also submit the estimates to his staff and ask if the projected costs are reasonable. The manager may find that the market has changed somewhat (because he is not always directly involved with the day-to-day issues), and that the estimate is either too high or too low.

A second method—even more important than the first—is to make sure that the business applications are clearly understood. If a new site is being planned, the telecommunications manager cannot arbitrarily accept the information from the end-user. If the new site is conveyed as a manufacturing facility with approximately 200 stations, it could be interpreted as a basic administrative system requiring only call coverage and voice mail. What if there was an order entry function within the plant that required ACD? ACD can increase the per-port cost of a PBX significantly. A surprise such as this would probably cause an overage. These are the types of unpleasant surprises that a system of checks and balances will help to prevent.

Whenever the telecommunications manager receives a business plan that will require a capital expenditure, he or she needs to review the plan with the staff. In addition, there will often be a need to set up meetings with the end-user. That is because the end-user may not know exactly what is needed. Consider the example of the PBX that might need an ACD. The end-user may not even know what an ACD is. Consequently, he or she does not understand what to ask for. That is why a staff member must review the business plan with an end-user in order to ask qualifying questions. The staff member will then report back to the telecommunications manager with any recommendations for adjustments to the estimate.

Systems that need to be replaced may need a justification from the telecommunications manager. The end-user may be frustrated with the system and want it replaced. However, a capital expenditure will not be approved simply because somebody does not like their system. There needs to be a formal documentation of the technical deficiencies of the system, and how the problems would be solved with a new system.

Technical deficiencies need to be identified as the source of the problem. Before senior management will approve a major expenditure, it must be proven to them that the money will solve a problem or support the overall business objectives. Continuing with the example of the ACD, assume that the site was already in existence, and that the order entry function was being supported with a crude hunt group. In this example, the inbound calls always hit the first extension of the hunt group. This creates numerous problems in a call center environment. First, the calls are not distributed equitably. Rather, one extension receives the lion's share of the calls while the last extension in the hunt group receives the least amount. This design creates managerial issues. A second problem is that calls are offered to unattended stations. In an ACD environment, the agent must log onto the telephone. Theoretically, no calls will be offered to a station that is unattended. This is not the case with a PBX where a call can be offered to any station. When revenue-generating calls are involved, this is unacceptable. Finally, the system does not provide true management reports that allow the call center manager to evaluate how the department is servicing the customer. The telecommunications manager, in the capital budget, must convey these issues as supporting documentation to the finances. In this case, revenue generating calls are being lost and the only solution is to upgrade the system to an ACD.

Another method of cost-justifying a system is to address savings that will produce a payback period. For example, a new PBX system that costs $100,000 may offset the monthly lease cost of an existing PBX. Assume that the savings will be $10,000 per year if the new system is purchased outright. In order to determine the payback period, the capital cost is divided by the savings to yield the payback period. In this case the payback period is five years.

It is not the purpose of this book to educate the reader about advanced accounting techniques. In fact, many telecommunications managers rely on their accounting departments to provide advanced financial analysis regarding concepts such as net present value. Although such analysis does have its place, telecommunications spreadsheets typically do not have to contain large and complex formulas. Executive management and end-users alike are usually only interested in three things: capital costs, one-time charges, and operating costs. They also expect the telecommunications department to provide supporting data. For example, they will want to know what the life cycle of the product will be and if it will meet the needs of business for the duration of the life cycle.

Beyond capital costs are one-time costs which may be large, but they are not related to capital. These charges are usually for installation or labor. The same sources that provided information to support the capital plan may provide information related to one-time charges. For example, it may be determined that there is a need for X number of trunk lines at a new location slated for the forthcoming fiscal year. The installation charge for these trunk lines is $150. Therefore, the cost of installing 40 trunk lines is $6,000 (see Table 5.3). This example applies to a new location where it may be relatively easy to forecast the one-time charges.

There may be additional one-time charges that will be more difficult to predict. Consider that there may be a number of cable additions in a new site, which will cost approximately $200 for each new cable drop (i.e., one station that will support voice and data). The end-user estimates that there will be 200 drops required to complete the project. Any telecommunications professional that has worked on cable installations knows that the final number often differs from the original count. Unfortunately, there may be no way to estimate this figure. Consequently, the telecommunications manager can only annotate the figure on his or her budget forecast; indicating that the figure was based on end-user input, and that unplanned additions will cost $200 per drop. As in the case of the capital budget, the telecommunications manager needs to initiate dialogue with the end-user in order to understand and refine the information.

Recurring costs will come from a number of sources. Any telecommunications service provided through a carrier, whether LEC or IXC, will carry a recurring charge. In addition, there will be charges for maintenance of various systems that are out of warranty, such as PBXs, voice mail, hubs, routers, or videoconferencing units. These costs must be estimated prior to the system being installed. Whether a system is purchased or leased, the telecommunications manager needs to project when recurring costs will be incurred by the end-user (see Table 5.4). Recurring charges are divided into two categories, fixed and usage based. Fixed charges are relatively easy to identify and budget. Usage-based charges will carry a number of variables. For example, how much long distance usage will a new location incur? Without any history to study,

Table 5.3
One-Time Calculating Charges

Type of Service	Quantity	Install	Total
DID trunk	20	$150	$3,000
Combo trunk	20	$150	$3,000
Total	40		$6,000

Table 5.4
Estimated Recurring Costs

Voice Services	Quantity	Monthly	Total
DID trunk	20	$50.00	$1,000.00
Combo	20	$50.00	$1,000.00
Long distance*	5,252	$0.08	$420.16
Local†	7,842	$0.03	$235.26
Data Services	**Quantity**	**Monthly**	**Total**
Frame relay	1	$750.00	$750.00
ISDN BRI	1	$68.00	$68.00
Equipment	**Quantity**	**Monthly**	**Total**
PBX maintenance	200	$5.00	$1,000.00
Router	1	$200.00	$200.00
Total			$4,673.42

* Expressed in minutes and cost per minute
† Expressed in message units

there may be no way of knowing. Consequently, the telecommunications manager can only provide a cost per minute estimate to the new location.

Regardless of whether costs are charged back to the end-user or the direct responsibility of the telecommunications department, the telecommunications manager needs to maintain a system whereby all costs are tracked. He or she should also be able to generate reports based on any number of parameters. For example, if an end-user thinks that he or she is paying too much for long distance, the manager should be able to extract long distance usage and costs by location, department, or individual. The same should be true of any telecommunications cost that might be associated with a site. Not only should the database allow the telecommunications manager to look at each individual site, he or she should also be able to extract summary reports that relate to overall costs per service, product, or vendor.

Keeping a master system of all costs can be labor intensive. However, there is no alternative to this. There must be staff dedicated to entering information in a master database. This includes entering information regarding new sites or applications as they occur. Even if the telecommunications department is not financially responsible for the bills, they are responsible for providing technologies at a cost-effective price. Consequently, the telecommunications manager needs to ensure that processes are in place to maintain the accuracy of the database. In addition, he must extract reports that will allow him to assess the costs of the corporate telecommunications infrastructure.

As bills arrive on a one-time or recurring basis, they must be directed to a central point that will check the bill for accuracy and log the costs into the master database (see Figure 5.1). Once this is accomplished, the telecommunications manager needs to extract reports that will be used for discussion documents with staff. For example, long distance charges may be examined to see if the cost-per-minute is in line with current market trends. The reports may also indicate that usage has changed. Consider that a site may use switched access long distance, meaning that local lines are PICed to an IXC. Assume that the site has experienced a substantial increase in usage and can now justify the use of a T-1 access to reduce the cost-per-minute. The telecommunications staff should be proactive in approaching the end-user about the proposed change and the resultant savings. The management reports should cover all aspects of one-time and recurring costs similar to the spreadsheet depicted in Table 5.4. However, the exercise does not have to be terribly complicated. For example, there is no need to scrutinize the reports unless there is a change of costs, plus or minus, of 10 percent. There may also be charges, skewed to the original budget, which would require an investigation. Ultimately, the telecommunications manager wants to accomplish two things. First, he or she needs to have a standardized process in place for logging costs in a central database. Second, the manager must extract master reports that will allow him or her to

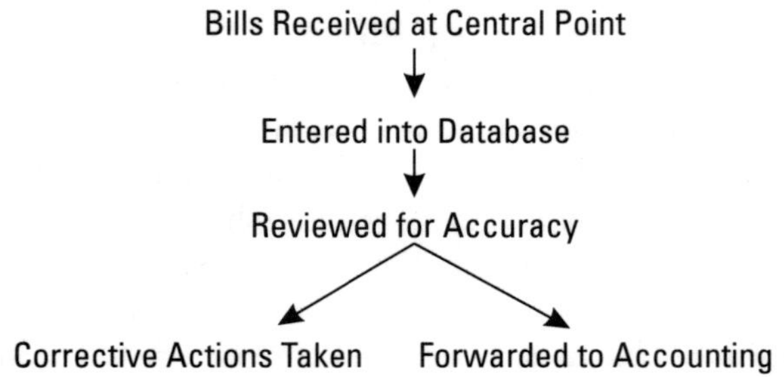

Figure 5.1 The bills must be directed to a central point that will check them for accuracy and log the costs into the master database.

assess the budget throughout the fiscal year. This can be done on a monthly or quarterly basis. This sets the stage for a system of checks and balances that will help to keep telecommunications costs in line.

5.3 Charging Back Telecommunications Products and Services

If the individual costs of the telecommunications infrastructure have been identified, it has to be determined who will pay the bills: the telecommunications department or the end-user. There is no standard that must be followed, however, there are compelling arguments for charging back all telecommunications costs to the end-user. The reason for this is very simple; the end-user should pay for what is used. This is no different than having to pay for utilities or office space. Telecommunications products and services are necessary to conduct business, and they are a cost of doing business. Beyond this simple concept, there are also public relations benefits to charging back services.

When the telecommunications department assumes complete financial responsibility for telecommunications products and services, the end-user is often not very discriminating about what products and services should be used. He or she will demand services, and will not necessarily be concerned with the cost, or even try to understand if they are actually needed. When the costs of all products and services are charged back to the end-user, he or she will act more as a consumer. Under a charge-back system, a determination concerning the true need of the product will be made, and the end-user will also want the best buy for his or her money.

Conversely, a charge-back system ensures that the telecommunications department provides the best possible products and services for a fair and reasonable price. Note that the concept does not mean "least expensive" cost. Ultimately, the end-user should view the telecommunications department as a source of knowledge that provides quality support and solutions that work at a fair price. If the telecommunications department can achieve this, the charge-back system will be successful. If the end-user does not feel that the solutions offered by the telecommunications department work, the charge-back concept will be in jeopardy, along with the reputation of the telecommunications department.

Ultimately, a charge-back system should create a system of checks and balances between the end-user and the telecommunications department. For example, an end-user may request a product or service that is expensive and perhaps inappropriate for the business application. A public relations problem may occur when the end-user is told that he or she cannot have the requested service. If, on the other hand, the end-user is told that he or she may certainly have the service—and here is what it will cost—the viability of the requested service will seriously be examined.

Of the many components that comprise the enterprise network, what exactly should be charged to the end-user? Although there are many companies that employ a hybrid solution, it is this author's opinion that anything used by the end-user should be charged back. Consider a remote site that is a completely separate department and charge center. There is a frame relay circuit used for WAN connectivity and an ISDN BRI used for dial backup to the WAN. A PBX and attached voice mail system provide voice services along with network services that include trunks and lines. In this simple configuration, the costs break down as such. The 200-port PBX carries a maintenance charge of $5 a port per month ($1,000). There are 20 trunk lines at $50 per month each ($1,000). The frame relay link is priced at $750 per month. Also, the router carries maintenance of $200 per month. Because the end-user is paying for all services, demands to the telecommunications department become rather simple. First, are the telecommunications products and services meeting his or her needs? Second, is there any way to reduce the costs?

5.4 The Politics of Charging Back

The charge-back system is not nearly as simple as it may seem. There will still be public relations problems, especially when costs go up, which they often do. If the telecommunications department upgrades a service, it will probably result in an increase for the end-user. End-users will typically not blindly accept an increase without some form of justification. There is also the issue of reliability. What if the WAN link keeps failing? How many telecommunications managers have heard the phrase, "You people are supposed to be the experts!" When the end-user is charged for telecommunications products and services, he or she expects value for his money. Consequently, the telecommunications manager must institute a number of policies in order to ensure that the charge-back system is effective.

Communications with the field is essential for maintaining good diplomatic relations. There is nothing more frustrating to an end-user than to have an unbudgeted increase thrust upon him or her. For example, if it has been determined that the routers will need to be upgraded to support a future computer application, it may mean that there will be a new capital cost, a one-time charge, and a higher monthly recurring charge. It will not matter to the end-user that the increase was a result of another department implementing changes. He or she will want to know why the charges have increased. As a standard policy, the telecommunications manager should provide notice to any department that will incur an increase before the change occurs. As basic as this may sound, it is often easy to forget about the end-user with all the responsibilities that the telecommunications manager must assume. If any change is going to be made that will affect an end-user, there should be an official policy of notification. This should also be done on a timely basis. The end-user will feel no better if he or she is informed two days before a change that there will be an increase of 20 percent in the telecommunications budget—especially if the change will make him or her go over budget.

The telecommunications department should also police the costs that are charged back to the end-user. There should be policies in place whereby end-user charges are audited on a periodic basis. This certainly does not have to be a full time job, but it may be advantageous to assign an administrative person within the staff to spot-check various divisions. There should be a systematic approach to the spot check. Consider that a staff member may find that a number of lines that have been cancelled are still appearing on a bill. He or she calls the LEC and has them credit the end-user's account. The staff member should then generate a simple audit report to the end-user, indicating the results and potential savings.

Although there is no way to prevent problems from occurring, the telecommunications manager also needs to be sensitive to the needs of his end-users. For example, if there is a change being made, such as a software upgrade to a router, the end-user should be informed of the potential for problems or downtime. He or she must also try to determine if there is a way of providing alternate service during the change. Perhaps there is a test router that can be used while the main router is being upgraded. Ultimately, the telecommunications manager needs to ensure that there is consistent and quality communications between the telecommunications department and the end-user.

5.5 Outsourcing

Outsourcing has been the bane of many telecommunications departments in the latter part of the twentieth century. Because telecommunications departments are not part of the core business, the cost of running these departments is viewed as overhead. Consequently, there will always be members of senior management who view the telecommunications department as a cost that must be reduced. In order to avoid being the victim of outsourcing, the telecommunications manager must provide quality service and keep the costs within reasonable boundaries. The danger of not implementing quality processes is that an executive manager may hire a consultant to assess and audit the telecommunications department for possible outsourcing. I have witnessed such situations where the consultant has been able to identify excessive costs and make recommendations for improving existing processes. The result is that the consultant recommends that the telecommunications department (or a portion of it) should be outsourced. Either way, this gives the telecommunications manager and his staff a black eye. If the entire department is outsourced, the telecommunications manager will lose his or her job. If it is only partially outsourced, his or her clout within the company will be severely impaired. Unfortunately, if the savings offered by outsourcing are attractive, executive management will have little choice.

There are some basic issues about outsourcing that should be noted. First, an outsourcing company can rarely provide the same quality as a well-run telecommunications department. That is because outsourcers usually cannot provide the same level of dedication as your own employee. Consider that a problem has occurred, and the outsourcer has been called. The outsourcer will make his or her best estimate of the situation, but may not take full ownership of the problem. In a well-run telecommunications department, the staff should be trained to take ownership of problems and to see them through to resolution. Although the same philosophy may be present with an outsourcer, the natural laws of business dictate that the outsourcing employee will not have the same level of commitment. Probably, he or she will be more concerned with following the rules defined in the outsourcing agreement. Also, an outsourcer supports multiple clients. Consequently, they may not be able to respond in the same manner as the telecommunications department. In fact, specific service levels are written into the outsourcing contract. The outsourcer is not obligated to go beyond the requirements defined in the contract. Many companies are disappointed once they change to an outsourcing company. They simply do not gain the same level of support. Unfortunately, outsourcing continues to be a major trend in the industry. The dollars are often too attractive to ignore.

The telecommunications manager needs to take several routes in order to avoid outsourcing. First, he must take any and all measures to ensure that the department provides the best possible service to the corporation. This is part of the value that the company is getting for their money. Second, the telecommunications manager must ensure that the costs of telecommunications services are provided at the best possible price. Last, he or she must be cognizant of outsourcing trends in the industry. Regardless of a how well the budget is managed, the department could still be the target of an outsourcing analysis. If the department is managed efficiently, it will be difficult to outsource the department. If there are efficiencies to be gained, senior management will give the proposal serious consideration.

The telecommunications manager needs to also consider using outsourcing to his or her advantage. That is, when there seems to be an advantage of using outsourcing for a particular discipline within the department, he should seriously examine the issue. There are instances where this could be a distinct advantage. Consider that the help desk may be experiencing a high degree of turnover. In this instance, it may not be an issue of the budget being mismanaged. Rather, the quality of support has been lacking due to turnover. New people cannot be trained fast enough and many existing staff are demoralized because they see no end to their stressful duties. Not only would the telecommunications manager have a problem with morale, he or she would also be receiving complaints about the performance of the help desk. Under these circumstances, it may be prudent to examine an outsourcing proposal.

If the telecommunications manager determines that outsourcing may serve a function within his department, a number of issues will need to be addressed. First, he or she will need to clearly define acceptable service levels for the function. Second, the telecommunications manager needs to write an RFP, just as he or she would with the selection of any system or service. This would include an in-depth examination of the bidders, and checking references with companies that are using the outsourcer for similar services. Last, the telecommunications manager needs to be cognizant of the morale issue that may surface within the department. Would staff members be losing their jobs? If so, would it be possible to offer them another position within the department? Also, would the staff interpret this move as a first toward total outsourcing? The telecommunications manager needs to maintain communications with his or her staff to apprise them of their future within the department and their job security.

5.6 Summary

The telecommunications manager needs to ensure that the budget is sufficient to allow for reasonable and customary raises. He or she also needs to ensure that they will be able to provide the education and tools that are necessary for the staff to efficiently perform their jobs. In addition, capital, one-time, and recurring charges must be logged and monitored for all aspects of telecommunications products and services. The costs must then be continually scrutinized in order to ensure that costs are reasonable and customary. The telecommunications manager must also be aware that failure to provide a cost effective solution may result in some, or all, of the department being outsourced.

6

Project Management

As stated in Chapter 3, projects are a constant in the corporate telecommunications environment. Only the most stagnant of organizations do not realize a constant influx of complicated projects. Projects can range from the very simple to the large and complex. They can be as simple as adding a few telephones to a department or they could encompass implementing a new, expensive, and untested technology that will be highly visible within the company. Regardless of the scope of the project, project management is one of the most critical functions performed by the telecommunications department. It is a set of skills that must be performed, to at least a minor degree, by all members of the department. Moreover, the reputation of the telecommunications department is often dependent upon how well telecommunications projects are executed. A failed project will damage the reputation of the department for years and individual careers could very well be ruined. Consequently, telecommunications managers need to aggressively develop project management skills and standardize processes within the department. In addition, there should be a policy whereby existing skills and processes are continually refined.

Project management skills, like all business skills, are learned and continually developed. The telecommunications manager, in the role as coach and mentor, should take concrete steps to ensure that all staff members have quality project management skills. This includes setting standards for how projects are planned and managed. Of course, each project will carry its own unique set of requirements, but there still must be general guidelines for managing each project. A telecommunications manager who ignores this duty simply asks for trouble. There will be no standards by which projects are planned, executed, or managed. In such an environment, each project manager would use different methodologies to manage their projects, with varying skill levels, and mixed results. When there is no central policy established, the best ideas and processes remain isolated with individuals; the worst ideas and processes go uncorrected. The result of such a situation is that department performance will be inconsistent, if not downright poor. An additional problem will be that the department's reputation will probably be based upon the failed projects, even though there may be many skilled and competent project managers on the staff who have had successful projects. The unfortunate reality of the telecommunications business is that failed projects gain much more attention than successful projects. Successful projects are not noticed as much because project management is assumed to be a core competency of the telecommunications department. In a nutshell, the telecommunications department "should know how to do their jobs." Perhaps this is not fair, given the uncontrollable variables that can affect telecommunications projects, but unfortunately, it is true. That is why every effort should be made to institute processes that increase the chances of successful project management. This chapter will examine many of those processes. The examples and techniques are by no means the last word on project management, and are intended to provide a fundamental understanding. The telecommunications manager through new courses, team meetings, and project reviews should continually solicit new ideas. Although there is no single or perfect way to conduct telecommunications projects, there are fundamental concepts that will transcend all business segments.

6.1 The Source of Projects and Managing the Project Load

Projects are generated from a number of sources. For example, executive management may develop new business applications or decide to invest more money in the IT infrastructure. These projects will carry the highest priority because they are dictated by executive management and are usually directly related to the business plan that has been projected to the board of directors and Wall Street. When projects of this nature are generated, they become key initiatives for the entire department, and carry precedence over all other projects. Unlike many other types of telecommunications projects, the high level projects that are key initiatives of the corporation usually are recognized when they are successfully completed. That is the good news. The bad news is that high profile projects that fail could jeopardize careers. That is why the planning effort must be carefully executed so that the proper resources (internally or externally) are allocated in order to assure a successful project. If the experience or resources do not exist within the telecommunications department, this must be quickly identified and allocated. I once knew a telecommunications manager who miscalculated the budget for implementing a WAN for his company's European sites. The miscalculation was so large that the CEO was quoted in the *Wall Street Journal*, citing the telecommunications department as one of the factors contributing to the company's disappointing financial performance. The telecommunications manager was inexperienced with international telecommunications, and this certainly contributed to his problems. Nonetheless, the story demonstrates how critical it is to properly plan projects that are generated from executive management.

Individual departments within the company will also generate projects of varying scope, complexity, and importance. Unlike the high level projects that are dictated by executive management, the importance of these projects varies greatly. Not all of them will be accepted, and the telecommunications manager often has to make a judgment call regarding each request. There are times when the end-user will request a technology that is essential to supporting a mission critical business application. Under such circumstances, it behooves the telecommunications manager to support such requests to the best of his ability. I have known managers who have tried to avoid doing work for the field, often citing their large project load and their limited resources as the reasons. Unfortunately, this strategy will only garner limited sympathy from the end-user. The end-user is probably just as strapped as the telecommunications department. In addition, if the requested technology is necessary to support his business, there will be negative repercussions throughout the organization.

Many problems surface when the telecommunications department cannot (or refuses) to meet the needs of the end-users. The most obvious problem is that the telecommunications department will gain a poor reputation. A second problem is that the end-user may try to perform the work themselves. This creates a number of difficulties. Very often the end-user will choose the wrong product or service, and then implement it incorrectly. The net result will be a technology that is nonstandard that will eventually have to be interfaced to the corporate network, or a botched job that will have to be repaired. This will probably be a more difficult project than if the telecommunications department did the project in the first place. Either way, this will reflect poorly on the telecommunications department. The end-user will complain that if the telecommunications department had helped him the first place, he wouldn't be in this mess. If the end-user does a stellar job (rare, but it does happen) of implementing the new technology, it could reflect equally as badly for the department. The logic will be, "Why does the company maintain this group when you could very well do the work yourself?" An additional complaint may be "Why bother to even call them, they always complain about their heavy project load, and they never do anything for you anyway."

Not all requests from end-users will be mission critical in nature. In the information age, where many people are "computer and telecommunications experts," there will be requests that do not make sense. A manager may read of a new technology and become enthralled with it, not realizing that the technology may be immature, untested, expensive, or simply unsuited for business applications. I once worked with an end-user that was very technically astute. Although he was not a telecommunications professional, he read heavily in the field. It was commendable that he became so knowledgeable in a field that was not his core business. Unfortunately, he became what many refer to as "magazine smart." In other words, he had no practical experience. He had no concept of cost, the resources required to implement and manage the requested technologies, or if the technology would even yield a tangible benefit. His requests were numerous and taxing, to say the least. He also had a tendency to design network applications that were incredibly complex, but yielded few tangible business benefits. If I did not carefully select the projects that he requested, I would have worked only for this end-user. I also had to exercise a great deal of diplomacy, because the requests often did not make much sense. I often had to convince him, as politely and as logically as I could, that his designs were poor. This was no easy task, given the emotional commitment this man had made to his design.

These examples bring up an important point regarding project load. The telecommunications department has limited resources and must use them wisely. Consequently, not every request can be honored. Requests that are denied, however, must be handled professionally and diplomatically. This will be covered in more detail later in the chapter.

A third source of projects is the telecommunications department itself. It is stating the obvious, but the telecommunications department is the center of telecommunications knowledge for the company. Executive management and end-users alike look to the telecommunications department for technological guidance and support. Therefore, it is often the department's function to assess the current array of telecommunications technologies, how they service the current business applications, how they can be improved, and what is required for the future. Very often, only the telecommunications department has the expertise to make such decisions. Accordingly, the telecommunications manager and his or her staff are responsible for ensuring that the business applications are properly supported, for the present and in the future. This responsibility includes proposing projects, many of which are major projects that are both complicated, costly, and highly visible. The telecommunications manager should solicit ideas from the staff, analyze the proposed projects, pick the best ideas, and present those projects to executive management when appropriate.

Projects that have quantifiable benefits in the form of cost efficiencies or enhanced productivity will have an excellent chance of being approved. Proactively developing projects that benefit the corporation also lends credibility to the department.

Regardless of the source of projects, all projects should have quantifiable benefits and results. The projects should improve efficiency, correct problems, provide completely new capabilities, or save money. Once again, referencing the example of the "magazine smart" end-user, the telecommunications department cannot participate in projects that are simply technology for the sake of technology. The telecommunications department will want to only execute projects that have quantifiable results to the company, and to communicate those benefits to executive management, individual departments, or the entire company once completed. The ultimate objective of the department should be to complete projects successfully and to communicate those successes to the most influential and affected departments. This helps employees outside of the telecommunications department understand the importance of telecommunications services and their critical importance to the business infrastructure. They should understand, at a high level, the scope of the project, and the tangible benefits realized by the company. The motive behind publishing and advertising successful results is to ensure that the worth of the telecommunications department is realized. There are many people who still look upon the telecommunications department as simply being overhead, a utility bill that has to be paid. Because business is now conducted on networks, the department should be viewed and funded as an essential part of the business infrastructure. It is the telecommunications manager's duty to convey this information to the corporation. This not only strengthens job security, it also strengthens the manager's position when lobbying for additional resources.

With so many sources of projects available, most telecommunications departments have more projects than resources. Inevitably, this means that some projects will have to be placed on the back burner while other projects may never be addressed if additional resources are not provided. The telecommunications manager must document, categorize, and prioritize the list of projects (see Table 6.1). When analyzing project loads, the telecommunications manager must quantify the benefits of each potential project. Projects that are dictated from executive management will, of course, require no justification. Projects that are requested from end-users—or from within the department—will have to be justified with quantifiable goals. Once documented, the telecommunications manager will have to evaluate existing staff and resources, the current project load, and extrapolate the project schedule into the future.

Table 6.1
Analysis of Project Loads

Project No.	Description	Benefit	Priority
98-0023	Replace 3 ACDs—catalog operation.	Enhance system to support order entry function.	1
98-0056	Provide voice and data communications to new plant.	Support new business application.	1
98-0027	Upgrade European WAN links to frame rela.y	Improve response time and reduce budget.	2
98-0015	Design and implement fiber optic backbone for corporate campus.	Prepare campus for future high bandwidth data applications.	2
98-0041	Add networking software to voice mail units in manufacturing division.	Improve productivity.	3

1 = Project in progess—top priority
2 = Priority project, but not mission critical. Will be executed when resources are available.
3 = Low priority, not mission critical. Will be executed when resources are available but not before priority 2 projects.

Creating a timeline in the form of a Gantt chart (see Figure 6.1) can do this. In this example, there are a number of projects that are pushed far into the future because of limited resources.

This data is backed up by the information provided in Table 6.1. Also note that the projects are presented with a requested date and a projected date. Based on this information, the telecommunications manager can prepare a recommendation to executive management to request additional staff or resources (see Figure 6.2).

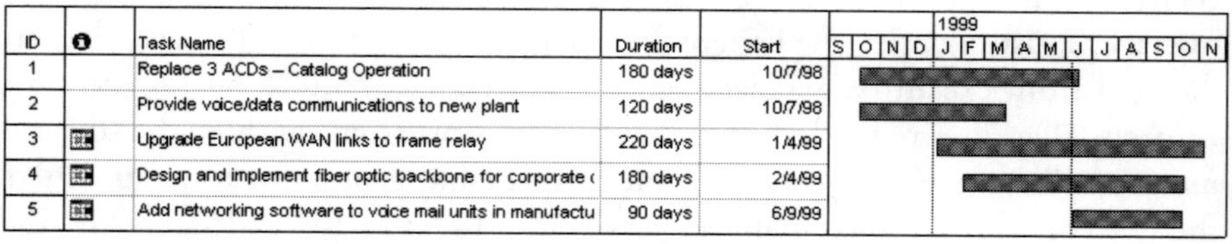

ID		Task Name	Duration	Start
1		Replace 3 ACDs – Catalog Operation	180 days	10/7/98
2		Provide voice/data communications to new plant	120 days	10/7/98
3		Upgrade European WAN links to frame relay	220 days	1/4/99
4		Design and implement fiber optic backbone for corporate (	180 days	2/4/99
5		Add networking software to voice mail units in manufactu	90 days	6/9/99

Figure 6.1 Gantt chart.

Name: ______________________________

Department: ______________________________

Telephone No.: ______________________________

E-mail Address: ______________________________

Requested Project: ______________________________

____

____

____

____

____

Location of Project: ______________________________

Justification: ______________________________

____

____

____

____

Requested Date: ______________________________

Figure 6.2 The telecommunications manager can prepare a request for additonal staff or resources based on the information presented in the Gantt chart.

As any manager will attest, justifying additional personnel is an arduous and challenging undertaking. It is less difficult to justify monies for consulting fees or contract workers because the commitment is short term. Justifying new personnel, however, is a different matter. Executive management will want justifications that are quantifiable, so that the investment carries well into the future. That is why the justification and quantifiable benefits for each project must be carefully calculated and documented. Executive management will then be able to make informed decisions regarding monies allocated to the telecommunications department. For example, assume that the company is starting a mail order catalog division. This business will be reliant upon three call centers that will employ advanced call center technologies. Since the project is a key initiative set forth by executive management, it carries the highest priority on the project list. The telecommunications manager, however, has limited call center expertise on his or her staff. He or she feels that there will be considerable call center expertise required after the initial installation in order to continually support the complicated design. The argument is logical and compelling that additional staff should be hired.

The list of projects is a critical document that will be used to augment status reports to executive management. If the telecommunications department has limited resources and cannot meet the many requests that come from the field, it is inevitable that complaints will filter up to executive management. When this happens, the telecommunications manager should be prepared to offer documentation that logically explains why the work cannot be done. As executive management examines the list of projects and the benefits, they make a determination about postponing projects or hiring additional personnel. If the telecommunications manager (and his staff) has been professional in their dealings with field requests, there will be more understanding as to why specific requests could not be addressed.

At this point, interdepartmental communications bears some closer examination. Telecommunications professionals should strive to be professional and courteous when dealing with end-users. As previously stated, it is not always feasible to immediately service every request from the field, nor does every request necessarily make sense. The end-user may be angry and frustrated that his or her needs cannot be met immediately. A staff member who is perceived to be arrogant, insensitive, or uncaring will exacerbate the situation, and this may certainly reflect upon the entire department. The end-user is not concerned with the overwhelming project load of the telecommunications department. What he or she does understand is that they have a business plan to support that is contingent upon a specific type of telecommunications technology. If support is not given, frustration will most likely escalate throughout the company, especially if the refusal or delay is not handled in a diplomatic and professional manner.

There should be a defined system by which projects are solicited to the telecommunications department. A project request form can be developed in hardcopy or electronically (see Figure 6.2). Each request should be assigned a control number and reviewed within a predetermined period of time that should be of short duration. A designated member of the telecommunications staff should respond, verbally and then in writing, conveying the time frame about when the project can be finished. During this process, however the request arrives, the staff should be courteous and diplomatic, especially if the project will be rejected or delayed. The process of requesting and assigning projects should not be perceived as a bureaucratic roadblock, but rather as a systematic method that is efficient and accurate. It is a system that has been developed to make the most efficient use of an expensive corporate resource, the telecommunications department. Under this methodology, all requests are documented and acted upon. When the requests are documented, there is an audit trail created that can be provided to both executive management and end-users. If a complaint is generated, the telecommunications manager has ready access to accurate information. This is a more efficient method than the telecommunications manager receiving a complaint, then trying to chase down a staff member who fielded the initial inquiry, and becoming bogged down in a "he said-she said" situation.

The master project inquiry report (see Table 6.2) also yields salient information regarding department response and efficiency. Each inquiry is assigned a control number along with the date and the results of the inquiry (whether a project was generated). Staff members should always adhere to the policy of using the project request form, even if they receive a verbal request. In such a case, even if the end-user is reluctant to fill out the form, the staff member should fill out the form himself or herself, with the end-user on the line. The point behind assigning control numbers and reports is to create an audit trail and record data so that the telecommunications manager and executive management can understand the number of inquiries, the scope of work, and what resources are available to address the workload. It also provides a tool with which to manage the workload. The telecommunications manager can see how many projects are requested, the nature of the projects, and how quickly the department responds to requests. If complaints filter to executive management, the project request report is an instant justification for the slow response. There are usually many more requests than actual projects. Many of these requests are not project related and do not need to be logged. The request may be satisfied by answering a question or through the assignment of a simple task. The difference is that projects are actually large-scale plans that are comprised of a number of tasks, multiple resources, and a timeline that usually stretches beyond a day. Therefore, individual tasks are not registered on the project request report, only potential projects. When an actual project is requested, it is

Table 6.2
Project Inquiry Report

Inquiry No.	Date	Proposed Project	Result
98-001	1/5/98	Replace ACDs—catalog operation.	Project accepted.
98-002	1/9/98	Provide remote access from home for accounting employees.	Prohibited by company policy.
98-003	1/15/98	Recable renovated floor in administrative center.	Project accepted.

assigned a project request control number. Upon review by the telecommunications manager (or a project review team), if the project is accepted, it is assigned an official project control number. This is logged on the project request form and is an indication that the project has been accepted and is placed on the telecommunications department agenda.

Beyond the projects that are directed via executive management or other departments, the telecommunications manager needs to continually solicit ideas from his staff. These ideas are also placed on the project request form. The project ideas may range from the relatively simple (e.g., clean up the patch cables in the telephone closets) to the large and complex (e.g., network the corporate voicemail systems together). Of special interest may be projects that save considerable amounts of money. The telecommunications manager may present the projects to executive management citing that, "We are so involved with implementing projects that are critical to the business plan, or solving problems, that we can't address some of these cost-saving projects." If the savings are significant, it could motivate executive management to approve monies above the budget, which could be additional staff or consulting or contract help. A second advantage of continually soliciting project ideas from the staff is that there is a constant backlog of projects that offer tangible benefits to the company. Included in the benefits are cost-saving ideas. This is especially beneficial during times of recession. If directives are given to cut costs, there is a ready-made list to meet the demand.

6.2 The Components of Project Management

For as complex and daunting as telecommunications projects can be, the basic components of a project are actually very simple. The most basic element is a list of tasks (and subtasks) that need to be completed. These tasks must be performed in a predetermined sequence. The tasks are then assigned to specific individuals along with start and completion dates. Experienced telecommunications professionals know that hidden within these fundamental components are many landmines, some foreseeable, some not. That is when the skill and experience of the project managers within the department becomes critical to department success.

One of the first, and most critical, steps in project management is to assign a project leader or manager. The selection of this person is critical because his or her skills will be instrumental to the success of the project. There will be a project team created for most projects, although small-scale projects might be assigned to a single person. Most other projects will require a team leader, of which the project manager will be the leader. It must be clearly understood, by all concerned parties, that the project manager is the central authority for the project. Members of the telecommunications department will understand this, but it may not be apparent to other departments or vendors. This must be established so that there is one central point of control. Although project management is a team effort, there still has to be a final authority. There also has to be a clear set of rules. For example, an end-user cannot directly contact a vendor and make arbitrary changes to a system design or change a date without the consensus of the project team or manager. Because tasks and dates are often sequenced, and must be followed in a specific order, changes should never be made without going through the official channels. Decisions related to the project are to be made by team consensus. When the team cannot reach consensus, the project manager will be the final authority. Decisions made outside of this system will only disrupt the flow of the project.

The telecommunications manager will decide who the best project managers are within the department. He or she will assign them as leaders of specific projects based on their knowledge and experience. For each project, there will be a kickoff meeting with all concerned parties. Many issues will be discussed, but one critical objective will be to establish who the project leader is with all affected departments. At this point, it should be noted that the telecommunications department does not always provide the head project manager. In fact, very often, the telecommunications department will be one of many departments participating in a master project. For example, the facilities department often provides the project leader for corporate projects that include the telecommunications department. Consider that a company may be constructing a new building to support a new business application. The telecommunications department will be responsible for overseeing the cable design and installation; providing a PBX, the data hubs, and WAN connectivity. The telecommunications department will be given sole responsibility for these technologies, but all tasks will have to dovetail in with the main list of tasks that involve building construction. The facilities department will have a master schedule with specific dates. There must be a coordinated effort between the facilities and telecommunications departments or else costly mistakes will happen. There will also have to be consensus between the two departments regarding many of the tasks. For example, the telecommunications department will have to provide specifications for the telecommunications closet. During the design phase, specifications for all communications equipment are provided which includes electrical and environmental specifications along with physical requirements. The communications equipment cannot be installed before the room is completed according to the specifications provided by the equipment vendors. If the equipment is installed in a hostile environment, there could be component failures, and the warranty could be nullified. In such instances, it is imperative that the equipment is only installed when the proper specifications have been met, even if there is pressure to install it prematurely. This is only one example of how tasks are linked and timelines affected because of those links.

Another essential component of project management is some form of documentation. There needs to be a central document that meticulously documents all aspects of the project. This includes start dates, end dates, tasks, subtasks, responsible parties, and a budget. The project documentation can be developed in a number of ways. Many organizations use word processing documents, or spreadsheets. Reports generated in this manner will offer some capability, but they are not software programs specifically developed to manage projects. Because of this limitation, the process of documentation becomes more labor intensive. The preferred method is to use a software package that is specifically designed to manage projects. For example, a common software package is MS-Project. A master file is established on a server that documents all tasks, along with responsible parties and timelines.

The master project file should be developed from a standard template. Each technology should have a standard template that can be used as a developmental platform. For example, if the telecommunications department has been charged with implementing a structured cable system, the project manager should be able to pull a master template that outlines the tasks and issues that are critical to such a project. If the department has already performed this type of project, there should be no need to reinvent the wheel. Of course, every project will have its own unique criteria, and these can certainly be modified to the master template. When a master template is created, the project manager needs only to copy the file and rename it.

Beyond the documentation, there will need to be timetables and procedures for meetings. Many project managers find success in establishing a standard time for meetings. There are, of course, pros and cons to this methodology. A standard meeting time ensures that all parties mark their calendars ahead of time. There is always a better chance that they will be able to attend the meeting if this happens. Conversely, there may not be enough activity or information to warrant a weekly meeting, and the valuable time of the participants may be wasted. This is a judgement call on the part of the project leader. He must determine what is the most effective method of meeting and when those meetings should take place. The project may be served by conducting an audioconference once a week and then having an on-site meeting once a month. Much of this is dependent upon the scope and complexity of the project. The project manager has to balance two things. First, has sufficient meeting time been established so that the requirements of the project are served? Second, are the meetings efficient and productive?

Project meetings should be conducted so that they are efficient and documented. There should be a written agenda provided to each team member before the meeting. The agenda should be provided so that team members have sufficient time to prepare for the meeting. Each meeting should have minutes taken, and be distributed to all concerned parties. Because of the critical nature of telecommunications projects, the project manager cannot feel safe with verbal instructions that rely on somebody's memory. The minutes should include a review of issues discussed in the meeting along with action items and assigned responsibilities.

After each meeting, there will probably be changes that will be required to the project documentation. The project manager needs to assign a central authority to manage the project file. This may be the project manager or a project team member. By assigning a central authority, the project manager is assured that a standard format is adhered. When the project file is designed in a standard format, all participating parties know how to read the document and how to extract specific information.

Part of the ongoing process of managing the project will be to issue progress reports. Reports can be generated from the master file on a number of levels. There can be a simple list of tasks, along with responsible parties and start and completion. There can also be Gantt and Pert charts to provide a visual status of project progress. Once again, there should be standards, and the project manager should refrain from overwhelming project team members with reports that will not be read or are simply looked upon as a nuisance. Reports should only be issued when they convey pertinent information that has changed from the last report. The reports should also be formatted to be user-friendly so that information can be accessed quickly and easily. Telecommunications projects can become very large and complex, which can make for some daunting reports.

6.3 The Planning Process

The planning process is one of the most critical components of a telecommunications project. This is where budgets are determined, tasks are assigned, timelines are defined, and technology is evaluated. A project team will be established that will define the participants, their lines of responsibilities, and the procedures under which the team will operate. When it comes to project planning, it can be safely said that over-planning is nearly impossible. There is an old adage in the carpentry business that states "measure twice, cut once." Certainly this wisdom carries well into the telecommunications business, but the stakes are a lot greater than a wasted board. Failed telecommunications projects can mean thousands (and sometimes millions) of dollars in lost revenue.

The planning process is where the success of a project is often determined. Planning meetings are not necessarily kickoff meetings. In a kickoff meeting, the project has been planned, the participants, timeline, and budget established. The planning process is the earliest part of the project where team-oriented brainstorming establishes many of the project objectives, and establishes the platform for how the project will operate.

One of the first questions that must be asked during the planning process is whether or not the project has ever been done before. The answer to this question will have a dramatic impact on how the project will be executed and what people will comprise the project team. If the telecommunications department has done many similar projects, there may be a wealth of experience and reference material available to the project manager. If this is the case, it may be a simple matter of copying a software template, and then adjusting the project file to fit the unique criteria of the new project. It is then easier for the project manager to establish the project team, assign tasks, establish a budget, and estimate timelines. However, if the project encompasses new or unfamiliar technology and applications, there is much more work that must be done.

A first step in planning the project is to establish who within the telecommunications department should be part of the planning team. Then the team (which may not be the final project team) will address the following questions:

- What is the objective of the project?
- Who will pay for it?
- What is the expected completion date of the project?
- What are the major constraints?

The most obvious choices for team members are the people who have the most experience and knowledge of the proposed technology and application. If the project addresses a completely new application, new variables are introduced into the planning process. For any project that entails a new technology or application, extensive research will need to be conducted. Telecommunications departments should have corporate libraries and, however small, they should have some information about the proposed technology. Within the library should be a systematic method of subscribing to periodicals and cataloging them for at least a year. This will be covered in more detail in Chapter 10. The initial project team should be able to find information in these periodicals. In addition, there should be a library of books and reference works on file. Project team members will also want to contact colleagues and surf the Internet for information relating to similar projects. Regardless of how much research is conducted, the project manager should consider hiring outside expertise if there is a lack of experience within the staff. There is a big difference between textbook knowledge and practical experience, and neither the telecommunications manager nor the project manager should assume that the former is a substitute for the later. The hiring of outside expertise may be an added expense, but it could also be the one factor that may assure that the project is successful. This is one of the judgement calls that the initial project team will make, answering the question, "What are the major constraints?"

At this point, the use of consultants and contract workers in projects bears closer examination. Many telecommunications professionals, both management and staff, look upon outside expertise as a threat to their position and professional integrity. They reason that the hiring of consultants is a sign of weakness, and that perhaps they are perceived as being incompetent or lacking critical knowledge. Contrary to this philosophy, consultants are often a very valuable part of a project, and can very well be the primary reason that a project succeeds. However, there are some basic guidelines that should be followed so that the use of a consultant is complimentary to the project, rather than adversarial. The consultant should be selected via team consensus. If the telecommunications manager or project manager makes an arbitrary selection, the project team could feel as though the consultant is being thrust upon them, or that the consultant is there to "grade" them. They will certainly be wary of such a situation and this could undermine the morale of the team, which could affect the success of the entire project. The consultant will be working closely with the team members and they should feel comfortable with his knowledge and style. More than simply trying to pick somebody from a stack of resumes, the project team should interview potential candidates and check references. The quality of consultants varies greatly. There are many that are experienced and well respected, and still others who can only be described as fly-by-night. There are even some that will try to infiltrate their way into the department after the initial project is completed, obtaining additional work by attacking the skills and knowledge of the existing staff. Any consultant who indulges in such practices is unethical. Checking references will go a long way toward eliminating such people. The project team will want to interview the best candidates and discuss the results of the interviews as a team. This method will create an immediate bond between the team and the consultant, once the successful candidate is selected. The consultant will immediately know that he will be working for (and with) the team and not exclusively for the telecommunications manager.

The preliminary project team will also evaluate the project objective. Once again, the project objective is conveyed from executive management, an end-user, or from within the telecommunications department. While the objective may have been accepted as a project, it does not always convey exact objectives. This question may open much debate and discussion within the team. The results of these discussions can have a large impact on the quality and scope of the project. Consider a directive from executive management to provide a cable system for a new corporate campus. The objective may come to the telecommunications department as "provide a cable system to support all business requirements with the most current cable technology that will serve the company for the next 10 years." This is actually a vague statement and requires further clarification. The project team cannot guess about "the most current technology." If the campus supported mostly administrative employees who only needed access to e-mail or applications that pass little data, the requirements of each user could be met with category 5 cable. If there were high bandwidth applications such as CAD/CAM or video, a fiber optic backbone would be required. Moreover, the objective only specifically states the physical layer of the OSI reference model. It makes no reference to the technology that will actually carry the transmissions. For example, video and CAD/CAM may warrant consideration of ATM, which offers high bandwidth capacity and the capability of carrying multiple types of traffic. This may have a dramatic impact on how the project is executed and what will be required from the telecommunications department to support it after the project is completed. The project objective must be clarified and clearly understood so that the proper resources, time frames, technology, and budget are appropriated. In addition, the requestor may be contacted for further clarification and possibly as a participant in the planning meetings, when the objective requires close scrutiny, and possibly changes.

The role of the project requestor bears closer examination at this point. The requestor may assume that his or her department will not need to be involved with the project to any great degree. While this is often true, there will also be many instances where the end-user (the requestor) will need to be heavily involved in either the planning, the implementation, or both. A classic example in which the end-user must participate heavily is a call center project. Call center technologies are heavily intertwined with business applications. A call center manager will work with an ACD, and the ACD reporting system, on a daily basis. The reports will tell him or her how many people are available to answer calls, when they are available, and how many calls are being received. The manager may often make adjustments to the ACD software, based on staffing and business condition. If the system is cumbersome to use, or the reports do not yield the information he requires, the project that provided the ACD may be construed as a failure. One of the reasons that the project failed may have been that the end-user was reluctant to participate in the project, or simply refused. As unfair as it may seem, the telecommunications department may still bear the brunt of the blame for the failure. That is why during the planning process it is imperative that the project manager clearly identifies who should be the key players on the project team.

There are times when diplomacy and discretion are required when planning projects. What if the end-user refuses to become involved? "You people are the experts ... just find me a system that works!" However busy the requestor may purport to be, it still does not absolve him or her from participation. The end-user's participation in the project team will provide critical information to the planning process. Once again, examining the case of the ACD, there are questions that only the end-user can answer. Do you require a true real-time screen for the ACD, or will a refresh rate suffice? Do you require simple ACD reports, or do you want the ability to manipulate the data to extract other information? If the project team makes assumptions, the chosen system may fail to meet expectations when it is finally installed. That is why the project team must bring the end-user onto the project team when his input is critical. However, participation of the end-user may mean that he or she will be heavily involved, at the expense of regular duties. This is something the end-user will be reluctant to do, and diplomatic pressure may be necessary which could mean applying pressure to his or her superior. Such decisions should not be made lightly. However, when deemed necessary, all concerned parties should understand the jeopardy areas. In fact, a special report should be prepared if the proper resources are not granted. In the event of a failure, there will be clear documentation that efforts were aggressively made to procure the proper resources.

With the proper participants on the project team, the objective(s) can be more clearly defined. This is where the project team needs to thoroughly examine what the true objective is. As previously stated, telecommunications projects, at a high level, serve very basic requirements. Unfortunately, both executive management and end-users alike often fail to understand the scope of what they are requesting. They may not realize how complicated their request truly is, or that they have grossly over- (or under-) estimated the technology that will support their request. Regardless, the project team must examine the objectives and make recommendations for the proper technology and design. If an end-user requests 100 Mbps to the desktop for users that will only be sending text-based e-mail, the technology would be expensive and under-utilized. The project should recommend a more practical technology such as Ethernet 10BaseT that will still serve the end-user's requirements at a reasonable price.

Once the project team has examined the objective, all recommendations should be sent to the requestor in writing. If the recommendations are challenged, there should be a meeting to discuss the philosophical differences. This presents a common problem that all telecommunications departments face.

Eventually there will be a time when a request is made, but the telecommunications department disagrees with the objective. The telecommunications department should never lose focus that it is a service organization. If the requestor is demanding technology that is inappropriate or expensive, ultimately the department can only document its own recommendation, and then pursue the project to the best of its ability.

Once the objective is clearly defined, the finances must be examined. At this point, there may be an overall objective, but no financial plan. There are two high-level ways to analyze the finances. First, how much does it cost and second, who will pay for it? The telecommunications department can never assume that the requestor has a clear idea of how much a project will cost. In my own experience, I have seen end-users project budgets that were only a small percentage of the actual cost. Conversely, I saw budgets that were over-estimated by as much as 200 percent. Part of the project planning team's job is to assess the financials, make adjustments, and report the changes to the requestor and executive management.

Before financial responsibility can be determined, the total cost of the project must be projected. The first costs that are projected are those of the telecommunications staff itself. Consider that a new division is being developed. It will be located in distant city and a new building is being constructed to support the business venture. The telecommunications department will be responsible for designing the cable, choosing the cable vendor, and overseeing the installation. A telephone system will need to be selected that will require 300 stations of which 50 will need to be ACD. The department will also have to provide WAN and LAN connectivity which will include the selection of a router and hubs.

Staff resource requirements are listed according to project requirements. In this case, the project team will be comprised of the following people:

- Network Designer (LAN and WAN applications);
- Senior Voice Analyst (PBX and ACD);
- Telecommunications Analyst (cable system);
- Senior Data Analyst (hubs and router);
- Project Manager (overseeing the entire project).

Each of these people will have a role that will carry them through the entire duration of the project. In many companies, the time and expenses of the telecommunications department are charged back to the business unit that has requested their services. If this is the case, an hourly rate is pro-rated against a projected number of hours for each team member. The number of hours projected for each staff member can only be an estimate. Even if there is extensive experience on the staff regarding this type of project, every project has its fair share of unforeseen variables that may skew the final results. With this in mind, it is always better to overestimate than try to be exact. If experience is a teacher, whatever buffer is built into the projected finances, they will probably be consumed with unforeseen variables.

In addition to charging back time, travel expenses will be incurred for the project team members. If the company employs a charge-back system, this will be charged to the end-user and will need to be reflected in the initial financial analysis. If costs are not charged back, the project manager will still need to project travel expenses as part of the overall project budget. Once again, good project managers allow themselves a certain amount of buffer. If the team decides that twelve three-day trips will be required from various team members, the project manager should add 10–20 percent above the projected amount. In this case, the average trip consists of the following (see Table 6.3).

Twenty percent above the estimated budget is $3,360, or approximately two additional trips. Of course, the project manager will not want to advertise the fact that he has buffered the estimated costs. There is nothing ethically wrong with this, since the money is not charged to the end-user if it is not spent. However, if the end-user budgets 12 trips at $16,800 and the final tally is $20,160, it will reflect poorly upon the telecommunications department. Conversely, it will be a positive if the budget comes in at 20 percent less than projected costs.

Once the staff and travel expenses are estimated, costs must be developed for the technology that will be used. Once again, this will not be an exact

Table 6.3
Estimate of Average Trip Expenses

Average Trip	
Airfare	$750.00
Hotel × 3 days	$300.00
Car rental × 3 days	$150.00
Meals	$150.00
Miscellaneous	$50.00
Total for 3 days	$1,400.00
Est. no. of trips	$16,800.00

science. Unfortunately, budgets are often prepared before the formal process of issuing an RFP. Therefore it is always better to estimate higher than lower. It may appear that this process is more art than science, but experienced project managers can often readily project costs in their head. For example, they may know the general cost-per-port for hubs and PBXs, and the basic cost per cable drop. These base charges are multiplied times the number of ports and cable drops that are projected for the new building. This type of estimating will bring the budget to within a workable number. Again, it is still important to allow a certain degree of buffer. Inevitably the objective will change during the course of the project. Perhaps not dramatically, but there will be additional cable drops, new devices, and even new applications. When this happens, the budget has already been set, and the project manager is expected to meet it. Estimating and buffering are based on industry standards (e.g., $500 a port for a basic PBX) and project management experience (i.e., our history has always shown that we go 20 percent over the original projections).

The cost estimates of the planning team must project the total cost of ownership for the end-user. This includes one-time costs and ongoing costs. One-time costs may be consulting fees, charge-back for use of the telecommunications staff (time and expenses), installation costs for network services, and installation costs for cable and various devices. For all equipment purchases, there may be an analysis comparing purchase vs. lease. This decision is normally based upon the availability of capital funds. It is always less expensive to purchase equipment rather than lease. This decision does not have to be made immediately, but the end-user needs to understand what each financial option is and what the costs will be. Also included in the one-time costs may be education (e.g., PBX management class) and the hiring of administrators (e.g., PBX or LAN). Ongoing costs include maintenance for hubs, routers, and PBXs. There are also ongoing costs for network services.

There are many factors that can affect the finances of a project and a new technology can bring many variables. Beyond the actual cost of the technology there may be consulting fees or cost variables that are a result of a learning curve. Regardless of the variables, there must be a concerted effort to establish financial objectives as accurately as possible, combined with a reasonable amount of buffer. Once established, the project manager will want to stay within the projected budget. This will be a critical barometer of a core competency of the telecommunications department.

The requested completion date has a major impact on many aspects of the project. An aggressive completion date will require that more staff resources are assigned to a project, making it more expensive. An aggressive date may also mandate the use of consultants or contract workers. More importantly, an aggressive date may also be unreasonable. The telecommunications manager will probably have a number of projects on his or her plate when a project is requested. He or she must, with the input of the staff, determine how to accomplish the requested project and still complete the existing projects. This may simply be impossible with the current resources. Also, there may be constraints from the vendors. For example, most PBX providers require twelve weeks from the time an order is received to deliver and install a PBX. In many instances, this interval can be shortened, but there is only so much a vendor can do. Once a project is accepted, the project manager must determine how critical the requested completion date is. If the date is not negotiable, there will be no choice but to deal with it. If the date is negotiable, then it will have to be factored into the overall project schedule for the department.

A final aspect of the planning process is understanding the major constraints. Every telecommunications project has constraints. For example, there may be an elaborate move schedule established for the new building. Over the span of several weeks, new employees will be arriving. Each employee will require that they have a telephone, a voice mailbox, and a LAN port. They may also need Internet access or may be a part of an ACD group. But what if the construction schedule fell behind and the equipment room was not ready to house the communications equipment? This is a major constraint that could prevent the first wave of new employees from starting work. There are many components to a project and many tasks must be completed before another one can be started. It is imperative that the project team can identify the major constraints during the planning process so that they can make contingency plans. If there is no possible way that a contingency plan can address the major constraint, it should be flagged and publicized so that all concerned people understand possible roadblocks to progress.

6.4 Tasks, Resources, and Timelines

With the project team established, the completion date defined, and the finances projected, it is now time to establish the details of the project. Part of this is defining the many tasks that comprise the project. Tasks are defined as high level tasks and subtasks. A high-level task may be defined as large concepts that may contain many subtasks. For example, "Installing a PBX" is a high-level task. Underneath this heading are numerous subtasks such as:

- Conducting station reviews for accounting department;
- Conducting station reviews for audit department;

- Performing a traffic engineering study to establish trunk groups for CO lines and DID trunk group;
- Ordering DID trunk lines from LEC.

Establishing a list of tasks can be a daunting challenge, especially if the project encompasses a concept or technology that is new to the telecommunications department. Whether the project addresses a familiar technology or something completely new, it behooves the project team to solicit as much help as possible. Moreover, once a project has been successfully completed, it can be used as a boilerplate for future projects. Regardless of the source of information, the project team should always avoid reinventing the wheel. If a similar project has been completed in the department (or elsewhere), the project team should try to benefit from that experience. If there is experience within the project team, there should be documentation from prior projects that can be used as a starting point. This information should provide the major, high-level tasks. Minor adjustments can then be made.

The task list should be developed via team-oriented brainstorming. The subtasks are placed under the high-level tasks and then linkage should be examined. There are a number of ways that the project team can develop the task list. A team member can write each task on a whiteboard. Another way is to write each task on a post-it and attach the post-its on the walls of the meeting room. Regardless of how the brainstorming sessions are conducted, there should be somebody in the meeting who has a laptop computer and enters the data into the master file. Each task should begin with an action word. Install PBX, review station design, perform traffic engineering study.

Once the master tasks and subtasks have been defined, they need to be placed in some form of order. There will be a certain chronology to the tasks that will be obvious. There needs to be a business application before a technology can be selected. An RFP process needs to happen before a system is selected. A system needs to be designed before it is ordered. Beyond this obvious chronology, task dependencies are identified. A dependency is defined as a task that must be completed before a second task can commence. For example, there may be a master project list with a task that states, "audit equipment room to determine compliance with vendor environmental specifications." The PBX, router, and hub that will be installed in this room cannot be installed until the equipment room is finished with the proper environmental specs, such as air conditioning, humidity, and a dust free environment. This is a dependency where one task is contingent upon another.

After the tasks are defined along with dependencies, dates are then examined. The most obvious dates to define within the project are the start and completion dates of the project. Fixing dates to the individual master tasks and subtasks is a much more difficult exercise. The team must assess each task and determine how much time will be necessary in order to complete the task. Assessing task time can also be a useful tool in determining resources. Remember that the project team is not a fixed group of people that are part of the project from beginning to end. The project team may change during the course of the project. There will be people who are only involved during specific stages, such as a network designer who may only participate in the early stages. Once time frames are assigned to tasks, the number of required man-hours can be added and compared against the existing resources. If there are more man-hours than resources within the start and completion dates, more resources will probably have to be allocated.

Once the timeframes have been determined for each task, it is time to assign dates. Each task will have a start date and a completion date. There are many tasks that can be performed simultaneously while others will be contingent upon a dependency. Once the dates have been assigned, a Gantt chart can be created that connects the many tasks to form the critical path. The critical path becomes a useful tool in assessing the progress of a project and determining jeopardy areas. The critical path is also useful in presenting the scope of a large-scale project to executive management, giving them a quick and easy snapshot.

At this point in the planning process, the project team will be able to more readily assess the resources required to meet the project objectives. Because each task has been assigned a timeframe, a start date, and a completion date, the project team can add up the total hours and compare them to the available resources. They can also compare dates to see if there will be scheduling conflicts with the available resources. In either circumstance, additional resources may be required to meet the objectives set forth in the critical path.

One of the final objectives of the planning process is to compress the tasks that have been defined for the critical path. There are a number of ways that the schedule can be compressed. Are the time frames conservative or aggressive? If they are conservative, the project team may be able to cut time from selected tasks. There will be tasks that are dependent upon each other, but others may be performed parallel to each other. The project team needs to identify these tasks and determine if they can be performed at the same time. Additional resources may also be allocated so that tasks can be performed in parallel fashion.

6.5 Executing and Managing the Project

Project execution begins with the kickoff meeting. This is a meeting that all project team members should attend, regardless of when they will participate in the project. The project manager will communicate all aspects of the project at the kickoff meeting. He or she will review the project objectives, lines of responsibility, and time frames, and will also distribute project documentation and establish the methods by which the project team will communicate.

The kickoff meeting will be the first of many project meetings. In most telecommunications projects, there are more frequent meetings at the beginning and at the end of the project. But this is dependent upon the nature of the project. There will be times when progress is slow and there will be little to report. Therefore, there may be times when it is unnecessary to conduct meetings. The project manager should only schedule meetings when they are necessary. Moreover, each meeting should be meticulously planned so that maximum benefit is extracted for the entire group. If there are project team members who will only participate in 10 percent of the meeting, it would be more efficient to place their agenda items at the beginning of the meeting so that they may leave early. A bored or irritated participant may be disruptive to the productivity and morale of the meeting, and perhaps the overall project.

A specific agenda should be developed for each meeting. The project manager is responsible for developing the agenda, which will have action items and assignments for team members. The project manager should also designate an assistant project manager who will act on his or her behalf when the project manager is absent. Because of the critical nature of telecommunications projects, the project cannot be crippled in the event that the project manager may become ill or resign. Somebody within the project team should be able to step into the project manager's shoes in his absence. Minutes should be taken at each meeting, reviewing all subject matter and assigning action items.

The project manager should establish means of communications early in the project. A list of team members should be developed along with methods of contact: business and home telephone numbers, fax numbers, cell phone numbers, pager numbers, e-mail addresses, and street addresses. All meetings that can be scheduled in the early stages should be conveyed to the team so that they can mark their calendars. This will ensure that meetings will be well attended. The project manager may also want to establish a standard time for an audioconference meeting set at a time when most team members will be able to attend. For example, an audioconference set at 7 a.m. every Monday morning may be the most convenient time for all team members to participate. Because it is an audioconference, team members may even participate via cell phone while commuting to work.

Documentation will be an important part of managing the project. It will also be a critical method of conveying progress to the project team and other concerned parties. While there are many ways to develop project documentation, the most efficient method is to use a software package that has been developed specifically for the purpose of project management. One of the most popular programs in use today is MS Project. Project software is specifically designed to provide reports, document tasks, monitor costs, and chart progress. These software packages provide built-in charting capabilities that will provide multiple ways of reporting project progress.

There are several issues that the project manager should consider regarding documentation. First, there should be a keeper of the main document. This is a person who is responsible for all changes and updates to the main file. If the main file is opened to all team members, there will be little control and too much opportunity for mistakes to be made. Perhaps one person may be too restrictive for larger projects, but the project manager must maintain control so that the integrity of the document is maintained. Regardless of whether there is one, or several, people responsible for the project file, the file should only be maintained by people who are familiar with the software and the format of the file. A person who is unfamiliar with project software could well make a mistake that could take hours to repair, or even lose critical data. A second issue is to understand the role of documentation. I have seen both telecommunications managers and project managers become consumed with documentation to the point where the actual project work suffered. Project documentation is important, but it should not consume the project team to the point where it affects the actual progress and quality of the project. It is simply a means to an end, and is not the sole objective of the project.

The project file should be updated whenever change has occurred in the project, but not necessarily distributed. Periodically, a printed version of the updated file should be distributed to project team members. Distributions should only occur when there has been a substantial update. The project manager needs to avoid inundating the team members with a paper blizzard. Perhaps it may only be necessary when a major milestone is achieved or perhaps once every six weeks would be more prudent. The project manager should be careful about what parts of the project are distributed. Is it necessary to distribute the complete file? When the documentation is formidable, there could be a momentous task. The portions that are distributed should be simple to read and the reader should be able to identify progress easily. I have seen project files that contained thousands of items that were squeezed with the smallest possible font onto an 8½ by 11 piece of paper. Each update would literally contain hundreds of pages. When project team members received an updated version, they seldom bothered to look for the items that had been changed. The project manager should provide user-friendly reports that are easy to read and provide a clear snapshot of project progress.

When the project team is established, the project manager will establish the methodology by which decisions are made and problems solved. This means that the project team should be given the opportunity to work as a true team. Issues and problems are addressed through team interaction and resolved via team consensus. However, there will also be times when the debate may become heated and there is no clear resolution in sight. When this happens, the project manager will become the final decisionmaker. However, the project manager should make every effort to resolve issues through the democratic process of team consensus.

During the course of the project, it will be necessary to monitor and assess progress. If key dates within the critical path are in jeopardy, the project manager needs to call special meetings to address the potential problems. If the special meetings yield no solutions, a report will need to be made to management, assessing them of the potential delay or problems. There are times when problems will occur that will affect the critical path and there is no solution, other than to delay the project. In times like this, it is probably better to delay the project than to risk an all-out failure.

When problems occur during a project it can be an exasperating experience. Unfortunately, problems are often a rule rather than an exception in the field of telecommunications. When problems surface, this is one of the times that the project manager will prove his worth. Problems can be related to almost any aspect of the project. Consider that there may have been a miscommunication between the chosen vendor and the project team. The PBX will not operate in the manner that the end-user expects. Not only will it not work in the expected manner, the feature in question is the primary reason the system was chosen. Situations such as this are not uncommon. The project manager must call in the end-user and the vendor to discuss alternatives. Perhaps the software can be programmed in a way to offer a portion of the desired functionality. Can the end-user adjust the business application or his or her internal practices to meet the capabilities of the new PBX? Regardless of the outcome, the situation will not be easy to resolve. The vendor will be defensive and the end-user will be angry. The project manager will become an arbitrator trying to find a solution that is satisfactory to all parties while still keeping the project timeline on track. Many problems will occur during the course of a project. Dates will slip and technical problems will happen. Circuits will not be installed on time and vendors will point fingers at each other. When these types of problems arise, it is imperative that the project manager demonstrates quality leadership and problem-solving skills. This includes maintaining an even temper and diffusing anger. This also means that snap decisions must be made in order to meet the project objectives. The project manager may need to approve overtime costs, approve additional expenses, or bring in costly consultants or technicians at the last minute. Very often, this will affect portions of the project such as timeframes or the budget. Regardless, the project manager must understand what is necessary to meet the project objectives and must be willing to make snap decisions in order to do so.

Just as there are problems with every project, there are also lessons learned. While many problems are unforeseeable, each problem should be documented and examined in a final project meeting, once the entire project has been completed. If there is a process that can be avoided in similar future projects, the project manager needs to publish the procedure.

As the project progresses, there should be methods of monitoring quality. For example, the project manager cannot assume that all circuits have been installed and tested or that all telephone sets have been placed and tested on a PBX installation. There has to be some system put in place to quantify the completion of tasks and subtasks. The project manager will want to set up procedures during the planning process to ensure that work has been completed. Many PBX suppliers may not have a system of documentation that verifies such things (although most do). It would be the project manager's job to develop a checklist with the vendor as a stipulation of awarding business and of making milestone payments to the vendor. The project manager will not check all stations or trunk lines, but he will inspect the checklist and perhaps spot check certain items. The fact that the project manager (or a representative of the team) is inspecting work ensures that the participating vendors will proactively check their own work. While there must be a quality process in place, the project manager must also take precautions to not get too involved in the details. I knew of one telecommunications manager who insisted that a staff member carry a PC to every cable drop that had been installed in a new building, and test connectivity after the lines had been tested with category 5 test equipment. The staff member did not find a single bad line, but the directive did cause the manager to lose credibility and lower morale within the group.

When the project is finally completed, there will be a final review meeting. This wrap-up meeting will be to tie up loose ends such as project documentation, general paperwork, and remaining bills. It will also be a time to reward success and recognize achievement. This is covered in some depth in Chapter 9. Successful project results should also be published. This is done via memorandum to executive management and affected individuals and departments within the corporation. The telecommunications manager, not the project manager, should do the publishing of positive results in order to lend credibility to the testimonial. If the company publishes a company newsletter, it might also be used to publicize the success of the project and the importance of the telecommunications department to the company.

6.6 Summary

Project management is one of the important functions performed by the telecommunications department. Project managers should be developed within the staff along with a systematic approach so that all projects have standards for project management. Project managers are the leaders who will guide the project through the planning process, the documentation, project team communication, problem determination, and quality control.

7

Maintaining the Corporate Telecommunications Infrastructure

The corporate telecommunications infrastructure comprises a myriad of technologies, products, and services. It is a complex collage of services and components, designed by the telecommunications department to carry a variety of transmissions that are critical to business functions. The enterprise network is one that is in a constant state of change. It must constantly be monitored and evaluated, modified and upgraded. The importance of the enterprise network cannot be overstated, because it touches virtually all employees and business objectives. In the modern world, business is conducted on networks, and when the network fails, it is often devastating to the performance of the business.

There are fundamental support requirements that apply to all technologies. Therefore, there are certain skills and procedures that will need to be established for each discipline within the department. For example, each employee within the company will require a telephone and PC. Because the workforce within any company is usually one that is in a constant state of flux, these devices, and the technologies that connect them, will always have to be added, modified, or deleted as the workforce changes. The basic tenants of support can be divided into fundamental levels. They are additions, changes, deletions, and refinement of operation.

7.1 The Tenants of Support

Devices and services must be added for new employees or applications. This requires that somebody understand what type of device or service is required, how it should be configured, how it should be ordered, and how it should be installed. Such an undertaking requires knowledge and experience in order to execute the process properly. This holds true for the entire enterprise network, from the simplest of exercises to the most complex addition. There is no aspect of the enterprise network that should be given short shrift, and telecommunications managers should understand that even the act of installing a phone could have dire consequences, if not accurately executed.

To the novice in the field of telecommunications, the simple act of installing a telephone may seem an elementary exercise, requiring little effort or expertise. Although this is not true, it often appears this way to the end-user. Consequently, the telecommunications manager does not have the luxury of dismissing any aspect of support issues, however simple they may appear or be perceived. Consider the simple act of adding a telephone. There must be a process in place for ordering the telephone. Employees are normally provided a form that asks for salient information (see Figure 7.1) about the addition. Included in the fields will be employee name, location, department name, department number for charge-back, telephone type, line appearances, call coverage, voice mail, long distance authorization, and speakerphone requirements. There is also information that indicates an "in service" date. This tells the telecommunications department when the telephone needs to be ready for service. However, this form in itself still does not provide all of the information that will be needed by the telecommunications department. Is the new cube or office cabled with a jack for the new telephone? Just as important, is there capacity in the PBX to simply turn up the new extension in software, or will a new station card be required? If a new station card is required, there may be a standard ordering time that will delay installing the telephone on the requested "in service" date. Also, is there a DID extension available, and are there sufficient reserve numbers to support future installations? There must also be a stock of telephones kept under lock and key somewhere within the corporate facility. This inventory must be maintained so that new employees have telephones when they are ready to begin work. When quantities become low, there will be a need to assess future demands and determine how many phones will be needed to provide a sufficient stock.

As the reader can see, the installation of a simple telephone carries much more complexity than meets the eye. A new employee, however, requires more than just a telephone. He or she will most certainly require a PC with a LAN connection. Remote access into the LAN may also be required. There may also

❑ New Service
❑ Change
❑ Deletion

Employee Name ______________________

Location ______________________

Department/Division ______________________

Department Charge No. ______________________

Telephone Service

❑ 2500 Set
❑ Digital Set
❑ Multibutton Digital Set
❑ Voice Mail

Line Appearances:

1. ____________
2. ____________
3. ____________
4. ____________
5. ____________
6. ____________

Fwd. Target

❑ Data Access
10 Base T
100 Base T

Figure 7.1 Employees are usually provided with a telecommunications request form that asks for information about the addition.

be a need for a wireless phone. An e-mail address will have to be established and there may also be a need for an Internet account. Now consider what processes were required to support the installation of a simple telephone and relate that to all of these technologies. It now becomes apparent that many processes are required within the telecommunications department in order to support additions to the enterprise network.

Beyond additions, nearly all products and services will change or be deleted at some time. Once again, this will often appear to be a simple exercise to the end-user, not understanding the labor required or the logistics involved. When a change is requested, there must be a common reference point, such as a software program that will manage both inventories and processes. The staff member must be able to access information relating to the change. Continuing with the example of the telephone, the staff member needs to know the employee name assigned to the telephone, the number assigned to the telephone, the features programmed into the telephone, and the location of the set. The change could be a move from one location to another, a change in features, a change in call coverage, or an upgrade to a better telephone. As in the case of the addition, information will be required that relates to finances, inventories, technological capabilities, facilities, and time frames.

Supporting telecommunications products and services also requires measurement of service, documentation of actions taken, and financial accountability. While each technology will have its own unique set of idiosyncrasies, many of these basic actions and policies are uniform to the entire enterprise network.

7.2 Establishing and Managing a Work Order Process

There needs to be a logical and standardized process by which work orders are executed for all telecommunications technologies that support the enterprise network. A flowchart can be developed to aid the department in developing processes (see Figure 7.2) for the work order process. This chart can be applied to all technologies, whether voice or data. Companies may have a separate process for voice and data, or the functions may be combined. Regardless of how the telecommunications department is structured, there needs to be structure to the process. Otherwise, there will be no standard method of initiating a work order, or responding to one. The net result will be inconsistent support and disgruntled end-users.

The first stage of the process is to provide a form that the end-user will use to request a change. The change can be an addition, a change, or a deletion. The form may also be multifunctional, supporting both voice and data, and may also provide a section for trouble reporting. The form can also be either paper or computer generated. Many companies still use paper; however, this is awkward, messy, and difficult to control. A computer-generated form is preferable, and many corporate e-mail systems will allow the telecommunications department to generate a user-friendly form that can be used by almost any employee.

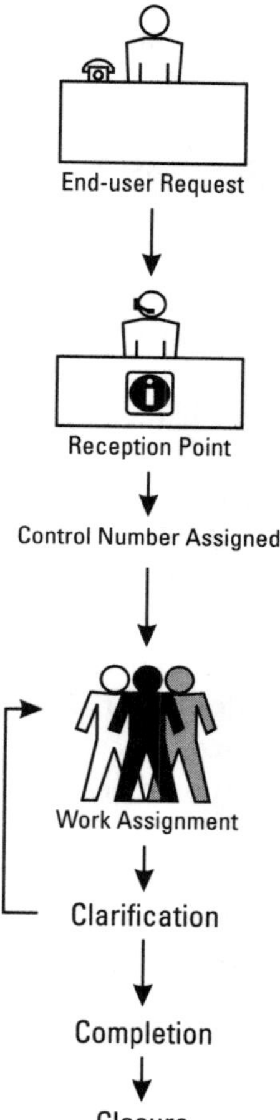

Figure 7.2 A flowchart can be developed to aid the department in developing processes.

The work order form can be an off-the-shelf package, or one that is created by the telecommunications department. The benefits to the off-the-shelf package are that they are turnkey, normally encompassing the fields required to generate a work order. The downside is that the package may be expensive and contain information that is not required for a company's unique situation. Should a telecommunications manager decide to create his or her own form, the process can be relatively easy, using a PC-based package such as a word processing program or spreadsheet. Once the blank form has been created, it can be stored on a central server to be retrieved by any employee, or sent to an end-user when requested.

There has to be a standard point where all work order requests will be sent. While it is preferable to have all requests made via e-mail, the end-users will also need telephone access to the department. The obvious point in the telecommunications department to receive the work order requests is the help desk. Because the help desk is designed as a support vehicle, help desk personnel are dedicated to acting upon requests from the field, rather than being involved with project work. Using the help desk as the reception point will ensure that all requests are received, recorded, and acted upon in a timely fashion.

At this point, it should be noted that the end-users need to be informed of the work order process. A system will do no good if nobody knows that it exists. Consequently, the telecommunications manager should take steps to ensure that all employees throughout the corporation understand the method by which work orders are requested. There are a number of ways this can be accomplished. One method is to contact the human resources department and ask that instructions be provided to all new employees as part of their orientation. Most companies will provide an orientation package, and this can be one method of disseminating the information. A second method is to periodically distribute a newsletter to reinforce the concept with employees. Last, all telecommunications staff members should be told to offer the correct information to end-users if they receive a request from the field that is not directed to the proper source. Of course, this should be done diplomatically, but a process cannot be successful unless all participants adhere to the standards.

When the request arrives at the reception point, it needs to be assigned a control number. This is where the advantages of a computer-based system become apparent. If a question or complaint is generated regarding a specific control number, any person within the telecommunications staff should be able to pull up the number and extract salient information. Since all staff members should have a computer at their disposal, this would certainly be easier than having to trace down a piece of paper.

The work order control number should be provided to the end-user after it is assigned. This will be used for future reference, should a problem occur with the work order. Of course, the end-user may lose the number but there will still be other ways to look up the information. However, the control number will always allow the help desk staff to zero in on the specific work when there is a question.

Once the control number is assigned, there needs to be an assessment of the request. Is the request clear? Is information missing? Is the requested in-service date unreasonable? Is there a proper department charge number? There may also be criteria for specific types of services. For instance, the process may require that a specific level of management approve a cell phone, Internet access, or remote LAN access. If corrections or clarifications are required, the form needs to go back to the requestor. This should be executed by sending back the form via e-mail. A staff member needs to call and clarify the issues over the phone. If the form is returned blindly, even with remarks or annotations, it still may not clarify the issue. The telecommunications manager wants to avoid a back and forth process that slows response time to the field. And while the form needs to be simple and concise, it also needs to be changed when it is apparent that end-users misunderstand the requested information. Still, there will be employees who will fill out the form incorrectly, regardless of how simple it is. When this happens, the problem needs to be resolved quickly. This function is better served by calling the end-user and resolving the issue on line.

Once the scope of the request has been clearly defined, there needs to be an assignment of work. Work assignments can involve a number of disciplines within, or outside of, the telecommunications department. The addition of a new telephone may go to a PBX technician. There may also be a need for new cable or a patch in the telephone closet. This could be a different technician or vendor. There could also be a need to order data connectivity, which would mean an assignment to a LAN administrator. This also could require a cable installation. The help desk becomes, in a sense, a dispatch center. The help desk staff has the expertise to understand who should be dispatched and under what circumstances.

The reception point will schedule all of the appropriate parties, whether in-house staff or outside party, and then wait until confirmation has been received that the work has been completed. This is where the value of the single point of reception becomes apparent. The help desk is actually assuming responsibility for the time frame of the request and the accuracy and completion of the work. There should also be benchmarks that are set within the department for response on work orders. Assume that the department has established a time frame of five working days as the required lead time to install either a new telephone or a LAN connection. Because the department has determined that this is a reasonable time frame, the telecommunications manager has informed the company of the requirement to place all work orders at least five days prior to the requested in-service date. Because this service level has been published to the corporation, there needs to be a process that monitors the flow of work. If a software package has been purchased, there may be a parameter that automatically flags all work orders that have reached a specific time frame. If the process were not automated, part of the help desk function would be to pull up all pending work orders and monitor progress.

Once a work order has been completed, it should be archived in a central file. The work orders should be stored in software for quick and easy access, but they should also be printed and stored in a file for a period of one year. There may always be a need to access the archived information and this is too valuable a source of information to be vulnerable to a computer crash.

Once archived, the work order files become a valuable source of information to the telecommunications manager. He or she can assess the volume and scope of the work, which helps him to continually assess the number of staff and support policies and procedures. For example, if the number of work orders begins to increase, there may be deterioration in service. Phones and computers are not installed within the guaranteed time frame of five days, and complaints begin to mount. The telecommunications manager knows this is because he does not have enough staff to meet the increased volumes. Justifying a new hire, however, requires documentation. By examining archived work orders, the telecommunications manager can document the type of work and the frequency. He or she can also prove that he has established service levels that are acceptable to the end-user, but cannot be met with the existing staff. Senior management would then have the option of hiring new employees or accepting the deteriorated service levels.

7.3 Infrastructure Support Issues

The work order process is only one layer of the entire support process. Underneath this process is a series of department processes and policies that must be in place in order to support the end-user (see Figure 7.3). Each of these categories must be managed for each technology in order to keep the enterprise network functioning smoothly.

Capacity management is understanding the capacity of each device and service within the enterprise network. Whether voice, data, or video, the telecommunications manager should be able to obtain, at any given moment, the capacity of any network component. There are unique capacity characteristics that are associated with each network device or service. For example, a voice mail unit that is attached to a PBX is measured in two ways: disk space and number of ports. If a complaint were generated that the voice mail unit was always full, a staff member should be able to access a database that would yield these two factors. Capacity management is achieved by three factors: port capacity, disk space, and bandwidth. Managing these factors and assessing them against future demand ensures that the enterprise network will be able to meet the growing and changing demands of the business.

Port capacity applies to any device where multiple ports are a factor in supporting end-users. PBXs, voice mail systems, and data hubs are examples of devices that require port management. There are three factors important in

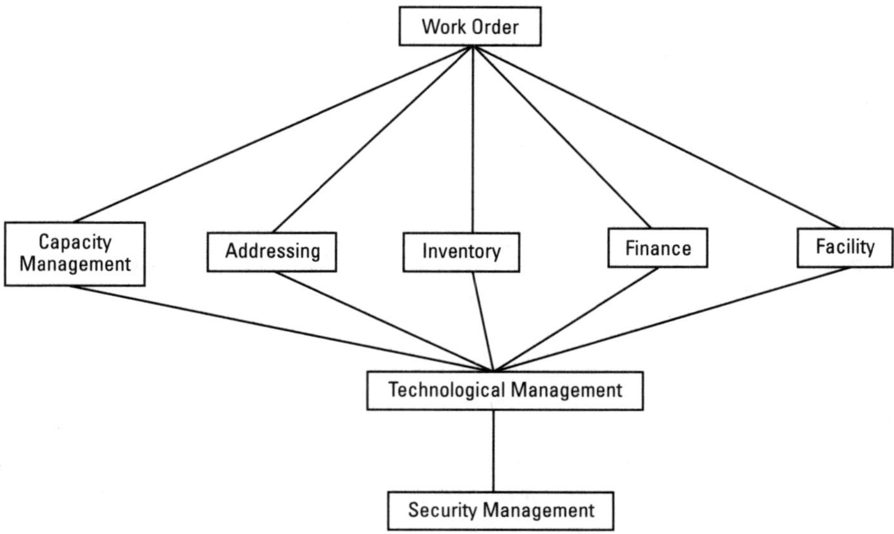

Figure 7.3 Underneath the work order process is a series of department processes and policies that must be in place to support the end-user.

port management. First, how many ports are currently available; that is, ports that are available and need only be turned up in software? Second, how can ports be added? This means that a circuit card is required to add X number of ports to the device. Last, what is the maximum port capacity of the device before the system requires replacement? The telecommunications manager needs to be aware of all three factors. Consider a hub that supports a number of users in a department. If a request is received to add 10 new users, it may be determined that there is only room for four more PCs. A new circuit card will need to be ordered and this may slow the process. Or consider that the hub may have reached its capacity and a new device is required. Depending on the purchasing and approval process within the company, and time frames for orders and delivery by the vendor, the service level established by the department may not be met.

Disk space capacity is important for devices that store data such as voice mail units. Whenever there is a growing business application, the storage capacity will become a critical factor for future growth. There are four critical factors in disk space management. First, what is the current capacity of the system in terms of supporting users? Second, what can be done to more efficiently utilize the existing capacity without upgrading the system? Third, in what increments can disk space be added? Last, what is the maximum capacity of the system?

Bandwidth is also a critical factor in capacity management. This can relate to devices or services. For example, a hub may support ports or varying bandwidth capacity (e.g., 100 Mbps vs. 10 Mbps). It is important to understand not only the number of ports available on the system, but also the bandwidth capacity of each port. A service is also measured by bandwidth capacity. A frame relay circuit has a port capacity and a committed information rate (CIR). Bandwidth needs to be managed by two factors: current and maximum capacity. Consider that a frame relay circuit may have a port speed of 64 Kbps but a CIR of 16 Kbps. Network congestion on the part of the carrier may cause frames to be discarded, thereby lowering the bandwidth of the circuit to 16 Kbps, which may be too low for the application. There may be a need to increase the CIR to the port speed in order to provide satisfactory response time.

Address management encompasses any form of addressing that is controlled by the telecommunications department. This would certainly include any addresses in the data network such as TCP/IP addresses. In addition, telephone addresses such as DID numbers are administered by the telecommunications department. Once again, the telecommunications department must maintain an inventory of addresses and manage the assignment and distribution of these numbers. Two very obvious problems are that there will not be enough addresses to support future growth or the numbers will be mismanaged.

Inventory management involves the management of physical devices that may be stored by the telecommunications department. Whenever a request is made for change, there should be sufficient stock to support the request. For example, the telecommunications department often keeps an inventory of telephone sets. It is a quick and easy task to install a phone by plugging it into the PBX and programming a new extension in software. Other items that may be kept in inventory are LAN cards, hub cards, patch cables, and modem cards. The telecommunications department can inventory items that are often needed immediately. Critical issues are to keep track of the number of items, to document the distribution, and to order new items when stocks become low.

Inventory management also involves the management of systems and services that are installed throughout the enterprise network. This can include almost any type of device, system, or service. For example, an inventory system should allow a staff member to pull up any router installed in the network and access the model type, serial number, and software level. It is important to understand this because the model type and software level may not support a requested change.

Facility management encompasses cable and wiring closets. There must be accurate and thorough documentation of the various categories of cable in each facility: entrance, backbone, riser, horizontal, and station. When changes are ordered, there must be a way to connect the requested devices. Installing new cable can involve a number of complicating factors. Not only is it important to have the right type of cable, there may be a long lead-time to install new cable. Management of cable can be either via paper or electronically, but it must be managed. A staff member should be able to access documentation and quickly determine if the site is properly cabled, or if new cable must be installed. Otherwise this can delay an installation.

Technological management includes the assessment of technological capability. This is perhaps a gray area, overlapping into capacity management, facility management, and inventory management. The objective, however, is to be able to assess the technical capability of any component of the enterprise network in order to support a change. The telecommunications manager, in assessing methods and procedures, must ask if his or her people can readily make this assessment if a change is requested. For example, a router may only be equipped to support specific types of data traffic. If a new application is required for a location, a staff members should be able to readily assess the current software level, and order an upgrade to support the new application.

Another concept that crosses into other areas is that of documentation. When a request is made of the telecommunications staff, there must be documentation in the form of network maps and text. A visual depiction of the network is worth a thousand words to the staff member who is trying to make an analysis. Documentation should be maintained for all aspects of the network. Text is important, but it must be kept to a minimum, otherwise it will be difficult to understand and maintain. The telecommunications manager should standardize on a graphics package that can be universally and immediately accessed by all staff members.

Timely service is not the only objective in supporting the end-user. The telecommunications department must also address security and financial issues. Budget has been covered in Chapter 5. Security is an issue that also must be addressed for both requested changes and ongoing support. Any requested change must address two fundamental issues. First, is the end-user authorized to use such services? For example, a new employee who has been issued a calling card usually needs to have authorization from a senior ranking manager. It is the telecommunications department's duty to check authorization and to then provide instructions about security procedures. Second, the department also needs to evaluate if the request will breach corporate security policies. For example, I once worked for a company where voice mail units were installed with toll-free numbers for remote access. Salesmen called into the voice mail systems remotely to check their messages. One salesman suggested to the site administrator that it would be great if he could dial in to retrieve his messages and then redial out, using the PBX outbound trunks. The PBX administrator complied and, within days, a hacker had breached the system and generated $60,000 in fraudulent toll charges. Such a request should have gone through the telecommunications department, which it did not, but the incident demonstrates that each request should be checked for security integrity.

If there is one essential aspect of support that every telecommunications manager must be aware of, it is to understand the structure and components of the enterprise network. This is, of course, true for strategic analysis, but it is just as essential for serving the immediate needs of the corporate end-user. The telecommunications manager needs to also assess what technologies should be directly supported and what technologies would be better served by a site administrator.

7.4 The LEC and Local Services

The local telecommunications environment is changing radically. There is now competition in the local arena and many companies now have a choice, but the basic array of LEC services has remained the same. There will be a mix of lines, trunks, circuits, and services that are used by any given company. One of the most challenging aspects of managing the local environment, especially in larger applications, is to keep a proper inventory of services and to manage the complex and cumbersome LEC billing. Most LECs are large and bureaucratic. Consequently, billing mistakes are common when services are changed or deleted. To further complicate the problem, LEC billing is not always understood very well, not even by experienced telecommunications professionals. Consequently, LEC bills often become a quagmire of complexity, riddled with billing mistakes.

From the financial perspective, support of local services encompasses managing the proper number of lines and trunks, data circuits (if any), local calling charges, and additional billable services such as central office-based voice mail or features. Most of the charges will be fixed, but there may be some variable costs, such as local LD or message units. The telecommunications department is charged with providing access from various systems and devices to the public network via the LEC. The goal is to purchase sufficient access so that business can be adequately conducted. This is to be accomplished without purchasing excessive service that may be expensive and unnecessary. For example, if too many trunk lines are purchased for a PBX, there may be trunks that are never offered a call and money will be wasted.

In addition to trunks and lines, local usage can add up to a considerable expense. The local calling area (typically within the LATA) can be flat rate or encompass message units. Special access lines may also be used to access discount services such as Feature Group B. There may also be special access services such as redundant cable entrance or various data services.

The array of services provided by an LEC may be considerable (see Figure 7.4). The problem is not usually in the design as much as the ongoing management. It is important to maintain an accurate inventory and to continually monitor bills for accuracy. This is accomplished by assigning resources to manage the LEC services. This can be paper intensive and will often require the staff member to assiduously plod through orders, bills, and customer service records (CSR) to ensure that the company is not paying for services it is not using. There will be services that were ordered removed that will remain on the bill. In an environment where there are literally thousands of lines, trunks, and services, this can be a full-time job.

It does not matter what type of LEC service is being ordered, changed, or deleted. It also does not matter what member of the telecommunications department makes this request, whether voice or data. What is essential is that all adds, moves, changes, or deletions are channeled to a central point within the department. The order needs to be checked and logged. The assigned staff member also needs to ensure that the proper discounts are applied and all salient information appears on the order. Just as important, they need to ensure

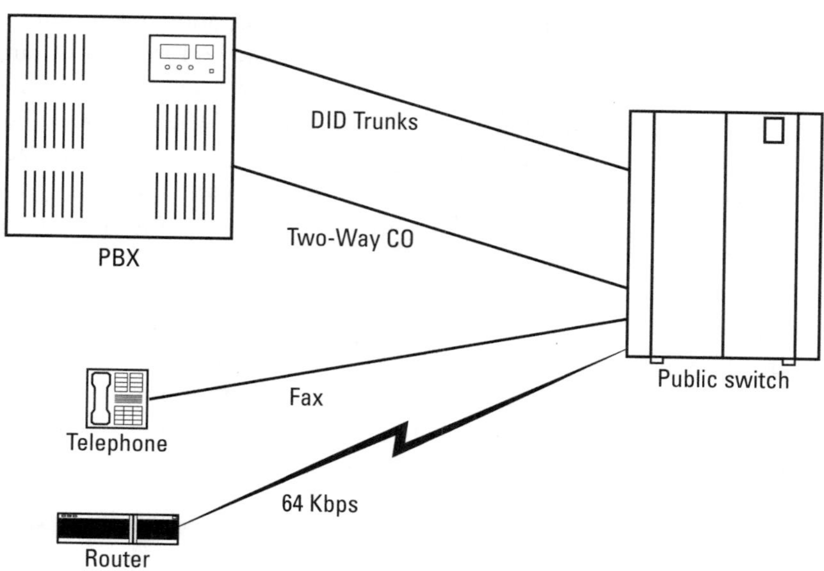

Figure 7.4 The array of services provided by an LEC can be considerable.

that the master bill reflects any changes. A master inventory file should be maintained to manage all existing services and changes.

Because LEC billing can be so complicated, it is sometimes advantageous to hire retired LEC employees. They have the experience to understand the esoteric coding schemes and the structure of the LEC organization. This helps them to facilitate quicker changes when there is a billing issue.

7.5 Long Distance Services

Long distance services come in three basic flavors: outbound, toll-free, and calling cards. While they can all be provided from the same company, they are certainly not managed in the same manner. Each one supports business applications in a different manner and must be supported as such.

Outbound long distance has become a commodity item. Since divestiture, the cost of long distance has dropped considerably, in addition to the pricing structure, which has become simpler. There is really only one main concern that a telecommunications department has in providing long distance services: to supply the lowest possible cost. Functionally, there are not normally any issues that should be of major concern. In most cases, it is simply a matter of providing enough access lines in order to conduct business. Because private corporate networks are no longer economically practical, most telecommunications departments concentrate more on the access method to the LD carrier, which drives the price. For example, a dedicated T-1 access to an IXC POP will offer a much more attractive rate than switched access (where the local line is PICed—pronounced "picked"—to a specific LD carrier). Dedicated access offers a price advantage for two reasons. First, the LD carrier does not have to pay transport charges to the LEC. Second, it is more difficult for the customer to change LD carriers when a T-1 or PRI is installed. Dedicated access, however, requires a higher volume of calling, because the lower rate must be offset by the cost of the T-1.

In addition to price, there are features that may be utilized such as seven digit proprietary numbers or speed dial numbers. These features are provided through the IXC switches and are a leftover from the days of private corporate networks. However, since these features offer no cost advantage, they are becoming less common.

If a company standardizes on a long distance carrier, the ongoing support issues encompass three areas. First, is each department, site, or division provided the most economical rate? Second, are all long distance services accounted for and subscribed to the master contract? Third, have the bills been administered so that each department is both aware of their costs and properly charged?

Staff members need to keep track of local services at each department, site, or division. Because local lines need to have a long distance carrier assigned, the staff members need to know when new local lines are assigned, and ensure that the proper PIC has been assigned. Not only do PIC numbers differentiate carriers; they also indicate various types of services. For example, there is a different PIC for basic service and a virtual private network (VPN). A VPN, such as AT&T's software defined network (SDN) or MCI's V-Net, provides a more attractive rate than basic service. This requires that the customer sign a master contract. Adding new local lines is not the only issue that must be addressed; there is always the problem of slamming. Because the change of a PIC is simply a software change at the CO, any carrier can call the LEC on behalf of the customer and have a PIC changed. This is known as slamming and is quite common with smaller long distance resellers who regularly telemarket their services to businesses. If a division or department is slammed, a staff member will have to issue a letter to the LEC to change the PIC back to the proper IXC.

Consider that a company may have thousands of employees and dozens of divisions. This could add up to literally thousands of local lines in various sites that need to be managed. Keeping an inventory of these lines is daunting enough. The problem is further exacerbated by end-users that add local services without informing the telecommunications department. The PIC may be arbitrarily assigned to the wrong carrier and months may go by where higher costs are incurred via the wrong carrier. There is no easy way to administer large quantities of telephone lines and PICs. The IXC master bill will reflect numbers, but this does not mean that all local numbers are represented. There must be a periodic action to reconcile the master IXC bill with local bills. This can be a labor intensive and arduous task, a problem that never seems to be completely solved. However, there must be a person, or persons, assigned to this task. Otherwise the problem will spiral out of control and the company could lose a great deal of money.

Management of long distance services requires that an inventory be established for each site. The database needs to list each line along with the application (e.g., fax, modem, CO, etc.). There must also be site-specific information that includes address, contact name, and department charge-back information. The person assigned to administration of the long distance charges must also monitor the calling volume at each location, and make a determination as to whether a dedicated access will be cost effective.

Support of calling card services entails cost management and the administration of individual calling cards. There are many telecommunications departments that do not support calling card services. Employees are given the directive to seek their own cards and then expense any business-related calls. Unfortunately, this means that the company will not be afforded a volume discount. Also, a master management report is not available, which will make issues such as toll fraud more difficult to police. When the telecommunications department does take on the duties of calling cards, it is—like the management of outbound services—a large and tedious administrative undertaking.

In a large company, there can be hundreds, if not thousands, of employees who require calling cards for business purposes. Each card must be issued, tracked, and accounted for. Beyond the issuing of the card, there are also a number of administrative tasks that must be executed. Employees who leave the company must have their cards cancelled. Employees will also change jobs and may no longer require a calling card or may change their department. In the latter case, it may mean that an accounting change is in order.

As in the case of outbound long distance, there needs to be a database developed so that calling cards can be effectively managed. In large applications, this will be nearly impossible if attempted with paper. The database needs to track employee name, address, telephone number, employee number, department name, department charge number, and calling card number. There must also be a field for in-service and cancellation dates. The purpose of the database is to administer the calling cards (i.e., adds, moves, and deletions). It is not necessary to track the costs on this program, although it could be used for this purpose.

Calling cards are a liability in that they are highly susceptible to toll fraud. Therefore, the costs must be more closely monitored than other forms of long distance. Consider that a salesperson must often use public pay phones to make business calls. Very often toll thieves will "shoulder surf" and steal a card number. They will sell the number for a flat fee, probably to a number of users, who will use it until the card is cancelled. If the fraud is not immediately detected, a great deal of economic damage can be inflicted in a very short period of time. An additional problem is that cards are lost. In this case, perhaps the card will not be "sold," but it may be passed out to friends. In my own experience, I once had the number of a lost card written next to a pay phone located in a high school in New Jersey. In a very short period of time, the student body had generated a rather large toll bill. Because of the financial liabilities involved with calling cards, the staff member should have procedures in place to monitor usage.

There are a number of methods that could be deployed in monitoring calling card usage. In many cases, the IXCs will offer services that will aid in this process. For instance, the IXC will set up parameters that will flag a card for possible abuse when specific types of calling patterns appear. For example, a call made in New York at 9 a.m. on Monday, and a call made at 9:15 a.m. on the same day from Los Angeles is impossible if the calls are registered to the same card. Unless the employee gave his or her calling card number to another person (something he or she should have been instructed not to do), the card number has probably been stolen. Under many IXC programs, this card would immediately be shut off, and the telecommunications department would be informed that the card had been flagged for possible abuse. There could be a number of other parameters that would flag abuse. For example, simultaneous calls from the same card, or calls to certain foreign countries that are known to have drug trafficking problems. Regardless of the problem, the card is disconnected and the telecommunications department is immediately informed. It is then up to the telecommunications department to address the issue with the employee. In most cases, they will simply issue a new card.

The dilemma of monitoring calling card usage is that there are usually so many cards accompanied by a plethora of call records. A master calling card report is usually provided from the IXC. This may be in hard copy or computer medium. There will be hundreds (or possibly thousands) of cards. Compound this with thousands of calls, and it can easily be determined that the meticulous examination of these records can be a frivolous undertaking. A more logical approach is to examine summary reports, which will provide information similar to the parameters flagged by the "hacker" program offered by the IXCs. Otherwise, the administration of calling cards can become a full-time job.

Support of toll-free services is not nearly as straightforward as the previously mentioned services. The reason for this is that toll-free services are usually more mission critical to corporate revenue objectives. For example, a toll-free service may service an order entry point in a call center, which means that revenue is generated via the toll-free number. An example of this is the catalog sales businesses that rely solely on customers calling the toll-free number published in their catalogs. In this instance, a failure on the part of the toll-free number could have a devastating impact on the core business. Whether the toll-free number is a direct sales line or provides a support function, the reliability of this service is critical to business. If a service line is rendered ineffective, it could mean that future revenue would be placed at risk.

Price is critical to supporting toll-free services, but it is often more important to establish reliability. In addition, modern toll-free services carry many advanced routing features that can make toll-free networks complex to manage. Whereas outbound long distance and calling cards may only require an administrative person to provide ongoing support, toll-free services often require more advanced expertise, such as a call center expert. In the case of toll-free services, it is no longer a matter of monitoring costs and managing a database. There are more critical issues at stake.

The telecommunications department needs to effectively monitor the traffic on toll-free numbers. A toll-free number is directed to a trunk line or group of trunk lines. The cost of the toll-free service is dictated by the access method (switched access versus dedicated access). If there are more calls than available trunk lines, the problem could result in busy signals. The problem then becomes potential lost revenue. The critical areas that need be addressed with toll-free services are: trunk availability, access method, advanced features, and disaster recovery.

Trunk availability is simply the ongoing effort to ensure that there are enough trunk lines to support the offered call volume. Monitoring usage (provided on bills and management reports) and performing traffic engineering studies achieve this. The call volumes that are analyzed are also evaluated for dedicated vs. switched access in order to provide a more favorable rate.

Beyond the number of trunks or methods of access, toll-free services are often deployed with advanced routing features. For example, a call center may close in Pittsburgh at 5 p.m. eastern time. The calls may then be routed via the time-of-day feature to another call center located in Chicago (central time zone) which is in operation for an additional hour. This simple routing feature offers an additional hour of coverage, which means possible additional revenue and good will. There are a myriad of other features offered with toll-free services. There is time-of-day routing, day-of-week routing, day-of-year routing, area code routing, exchange routing, and 10-digit routing. There are also many features offered with toll-free services, such as automatic number identification (ANI) and dialed number identification service (DNIS). Other features may include automated attendant, interactive voice response (IVR), voice recognition, and en-route announcement. As the reader can see, many of these features can be layered upon multiple toll-free services that can service a number of call centers.

In the case of advanced features, the telecommunications department needs to develop solid toll-free network documentation and continually fine tune the network features, making sure that efficiencies are provided to the business. Because IXCs do not provide toll-free documentation as a part of their service, the onus is on the telecommunications department to maintain accurate documentation.

Also critical to the support of toll-free services is the development of a quality disaster recovery plan. Since toll-free services are so critical to revenue objectives, there should be a thorough disaster recovery plan in place to allow for rerouting of calls to other sites. This can be done with a simple call to the IXC, or via terminal dial-up access. Toll-free services can be a very robust and versatile part of a disaster recovery plan. If local cable is cut, or there is an electrical failure, calls can be readily rerouted to another site.

7.6 Telephone and Voice Mail Systems

Basic telephone systems are often administered locally. That is, adds, moves, and changes are executed by a site administrator who has been trained on the system. However, this does not completely absolve the telecommunications department from supporting the telephone systems installed at distant sites. Local administrators can be very knowledgeable and well-versed on the systems they support. However, they still are not able to address more technically advanced or global issues. Perhaps an expensive software upgrade is proposed by the vendor, and is deemed "critical" to the future needs of the company. Or perhaps there are service-related problems with the chosen vendor, one that was recommended by the telecommunications department when the switch was installed. Or consider that a major upgrade is required in order to address a new application or a major expansion of the site. Would the site administrator be qualified to evaluate such issues? In most cases the answer would be no. In these cases, staff members such as analysts and project managers need to be available on a consultative basis.

In most cases, when a major change or upgrade is required of a basic PBX or key system, a project should be generated. As in the case of any upgrade, there will need to be financial and technical analysis performed to determine if the project should actually be done, what the proper technology should be, and what is the most economical way to approach the project. In the case of problems such as support issues, it will be necessary to document the problems and perhaps set up a task team to meet with the vendor and resolve the problem.

The same level of support can also be said of adjunct voice mail systems. These systems, just like site-based PBXs, are administered by the local PBX administrator. There is seldom a reason for the telecommunications department to become involved once the system has been installed unless there are performance problems or upgrades are necessary. The telecommunications department needs to take more of a direct role with telephone systems when there are advanced applications, such as networked PBXs, networked voice mail systems, or ACD systems.

There are key issues that must be addressed with networked PBXs or voice mail systems. The systems have to be connected with some form of network service. This is either via analog private line service or some form of digital service such as T-1. Traffic studies must be performed to determine the proper number of channels to service the call volumes. There is also software that must be programmed and continually administered to provide optimum performance. For example, is the traffic being sent over the most cost-effective route? Or could the traffic, as in the case of voice mail, be sent during nonpeak hours in order to obtain a more cost-effective dial-up rate, or not conflict with other traffic? Functions such as these need to be performed by a staff member with more advanced skills, usually an analyst.

ACD systems, as in the case of toll-free services, often support revenue-generating applications. ACDs have sophisticated software that routes calls to groups of employees (known as agents). Calls are routed to the least busy agent within an ACD group, or can be routed to another group, either on site or off site. Many companies use on-site expertise to support ACD programming, serving a similar function to the local PBX administrator. But because of the sophisticated nature of toll-free networks and the complicated designs of call centers, the telecommunications department is usually heavily involved in the on-going support of ACD systems. While local support may be able to serve the needs of the individual site, the local administrator will not be able to support the ACD environment globally. Such support requires the support of an analyst or call center specialist. The support includes understanding call volumes, types of calls, staffing, ACD routing capabilities, and making the appropriate recommendations for network and ACD design. Because ACD functionality is so critical to business plans, the telecommunications department is often expected to provide both expertise and ongoing support for the call center environment.

7.7 Data Services and Equipment

More than any other group of technologies that support the enterprise network, data equipment and services require the most direct support from the telecommunications department. An enterprise network can be composed of thousands of individual devices and services, and there must be a central, guiding force to design the entire network and continually manage it. There are many reasons for this, but one of the most basic is that data has yet to enjoy the universal standards that are applied to voice communications. Consider that any employee within the company may purchase their own telephone and business line. Once they receive their number from the telephone company, they simply plug in the telephone and dial anywhere through the public switched telephone network (PSTN). Conversely, anybody could call the employee if they knew the 10-digit telephone number. This scenario is not nearly as simple in the data communications arena. An employee cannot simply buy a PC and connect to a WAN or LAN. First, there are numerous types of data communications protocols. In fact, in large companies, it is not unusual to find multiple protocols being transmitted throughout the WAN. Many companies still retain legacy mainframe computers in addition to mid-sized and PC-based servers that must be connected. Consequently, a PC may need to be equipped with special network drivers or terminal emulation that provides the unique computer interface. Second, each computer and computing systems must have a network address. Then there are issues such as security and bandwidth requirements. Idealistically, data communications would be great if one could plug into the network and enjoy the same types of standards that voice does. This, however, is not the case, and the telecommunications department must address many variables in a field that is changing rapidly.

It should be noted that basic adds, moves, and changes are not always the charge of the telecommunications department. Many companies maintain a separate computer department that is responsible for the additions of PCs, which includes PC configuration, LAN addresses, LAN security, and HUB connectivity. Regardless of whether the telecommunications department supports this function or not, the department still needs to understand what is being installed and how that data traffic will affect network performance. For example, a workstation may need to be segmented on the LAN to a 100 Mbps segment in lieu of a standard 10 Mbps segment. This, however, only addresses the local traffic. If the high-powered workstation sends large files to another LAN within the building or over the WAN, the result could be a degradation of service. Consequently, the responsible party needs to maintain accurate network documentation. This is important because when problems occur, the telecommunications department will require accurate network documentation as part of the problem-solving process or if a change is requested.

LAN support is often the charge of a local administrator. As in the case of PBXs, voice mail, and cable, it is more logical and efficient to establish support this way. Because the telecommunications department is centralized, staff members need to understand the data applications and trends in the desktop environment. The day-to-day activity that may occur at remote sites should not be of concern unless operational problems should surface that relate to network design. The telecommunications department is more beneficial to the company when it serves as a consultant for general inquiries or problems that relate to local LAN support. Otherwise, the help desk will be burdened with installing LAN cards at remote sites, which they will be ill equipped to perform efficiently.

Where the telecommunications department takes on a more direct role is in support of the WAN. In viewing a simple enterprise network configuration (see Figure 7.5), each remote site requires a device that provides access to the WAN. In the modern corporate network, this is typically a router, but there are also devices such frame relay access devices (FRAD) or modems. Each device is provided from a number of manufacturers that produce diverse models to address various software levels, transmission speeds, and physical interfaces. For example, when the telecommunications department provides a router, the staff

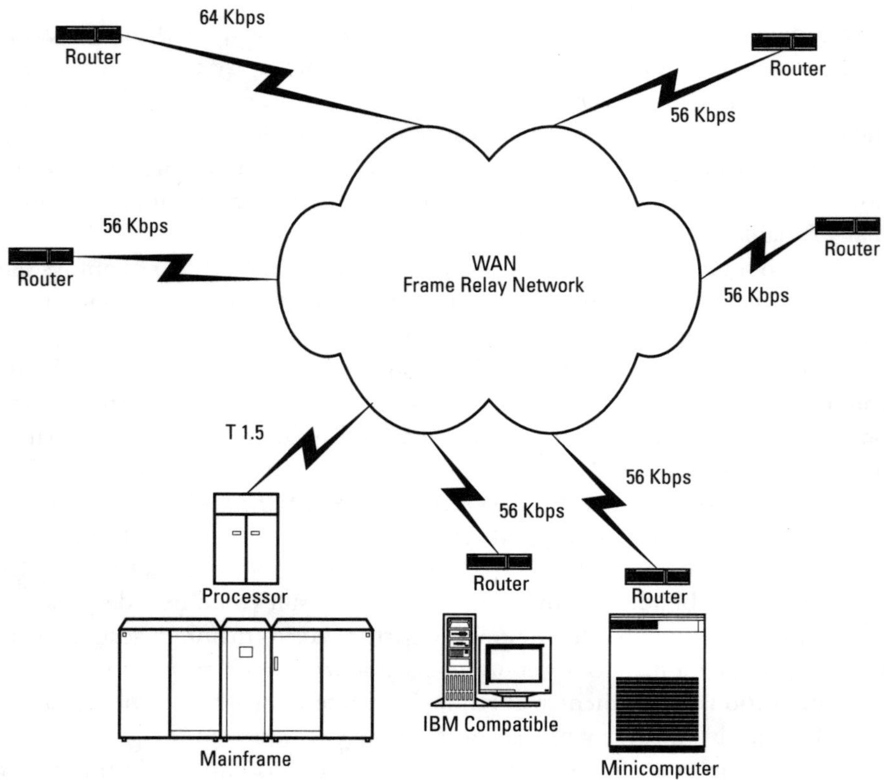

Figure 7.5 A simple enterprise network configuration.

member must understand what types of data traffic will be transmitted from the specific site, the bandwidth requirements, and the WAN service (e.g., frame relay) to which it is connected. Assume that a location has legacy IBM mainframe applications (i.e., 3270) and TCP/IP traffic. There is also a need to periodically send large graphic files throughout the network via TCP/IP. Because the IBM application supports an order entry function, response time is critical, although the bandwidth requirement is low. The network service that supports this site is a frame relay link. In addition, there is also an ISDN BRI link that is used for dial backup in the event that the frame relay link fails. Ongoing support of such a design requires that the telecommunications department execute a number of disciplines.

There must an inventory system that tracks the types of devices installed at each location. Staff members must be able to quickly understand what type of device is installed in the event that there is problem or a change is necessary.

If a new computer application was required for the site, it could mean that the router would have to be upgraded or replaced. Also, if response time became a problem, it could mean changing the network interface, or upgrading to a faster router. Regardless of the type of change required, staff members need to reference a master inventory file that yields salient information about the network device (see Figure 7.6). They must understand what is in place and what changes will be necessary in order to provide the necessary capabilities.

Because the enterprise network can contain thousands of devices, the only way to manage it is through a computer-based inventory system. The telecommunications department needs to track each device for its technical capabilities and financial status. Replacements or software upgrades will affect the budget, and they cannot be arbitrarily made without thorough documentation. One of the quickest ways that a budget will get out of hand is if the inventory is not properly maintained. This issue can sometimes become contentious in a telecommunications department. Analysts who are data communications experts often do not like to be bothered with administrative details; they are more concerned with the technology and managing the network. However, the network cannot be properly managed unless the myriad of devices and services that comprise it are accurately tracked. The telecommunications manager needs to ensure that inventories are properly managed. He or she must take measures so that changes to the inventory file are easy to execute, and not an administrative burden to his analysts who are often concerned with more complex issues. There are several measures that can be taken to ensure that this will happen.

There should be a person dedicated to executing changes to any aspect of the network inventory. In the case of devices, an analyst should not be able to call a vendor over the telephone and order a new router verbally. If this is allowed (and many vendors will do this), the analysts tend to place the paperwork at the end of their "To Do" list. Consequently, the inventory system will reflect a router that may have been removed and is lying on the floor of telecommunications closet, packed and ready to be shipped elsewhere. In my own

Device:	Router
Model:	XYZ
Software Level:	3.2
Protocols:	TCP/IP
Network Interface:	Frame Relay
Router Address:	
DLCI:	
In Service Date:	

Figure 7.6 Master inventory file.

experience, I have seen situations where the old router was simply forgotten. Such scenarios are not unusual. The results of such actions cause a bevy of problems. What if the analyst that ordered the router quits? Then consider that a new problem occurs and a new analyst looks into the inventory system and diagnoses the problem to be the result of an obsolete router. He or she then orders a new router, and the field engineer from the vendor discovers that the exact same router has already been installed. Obviously there have been monies wasted, but the true source of the problem has still not been solved.

Any change in network design should be logged as a project. Part of the requirements for project execution is to update the inventory system. An administrative person can make the actual changes in the inventory system, but there must be checklist that includes this action item as part of the project process. If the analyst does not make the changes to the inventory system himself, he must provide this information to an administrator. If the inventory file is not maintained, it will result in budgetary problems and support issues. There will be discrepancies between vendor bills and the inventory file. Staff members will also have a difficult time determining resolutions to problems, because they will not know what is truly installed. The net result is that the department performance will deteriorate and eventually there will be a need to create a major project to clean up the inventory file.

Understanding the components that comprise a network is only one aspect of the support issue. There must also be a system in place for monitoring data traffic throughout the WAN. This is not a simple issue because of multiple data communications protocols and proprietary network management systems. However, systems must be put into place that will provide information related to data traffic. Otherwise, it will be a matter of analysts throwing bandwidth at the problem. They will not know if they have ordered too large a circuit or if insufficient bandwidth was even the true source of the problem. Without network statistics, support becomes an issue of blindly groping about, and staff members must rely on dumb luck to solve their problems. If they are successful in solving a problem by throwing bandwidth at it, they will probably have ordered too much bandwidth, which may adversely affect the WAN budget.

The data communications device is always connected to some form of network service. The data link could be a dedicated analog private line, a frame relay link, or a T-1, to name some of the most common interfaces. As in the case of the network device, the network service also needs to be inventoried. Once again, there needs to be a computer-based master file. Many departments rely on vendor bills to manage their network services. This is fine in small applications, but large corporate networks have too many services and multiple vendors to contend with. Paper is not as efficient as a means of managing large volumes of data with a limited staff. There has to be a uniform method of storing and retrieving data for all staff members.

Certainly one of the reasons that an inventory is used is to track finances. However, there is also a need to understand what type of service is installed, which may have a direct link to network performance. The profile of each circuit will have information that relates to the characteristics of the circuit (e.g., frame relay, analog private line, T-1). It is necessary to have this information so that statistical information can be compared to circuit characteristics. This is done to monitor trends or to solve immediate problems.

Addressing also becomes a critical issue with data communications. Addressing is the means by which one device transmits to other points on the network. The addresses must be maintained in a file so that when a change is ordered, accurate information is available. Consider that a site must now transmit to two locations instead of one. The WAN links are frame relay, and it is a simple matter of programming the new data link connection identifier (DLCI) into the router.

There are factions of the telecommunications department that will be primarily involved in project work and there will be other factions that will be more involved in support issues. Data specialists, more than any other discipline, will be heavily involved in both. The data side of the telecommunications staff can be divided into support specialists and project managers. In certain cases (depending on individual ability and staff size), some staff members will carry both roles. The most advanced knowledge will reside in the analysts and project manager positions. The help desk personnel will probably not have as much experience or knowledge. Therefore, there should be a great deal of interaction between the groups when changes are requested from the help desk.

7.8 Videoconferencing Systems

Videoconferencing comes in two flavors: desktop and group systems. The telecommunications department does not always support desktop videoconferencing. Desktop systems are usually a simple matter of purchasing the PC-based hardware and software and transmitting over a basic TCP/IP connection. Group systems, on the other hand, require specific network connections, are of a higher quality, and normally require a video administrator to set up videoconferences.

Videoconferencing, like PBXs or voicemail, is often better serviced by a local administrator. Once a system has been installed, setting up a videoconference can be as simple as dialing a few telephone numbers and entering a few simple parameters into the video unit's software. This is, of course, under ideal circumstances. Very often, it is not quite that simple and there can be some very exasperating moments when trying to set up a video call. Still, the telecommunications department cannot be at every site to set up a videoconference every time a meeting is held. This is simply impractical.

The telecommunications department should have staff members that will be designated as video experts. These people will be responsible for selecting systems and setting up the network access. In most cases, the group videoconference systems will be capable of basic low (112 Kbps or 128 Kbps) or high speed (336 Kbps or 384 Kbps) connections. This will be at 15 or 30 FPS. The network access will be some form of digital dial-up, normally ISDN BRI. The software in modern videoconference units is quite user friendly and can be almost as simple as dialing a telephone number. Also, there is a suite of ITU standards that allows for universal connectivity. Given all of these circumstances, it should only be a simple matter to provide the units to a specific site and then train the administrator on the use of the system. In the ideal world, the video experts would only be involved in project work, not the daily details of setting up videoconferences.

What most telecommunications departments discover is that, in spite of their best efforts to train the site administrator, they are still drawn into direct support issues. The sites often need to set up a videoconference with a customer or business partner. While there are standards for connectivity, and basic telephone numbers are used to dial to another location, there can still be a number of issues that will surface. For example, the site administrator will need to enter two numbers into the software in order to dial the distant location over BRI at the low speed. The BRI at the distant end might have two different telephone numbers for each B channel or it might be the same number, depending on how the LEC is provisioning the service. Will the distant end support the ITU standards, or is it an older system with proprietary signaling? If it is the latter, then the conference will have to be set up through a video bridge so that a protocol conversion can be performed. What is the base bandwidth of the access channels at the distant end, 56 Kbps or 64 Kbps? If the distant end is a switched 56 access and 64 Kbps is used as the base rate, the connection may not happen. These small issues begin to loom large when a nontechnical person is assigned as the video site administrator. In my own experience, I have had site administrators who were secretaries and those that were managers. Technical skills or knowledge were not always present and, in many cases, they did not want the responsibility in the first place. More often than not, they will call the telecommunications department, rather than solve the problem themselves.

The problem may not seem that critical to the inexperienced person, however, consider a room filled with representatives of senior management. They have scheduled a videoconference with a potential business partner overseas, and they are expecting to conduct a face-to-face meeting. There is not a more stressful situation for the site administrator than to be in this room and not be able to make the connection. Moreover, they may try to place the blame on the telecommunications department for not providing adequate training or technical support. These types of problems will never go away for a number of reasons, but there are steps that can be taken to minimize the problems.

The first step is to develop solid documentation about most aspects of videoconferencing. The video manuals that are provided with videoconferencing systems are too generic to address the specific needs of a company. A video manual, developed by the telecommunications staff should provide enough information to address most videoconferencing situations. Consider the following subjects for a video manual:

1. Overview of videoconferencing
2. How to conduct a videoconference
3. How to dial up a low speed point-to-point connection
4. How to dial up a high speed point-to-point connection
5. How to set up a multipoint conference
6. Frequently asked questions
7. Videoconference scheduling form
8. Troubleshooting guide
9. Company locations, video systems, and video numbers
10. Glossary

Each section would be simple and concise, covering most situations that the site administrator may encounter. There are sections that will provide frequently asked questions, a scheduling form that will guide the site administrator through the details of the scheduling process, and a troubleshooting section that will address the common problems. While a video manual will serve as a good first line of defense against potential problems, the video experts will most likely still be called to assist. Although this situation will not completely disappear, there are steps that can be taken to counter this situation. First, the video experts on the telecommunications staff should always aggressively train any new site administrators. Second, the telecommunications department should recommend and lobby for a backup to the primary site administrator. Third, the video manual should be updated and distributed as changes occur.

In spite of all these efforts, there will still be problems with video support. There will be failed videoconferences for a variety of reasons. Some problems will be related to technology, others will be user related, and some will be problematic to other organizations. Consider that a site administrator has been charged to set up a videoconference with a major customer. Unfortunately, the video administrator for the customer is not well trained on the system, or well versed in telecommunications technologies. After lengthy discussions over the telephone, the conference is conducted via audioconferencing, but the meeting participants are angry and disappointed. The source of the failed connection could very well have been bad information provided by the customer contact. Such scenarios are common, and there will be nothing that can be done in such situations. The telecommunications department must make sure that training and information are provided in both quantity and quality, to reduce the direct support issue. Direct support, however, will almost always be required.

7.9 Cable

As in the case of PBXs, key systems, and voice mail, cable is better managed by a site administrator. The initial cable system is usually designed and implemented as a project of the telecommunications department. Once the system has been installed, it is normally not practical for the department to perform simple adds, moves, and changes for remote sites. When the installation is complete, the local administrator will be provided with documentation and a local contact, should additional cable need to be installed or a repair be necessary. Unless the cable is installed at the corporate headquarters, the role of the telecommunications department then becomes one of a consultant. For example, if a heavy data application is proposed for the site, there may be a need for fiber in lieu of copper. This would be a major upgrade to the existing system. The telecommunications department would then need to assign a project to an analyst and redesign a portion of the original system.

Supporting cable systems can be a major headache for the telecommunications department. If properly designed and supported, there should actually be relatively few problems. Unfortunately, the site administrators do not always take the time to document moves and changes or to keep the cable system neat. In my own experience, I have installed cable systems that were installed beautifully and with accurate documentation only to return a year later to find a virtual rat's nest with no documentation. The mismanaged cable system might then go on the books as a major project. This project is major only because it may be labor intensive. In essence, the telecommunications department will have to tear down the system (i.e., patch cables), document the connections, and reinstall the system neatly.

Of course, the telecommunications department cannot police everything. Staff members are not corporate police and cannot control everything that site administrators will do. However, there needs to be a concerted effort to provide a quality system and training during the initial installation. First, the telecommunications department needs to establish standards. There should be standard designs and systems that can be provided to any application within the corporation. In addition to a design, there should be standard naming conventions, documentation, and rules for executing changes. The site administrator for cable needs to be trained on the system documentation; how to perform adds, moves, and changes; and how to keep the wiring closets neat. The trainer should clearly point out the negative aspects of not properly managing the cable system, from wiring that becomes tangled to not understanding what devices are connected. If all of these efforts are taken, the telecommunications department will have done everything possible to support cable in the corporate environment.

7.10 Summary

The ongoing support of the telecommunications environment requires that there be a system established for requesting and executing adds, moves, and changes for all aspects of the enterprise network. In order for this system to be effective, the telecommunications department must manage various aspects of the technologies such as capacity management, addressing, inventory management, and facility management. In addition, the telecommunications department must establish distinct support requirements (local or central) for various aspects of support.

8

Setting Policies for Acquisitions, Support, and Strategic Direction

Setting policies for acquisitions, support, and strategic direction are among the most important functions that the telecommunications department serves. Unfortunately, many telecommunications managers often overlook these functions or only perform them on an ad hoc basis. The reasons for these oversights are often very logical, dictated by the demands of the business. The project load may be so overwhelming that there is simply no time to spend on these issues. Limited staffing may also place the department in a reactive mode. Setting policies and strategic direction means that the telecommunications department is proactive rather than reactive. This means that the project workload and overall department responsibilities must be manageable and staff sufficient in order to afford these concepts the time they deserve.

Regardless of the constraints faced by the telecommunications department, setting polices and strategic direction must be addressed because they do have an impact on the future of the department. When there is no consideration given to policies or strategic direction, systems are replaced prematurely, technology does not meet business applications, costs are unnecessarily high, and problems are not solved in the most efficient and logical manner. Moreover, the telecommunications department does not meet its maximum potential. If this is perceived by sources outside the telecommunications department, it may make the department a target for outsourcing.

A telecommunications policy is a process that is developed and then established in order to support the business environment in the most logical, efficient, and cost-effective manner possible. There will be many policies that will be implemented by the telecommunications department. Many of the policies will be guidelines that will only affect the telecommunications department. Other policies may be guidelines for end-users to follow. Regardless of who is responsible for supporting or executing the policy, the telecommunications department should be the primary source for creating telecommunications policies. If it is not, it may be an indication that the reputation of the department is suspect. There will also be many residual problems. The telecommunications department will find itself supporting technologies that it did not choose. Eventually, there may be many operational problems because various aspects of the technology have not been thoroughly planned. A company needs a central telecommunications planner and that planner is the telecommunications department.

Policies encompass a variety of concepts, ranging from minor tasks, such as the addition of a telephone to a PBX, to high level concepts, such as the acquisition of a major system. Established policies ensure that quality business decisions are made. Well thought out and meticulously defined policies bring structure to the processes that are performed by the telecommunications department. Because they are defined and documented, they ensure that a smooth transition can be made when there are staff changes, or when there are changes within the department. For example, assume that a company has decided to undergo a major restructuring. Part of the restructuring plan is to offer a number of experienced employees a buyout of their pension. This may mean that a number of people will leave the telecommunications department. It may also mean that the change will happen so quickly that there may not be time to allow for a smooth transition. Whether new people are hired or existing staff will take on additional duties, documented policies will help ensure that the transition will be less painful.

Policies define department processes, but they also lay a foundation for strategic direction. For example, the acquisition of any major system certainly merits the examination of the technological platform. Will it meet the needs of the business today and, more importantly, will it meet the needs of the business for the life of the system? A standard policy should dictate that each new system is evaluated on three levels: technological, financial, and strategic. This helps ensure that the technology will meet the needs of the business in the future and that factors such as capacity or bandwidth will not become inhibiting factors, perhaps requiring a premature system replacement.

As previously mentioned, policies are defined on a departmental and corporate level. Departmental policies encompass a number of disciplines. One of the most important is that of acquisitions.

8.1 Policies for Acquisitions

When a new system is acquired, there are many potential risks that the telecommunications department will assume. The system must be suitable for the existing business application, and yet versatile enough to meet the future needs of the business. This must be done in a cost-effective manner, but the methodology for acquiring the system must be logical, systematic, standardized, and documented. Also, the process for selection should be unbiased. There should be standard forms and procedures that are used for every acquisition. This will create an audit trail so that all parties can access the information and understand why a decision was made should questions surface. This should be a standard policy even if the telecommunications department is not required to use a formal process. The reason for this is very simple. All acquisitions have the potential of not meeting the needs of the business application. There may also be vendor-related issues, such as bankruptcy, a merger, or nonperformance. Whatever the source of the problem, the telecommunications manager may be asked to justify the decision that was made. If he can provide a formal, documented process that yielded decision matrixes, it will be determined that the telecommunications manager is a victim of circumstances beyond his control. If the decision was made arbitrarily, even if based on prior decisions that were well documented, it may cast an unfavorable light on the telecommunications manager and the department.

There are two situations by which acquisitions are made. First, there is a large-scale acquisition for a new system(s) that is applied to new or existing business applications. This acquisition may be the purchase of a single system or multiple systems, a method for establishing a standard, or a means by which the lowest cost is established. Second, there is the acquisition of a system for which a standard has already been established. Both situations require a formal process.

Large-scale acquisitions are executed via two processes, the request for information (RFI) or the request for proposal (RFP). The RFI is normally used as an analysis tool to prepare for the RFP process. The RFI is normally a relatively short document, concentrating primarily on technical specifications and vendor profiles. The purpose behind the RFI process is to access the technical viability of a particular product. Also, the vendor organization will be assessed at a high level in terms of size, experience, and financial stability. The project manager assigned to the acquisition will use the RFI process to eliminate a number of potential bidders. Because the RFI document is much shorter, a larger number of potential vendors are more manageable. For example, the project manager may evaluate 10 potential vendor organizations via the RFI, with the goal of selecting five finalists for the RFP process. Potential bidders may be eliminated for a number of reasons. The technology may be young and untested in the marketplace; it may not be a correct fit for the application; or the organization itself may present some limitations.

There is no right or wrong way to write an RFI. The size and complexity of the project dictates the length. However, the project manager should understand that many of the respondents will be eliminated and it is both unfair and unprofessional to expect them to expend a tremendous amount of effort. The RFI should be no longer than three or four pages, but preferably no more than two. Special care should be taken to only list the technical requirements that are most appropriate for the business application. Meticulous technical details should be avoided unless they are absolutely essential to the performance of the product. Consider the sample RFI in Figure 8.1. This sample RFI for an ACD system reflects basic capabilities that will be required to meet the business objective. In this particular example, a company is looking to install three large ACD systems that will be used to support a reservation system for a small airline. Because there are a number of unique criteria, the RFI examines these criteria at a high level, thereby eliminating the vendors that will not be able to meet the demands of the business objectives.

The RFI is not always required. When there are a number of players in a given technology sector, it can be an effective means to logically eliminate the vendors that will not be qualified. It is also an educational process for the project manager who will gain valuable knowledge of numerous products and technologies. It should also be noted that elimination from the RFI process does mean that the vendor will not serve a future need. The RFI responses are filed in the project file and may be accessed at a later time should a new requirement or technology be needed.

The RFP process is more involved than that of the RFI. This is the process by which a major system(s) will be selected, or a major policy established. The RFP itself defines the business application, the technology that will be applied to it, and how the technology will be designed and configured. Beyond the technology, there are also business, legal, and implementation issues that will be defined in the RFP. Any major acquisition, or newly defined standard, demands that these issues be defined and examined. There are two aspects to the RFP process: the development of the RFP document and the process by which the RFP is used to make a selection.

The actual RFP document can be quite long and detailed. There are many telecommunications professionals who take the concept to absurd lengths, developing documents that actually become self-defeating. As in the case of any process, there needs to be a degree of common sense applied, otherwise the project manager will find himself or herself managing the RFP process and not the actual project. The RFP is a means to an end, not the sole

Request for Information—XYZ Company

XYZ Company is currently in the process of gathering information in order to establish bidders for an ACD application for three major call centers. Please fill in the questionnaire and mail back to the address listed above by January 15, XXXX. Each call center will support approximately 300 agents. The three call centers will act as one, interflowing calls based on agent status, time of day, and priority of calls. The system will use CLID via ANI to prioritize customer calls. The CLID will be cross-referenced against a computer database via ANI.

---

1. Maximum number of agent stations per system = ______________
2. Maximum number of trunk ports = ______________
3. ANI capability (Y/N) ______________

---

4. DNIS (Y/N) ______________
5. Interflow (Y/N) ______________
6. Number of ACD routing tables = ______________
7. Priority queuing (Y/N) ______________
8. Priority lookback overflow (Y/N) ______________
9. CTI capable (Y/N) ______________
10. Realtime display (Y/N ______________

How many similar installations does your company have? ___________
How many years has your company been in business? ___________

Figure 8.1 A sample request for information (RFI).

responsibility of the project manager. Large and complex RFPs may have their place, but the more prudent course is to develop the smallest possible document while still addressing the critical issues.

The telecommunications manager needs to develop a boilerplate RFP that will serve as a basis for all projects. This document should be generic and stored in software so that project managers can access the document and then modify it to the specific needs of each new project. While many projects will require different technologies that are applied to varying types of business applications, there are some very fundamental aspects of the RFP document that will apply to all projects. Consider some basic concepts (see Figure 8.2).

1.0 Introduction
2.0 Present System—Background
3.0 Instructions to Offerers
    3.1 Critical Dates
    3.2 Pricing Criteria
    3.3 Confidentiality Clause
    3.4 Compliance to Requirements
    3.5 Rights of Rejection and Incorporation
    3.6 Questions—Contact
4.0 System Requirements
    4.1 Port Capacities
    4.2 Traffic Requirements
    4.3 Circuit Card Capacities
    4.4 System Architectures
    4.5 Redundancy
5.0 System Features
    5.1 Network-Related Features
    5.2 Software Questions
    5.3 Attendant-Related Features
6.0 Transmission Characteristics
7.0 Station Equipment and Features
8.0 Power Requirements
9.0 Vendor Background
    9.1 Company Profile
    9.2 Product Experience
    9.3 References
    9.4 Project Organization and Staffing
10.0 System Pricing
    10.1 System Cost Breakdown
    10.2 Itemized Unit Costs
    10.3 Payment Schedule
11.0 Switchroom Information
    11.1 System Footprint
    11.2 Drawings
    11.3 Environmentals
12.0 Service, Maintenance, and Training
    12.1 Maintenance Agreement
    12.2 Training Schedule
13.0 Supplemental Applications
14.0 Appendixes

Figure 8.2 Basic concepts of a request for proposal (RFP) document.

While this example is more specifically geared toward a PBX acquisition, the majority of these concepts will apply to all technologies. Each section of the RFP is numbered. Sections that address more complex issues will have subsections.

Section 1.0 should be a short overview of the company and the business application. As in the case of all sections, the objective is to convey a concept concisely. Normally, this section should only reflect a few paragraphs.

If there is a present system, it is described in Section 2.0 with background information. Once again, there is no need to go into elaborate detail. After all, the system is being replaced, and only information that is relevant to the business application needs to be conveyed.

Whereas Sections 1.0 and 2.0 are basically text, Section 3.0 reflects a more exacting format. This section dictates how the bidders will present information. Critical dates that reflect specific project milestones are listed. The dates reflect a timeline that is critical to the project, and vendors must understand that the dates are not negotiable and will not change. The only time that a date will change is if the project manager or end-user determines that the project timeline should change. Pricing criteria is also reflected in this section. Here the project manager will specify how the specific product or service is to be priced. It is important that instructions are clear and concise. One of the reasons behind the RFP process is to provide a standard method for bidding. The vendors should know how to break out all costs. For example, if equipment is being purchased, the project manager will want to see an itemization of each system component. Also, there may be a need to show purchase vs. lease, depending on the financial status of the company or department that is acquiring the product. When all bidders submit their pricing in the same format, it is usually a simple matter of plugging the numbers into a spreadsheet.

The Confidentiality Clause, Compliance to Requirements, and the Rights of Rejection and Incorporation subsections all relate to rights of the RFP issuer. Bidders must understand that the information contained in the RFP is proprietary in nature and cannot be disclosed to any other party without written permission from the issuer. The bidders must also understand that they must comply with the specific requirements of the RFP, or face arbitrary elimination. Many vendors, in spite of clear and explicit instructions, often chose to submit bids that are contrary to the format and specifications of the RFP. This creates extra work for the project manager and defeats the purpose of the RFP. Bidders should also understand that the issuer of the RFP has the right to reject any and all bids. Finally, a single point of contact should be listed, which will be the project manager from the telecommunications department. It is important that all questions are directed to this single point of contact. Otherwise, questions will be filtered to people who do not have the technical expertise or project knowledge. This will also introduce an element of confusion into the project. All bidders should understand that the project manager is the sole authority of the project. If bidders attempt to bypass the project manager they could face elimination from the process.

Section 4.0 defines system requirements, which includes system configurations and design. This section educates the project manager and the end-user about the capabilities of the systems that are being bid. This section also tells the bidders how to configure the system that they will be bidding. For example, redundant processing is a desirable feature but an added expense. The premise is to configure an additional processor that will act as a backup to the primary processor. If the primary processor fails, the backup will take over. By specifying such capabilities, all bidders will bid the same, or similar, systems. The system features will be incorporated into a decision matrix that will be used in the selection process. Other capabilities will also be detailed, such as circuit card capacity. The project manager will understand what the future requirements of the system will be and will factor this into the selection process.

System Features, Transmission Characteristics, and Station Equipment/Features, as outlined in Sections 5.0, 6.0, and 7.0, address the capabilities of the system to meet the business application. These sections should reflect requirements that are a culmination of end-user input and the expertise of the telecommunications staff. The project manager should thoroughly understand the business application and then recommend the appropriate technology to the end-user. Beyond the end-user's immediate requirements, the project manager will make sure that the specifications detailed in the RFP reflect future needs and also fit into the standard corporate network requirements.

Section 9.0 is important in evaluating the stability and capability of the vendor. In the-fast changing world of telecommunications, there are many start-up companies, mergers, and failures. Also, many telecommunications manufacturers constantly retool their methods of distribution. The best technology in the world will be of little value if the vendor does not have the resources to support it. There is nothing more exasperating to the telecommunications department than to have an angry end-user call about the poor support he or she is receiving from the vendor. Even under the best of circumstances, and with the most reputable of vendors, there will be problems. To not consider the experience and financial stability of a perspective vendor simply invites trouble. The RFP brings the end-user into the selection process and allows him or her to examine, with the project manager, the quality of the organizations that are being considered. If, or when, problems occur, he or she will be understanding rather than thinking that something has been forced upon him.

System Pricing, Section 10.0, provides a template for pricing detail. This section allows the project manager to assess how the system is priced in terms of hardware, software, installation costs, etc. It is important that each vendor adhere to the format specified in this section. As each bid is received, the project manager will place the unit costs into a matrix, in an effort to achieve an apples-to-apples comparison.

Switchroom Information, Section 11.0, defines the environmental specifications that will be required to support the system. This section is more for the benefit of the project manager and the end-user. Modern telecommunications systems, whether voice or data, are computers. As such they require the same demanding environmental specifications as any mission critical computer. The room must be dust free and kept at a specific temperature and humidity. For many projects, this may mean that the telecommunications closet requires expensive modifications. The end-user may not have budgeted for these modifications and the project manager must convey that system performance may be degraded and warranties negated if the environmental specifications are not met.

Service, Maintenance, and Training, Section 12.0, details the options that are available for general maintenance and trouble reporting. Most systems will have a factory warranty that generally ranges from three months to one year. After the warranty period has expired, the customer will have to pay for time and materials to have certain types of work performed. Signing a maintenance contract is sometimes considered to be a waste of money. In essence, a maintenance contract is a gamble on the part of the end-user. Under the terms of the contract, it is agreed to pay a flat amount that will be less than anticipated expenses for a given year. If there are relatively few changes required, and the system is reliable, it may appear that the money was wasted. The maintenance agreement, however, serves other purposes. For example, the equipment may be supporting a critical business function. Under the conditions of the contract, a major system failure may specify a guaranteed four-hour response time from the vendor. If there is no maintenance contract, the vendor is not obligated to meet a specified emergency response time. Although the vendor will respond to a trouble call, priority will be given to end-users that have signed a maintenance agreement with specific response criteria. Although there may be instances where a maintenance contract is not required, the end-user assumes a large liability if a contract is not signed.

Included in Section 12.0 are training issues. Training for end-users, whether as system administrators or users of the system, is listed as a separate price. The end-user must understand his or her responsibilities for supporting the system that is being installed. This may mean that a system administrator is designated. It may also mean that a backup is designated and that, in the event of employee turnover, a new administrator will need to be trained. Thus, the end-user should understand the operational changes to the department and the potential costs to support the new system.

Many systems require training of each employee. For example, a PBX requires that each employee is trained on the feature codes of the PBX and voice mail system. The cost of providing a trainer for a specified number of classes will be listed in the bid. However, as employee turnover occurs, there will be a need to train replacements. This may be done via a train-the-trainer program, videotapes, or hiring trainers from the vendors on an ad hoc basis. Regardless of the most practical way to approach training, the vendor will spell out in this section the training price for the installation along with any additional costs that may be requested after the fact.

Beyond the actual document, the steps of the RFP process are relatively simple. Consider the following steps:

1. RFP is mailed to participating vendors;
2. Vendor's conference is held to answer vendor questions and perform a walk-through of the facility (if applicable) approximately two weeks after receiving the RFP;
3. Addendum to the RFP is issued to all vendors addressing answers to questions that were raised during the vendor's conference;
4. Bid due date;
5. End-user and project manager may review the proposed systems via on-site visits;
6. Bids are analyzed and reviewed;
7. Decision matrixes are developed;
8. A finalist is selected;
9. Business and legal issues are reviewed;
10. Vendor is selected.

The use of RFPs brings structure and logic to the acquisition process. All bidders will submit their information in the same format and matrixes are developed by the project manager, which will allow, as closely as possible, an apples-to-apples comparison. The telecommunications department should always use the RFP process as a standard method for acquiring new systems or establishing standards for future purchases. Adherence to this policy ensures that products and services are selected in an unbiased manner. It also creates an audit trail, should the vendor not be able to deliver as detailed in the proposal.

Many telecommunications professionals frown on the use of RFPs, deeming them to be a process used by consultants to justify their fee. This is a disservice to the process, ignoring the benefits it brings to the department and corporation.

The project manager should try to involve the end-user in the RFP process. Certainly many end-users will not be interested in every detail of the process. However, there will need to be a certain degree of involvement. The end-user will need to make sure that the project manager understands the business applications and objectives. The project manager will want to make sure that the end-user understands the RFP process, what technologies and vendors will be considered, and how the chosen system will affect his or her business operations. The worst possible situation will be for the end-user to take a hands-off approach, not understanding the process, the system, or the vendor that will support it. This situation very often leads to a public relations nightmare for the telecommunications department.

Many telecommunications professionals view the RFP process as a tedious undertaking with questionable results. They interpret the process as the time-consuming development of a large document, filled with minute details and data that have little importance relative to the true performance of the proposed system. The problem is not with the RFP process, but rather with how the RFP is composed. For example, many consultants assume that they must earn their fee by developing large documents that encompass every possible feature and technical detail. I once worked with a consultant who had a question in his RFP that asked the "meantime to seize dial tone." In other words, how long would it take for a handset to seize dial tone from a telephone system once it had been taken off the hook. I argued that the question had little relevance, but the consultant was adamant about keeping it in the RFP. In reality, the question did not have much relevance; answers came back from the vendors in nanoseconds. Moreover, the question was viewed by the vendors as being a nuisance. In order to answer the question, all vendors had to contact engineers in their research and development departments. By comparative analysis, none of the systems reflected a deficiency that would have affected system performance. In a nutshell, this question was a waste of everybody's time. The RFP process can be time consuming and, if approached in the wrong manner, counterproductive. However, there are some measures that can be taken to make the process efficient and productive.

One of the most difficult aspects of the RFP process is the actual writing and composition of the RFP. Once this is accomplished, the boilerplate document can be kept in software, to be pulled up and modified for each new project. Writing the actual RFP need not be as difficult as it sounds. There are many sample RFPs available on the market from educational and research organizations, such as Faulkner Technical Reports or Datapro, that can be used as a basic guideline. These sample RFPs will be generic in nature, but they can be a good starting point. Also, if there is RFP experience within the telecommunications staff, this may be an additional resource. The objective is to do the bulk of the work one time, and simply modify all subsequent RFPs as specific projects dictate.

There may be a number of boilerplate RFPs that will be developed, depending on the technologies that are supported by the department. There may be RFPs for PBXs, key systems, videoconferencing systems, routers, hubs, cable systems, long distance services, or data services. While there may be other products or services that might necessitate an RFP, this list covers the general categories. There will be many sections of the RFP that will be universal across all technologies or services. Still, there will also be many aspects that will be unique to the specific technology. For each particular technology or service, the telecommunications manager will want to establish a boilerplate RFP. Some of these RFPs will be used quite often. For example, PBXs and key systems will usually be replaced every five years. In a company with a number of locations, there can be continual PBX replacement projects that require RFPs. Conversely, an RFP that is issued for a corporate long distance service will only be issued once, and will not be reissued until the existing long distance contract is about to expire. Nevertheless, there needs to be a reference point whenever a major contract or acquisition is pending. Therefore it is certainly more efficient to use a boilerplate document and customize it for the new project, rather than to start from scratch.

There should be a standard process by which the RFP is stored and named. The telecommunications department should have a server where intellectual property is stored. There should also be a number of directories related to department functions. One of these should be named ACQUISITIONS. Under the ACQUISITIONS directory can be a number of subdirectories such as RFI, RFP, RFQ, and PROJECTS. Boilerplates are kept in the RFP subdirectory and named according to application. For example, RFP_PBX, RFP_ROUTER, RFP_KEY, or RFP_LD. When a new project is assigned, the project manager simply accesses the RFP subdirectory and pulls up the appropriate RFP. The naming convention lists what type of document it is, and the application that it applies to.

The project manager will then rename the file to also indicate the specific project. For example, RFP_PBX_PHOENIX will indicate that RFP is for the PBX replacement for the company's Phoenix site. This will eventually be stored in a subdirectory. As new projects surface, project managers can benefit from similar projects that were done in the past. If the Phoenix site required an ACD, and similar requirements surface for a new site, the project manager can simply reference the RFP_PBX_PHOENIX file and modify it for the new project.

Not all acquisitions will require a document as large and all encompassing as the RFP. If a standard vendor has been selected, there may only be a requirement for a request for quote (RFQ). The RFQ is used in two situations. First, if a standard vendor has been selected and there is a need to order additional equipment or configurations. For example, it may have been determined that a standard router will be used in all corporate locations. While all routers will be supplied from the same manufacturer, there will most certainly be a variety of models, configurations, and software levels. The RFQ is a method of controlling the ordering process, while also instituting a system of checks and balances for inventory and accounting purposes. A second application is for controlling the ordering process when there is only one choice for the product. For example, there may be a need for monitoring software that is proprietary to the routers that are used in the WAN. Since no other system will work with the routers, there is only one choice. Still there needs to be a process that shows that the telecommunications department examined the costs before approving the purchase. If there is only one choice, and the cost is excessive, the most obvious question is whether the product is actually necessary.

The telecommunications manager will want to convey to all parties—end-users, senior management, and telecommunications staff—that there will be standard processes for all acquisitions. Regardless of how much freedom the telecommunications department is given for acquisitions, it is always advantageous to use a formal and documented process. Only in the most extreme of circumstances should an acquisition be made without the use of an RFI, RFP, or RFQ. If the timeframes are short and the pressures great, there may be no other choice. In these instances, the telecommunications manager will want to write a letter to the project file, copying all concerned parties, and informing them that the standard procedure was recommended and could not be used due to circumstances beyond the department's control.

8.2 Policies for Trouble Reporting and Maintenance

After a system has been installed, in almost every case, there needs to be some methodology for trouble reporting and maintenance. This usually entails the development of a separate discipline within the telecommunications department: the help desk. Beyond developing a help desk, the telecommunications manager must also ensure that all staff members develop quality problem- solving skills. That is because all staff members will be required to provide some level of support and must be prepared to be part of the process. The help desk is considered first level support and various staff members are often considered second level support.

Problem determination is a critical function of the telecommunications department and an essential skill set for all telecommunications professionals because problems are endemic to the field of telecommunications. There will be installations that go awry, technology that does not live up to its expectations, poor performance on the part of the vendor, software bugs, disasters (both due to human error and acts of God), or devices that simply fail. Seasoned telecommunications professionals take concrete steps to prevent problems from happening, but in spite of the best efforts and planning, problems do happen. Therefore, all members of the telecommunications department should be prepared to face numerous problems during the course of their career. There is no way to avoid problems, because this is in the nature of the business.

The telecommunications manager needs to understand that a problem-solving policy is one that will undergo continual evolution. In addition, there are a number of components that need to be in place encompassing organizational structure, policies, and individual staff skills. The first component of the plan is to provide a standard method for end-users to report problems. This is a method by which problems are reported, logged, and acted upon. What must be avoided are calls randomly filtering through to staff members who are involved in critical project work. When a staff member receives a trouble call, acts upon the problem and solves it, it is good PR for the department. Unfortunately, when this happens, there is no methodology in place for addressing all trouble calls that may arrive at the telecommunications department. If the end-user is helped, he will make a note of the staff member's name, and probably pass it on to fellow employees. This "helpful" staff member might then be inundated with trouble calls. Inevitably, if the staff member is involved with a heavy project load, he will not be able to effectively service trouble calls. What started as a good deed becomes a bevy of angry end-users who have not had their calls returned or their problems solved.

There are a number of ways that trouble calls can be addressed. The preferred method is to establish a help desk to service most trouble calls. The reader should note that not all trouble calls can, or should, be serviced. The reason for this is that trouble shooting for some devices (such as PBXs) is often better served by having the end-users perform it themselves. Many companies establish a policy that mandates that PBXs, cable, and some LAN functions be administered locally.

The help desk is essentially a call center staffed by specialists who are trained in a number of technical disciplines. The call center should be operational as business dictates. For instance, if a company only has sites located in the eastern time zone, and is only operational during regular business hours, the help desk might only need to be operational from 8 a.m. to 5 p.m., perhaps with a person on call in the off hours. Many other businesses will need to cover additional time zones and perhaps international sites. Under these circumstances, the help desk will probably need to be available 24 hours a day, seven days a week.

Help desks are not always economically feasible. As in the case of all call centers, there needs to be a room that houses the staff, an ACD, and computer system access. There may also be a variety of network management and diagnostic software that the staff will require in order to trouble-shoot various technologies. This design entails a rather large capital investment and considerable operating costs. The staff also needs to be managed so that vacation or routine absenteeism does not impact the performance of the help desk. In addition, the help desk staff will be comprised of expensive telecommunications professionals who have considerable technical knowledge and require constant training. The facility, the ACD, and the toll-free number will also contribute considerably to the expense. Provisioning all the essential parts of the corporate telecommunications help desk is indeed an expensive undertaking.

Even the smallest of telecommunication department help desks are expensive and difficult to manage. If the help desk does not exist, it will be a rather arduous task to convince executive management that the expense is justified. However, there are many compelling reasons why an in-house help desk is the best method of supporting the corporate telecommunications infrastructure. But before one can justify the implementation of a help desk, it must be understood what the true function of the help desk is.

The corporate telecommunications help desk provides first level support for all problems that may occur with some (or all) technologies that comprise the corporate enterprise network. It is not always necessary for the help desk to be responsible for all technologies. As previously mentioned, certain technologies are better served by the end-user. One of the most common examples of this is the PBX. For example, in a branch location, PBX technicians are usually dispatched locally. Reporting trouble for a basic PBX may be a relatively simple matter of calling a number and reporting the source of the problem. For instance, there may be several telephones that are out of commission. The PBX administrator for the site would call the vendor, who would issue a trouble ticket. The problem may be resolved by dialing remotely into the PBX or by dispatching a technician on site. If the problem is seemingly more complex (perhaps network related), the PBX technician may initiate a trouble-shooting procedure with either the LEC or the IXC. All of this can be accomplished independently of the corporate telecommunications help desk. If the problem relates to a network design issue (perhaps there is an advanced toll-free network), then the PBX technician may have to call the help desk in order to understand how the PBX is designed to interact with the corporate network.

The reason that some trouble reporting functions should not be supported by the corporate help desk is that resources will be limited, expensive, and therefore must be managed to provide maximum effectiveness. The telecommunications manager must define what should be supported and what can be delegated to the end-user. Otherwise, the help desk will be burdened with simple, mundane problems while more serious problems are held in the queue. Consider the examples illustrated in Table 8.1. WAN services are almost always supported by the help desk. Once data leaves a remote site and is transmitted onto the WAN, responsibility is handed off to the telecommunications department. The end-user will generally not know how the network is designed or why the problem is occurring. Moreover, he will not have the diagnostic tools necessary to determine the source of the problem. The same is true of CPE that supports the WAN. If a router fails, the end-user will seldom understand how to approach the problem. As first level support, help desk personnel will dial into the router to determine if the problem is related to addressing, equipment failure, or an unexpected change in data traffic. Based on the nature of the problem, it may be resolved on line, or a technician may be dispatched.

LAN support, on the other hand, is generally supported locally. Just as the PBX should have a site administrator, there should be a designated LAN

Table 8.1
Help Desk Support Needs

Technology and Function	1st Level Support
WAN network services	Help desk
WAN CPE	Help desk
Toll-free services	Help desk
Long distance	Help desk
LEC voice services	End-user
LAN	End-user or help desk
Cable	End-user
PBX	End-user
ACD	End-user or help desk
Cellular telephones	End-user or help desk

administrator. Many LAN connectivity problems are related to the PC configuration, local traffic, or cable issues. Many problems are commonly known and can be readily solved with a certain degree of experience. This does not mean, however, that all LANs can, or should, be supported locally. A simple LAN configuration comprised of low speed Ethernet or Token Ring connections can be administered by an employee with a basic knowledge set. If there are high-speed workstations involved, multiple types of data, or complex cable configurations, problems could be much more involved and will need to be escalated to a higher level of expertise.

Cable is almost always served better with local support. Once the cable system has been designed and installed, there should be a local administrator to support adds, moves, and changes. If the cable system is properly designed and administered, little can happen. If a connectivity issue should occur, it normally means that some office mishap has damaged the cable. In this instance, the end-user should have been provided with the number of the local vendor. Resolution of the problem is to simply call the vendor and have the cable repaired.

ACD, depending on complexity, design, or business application, may require first level support from the help desk. ACD is an enabling technology of call centers, which often have a critical relationship to corporate revenue plans. For an order entry point, there is a dollar value carried for each telephone call. This is calculated by taking the total dollar value of orders entered in a day and dividing it by the number of daily calls. If there is a failure in any component of the call center, there could be potential lost revenue. Therefore, it is imperative that call centers are given first level support by the corporate help desk. There are usually elaborate disaster recovery plans in place with call centers that may encompass a number of complex plans. Execution of these plans will require in-depth knowledge of the network and call center design. This should seldom be entrusted to the end-user, who is usually more involved with call center business issues rather than technology.

Outbound long distance and toll-free services are usually provided from a single carrier because a corporate contract has been negotiated. This means that the telecommunications department has records of all corporate locations, what types of services they have, and what types of features are being used. Because the end-user will not be schooled in all aspects of telecommunications trouble reporting, he will probably call the PBX vendor when there is a problem. This would be the best course to follow, but the end-user should also be instructed to call the corporate telecommunications help desk if the problem proves to be unusual or debilitating to the business. For example, if long distance is suddenly not available, and the site houses a telemarketing operation, there could be lost revenue.

Toll-free services, especially advanced services that support multiple sites, will almost always require first level support from the telecommunications department. Because of the critical nature of toll-free services, there is too much revenue at risk to trust this function to the end-user.

Local services are normally best supported by the end-user. Since each site may be supported by a different LEC, it is usually better for the end-user to trouble shoot local service. Once again, trouble reporting usually begins by calling the PBX who will determine the nature of the problem and escalate accordingly.

The number of users and the scope of contracts determine support for services such as local cell phone usage. Because many cellular contracts are negotiated locally, it is usually more efficient for the end-user to support this. Also, most cellular issues relate to broken phones or billing issues. However, this is not a steadfast rule. With the advent of PCS, some carriers can offer a national contract. It is also possible to have a large-scale contract on a local basis whereby a company requires hundreds (if not thousands) of phones. In these cases, it may be more logical to support the wireless contract from a central point.

In reviewing the various technologies and services, the aforementioned scenarios are only provided as a rough guideline. Whether first level support should be provided is dependent upon the business application, the scope and complexity of the technology, and the visibility and perceived importance of the business application.

Critical to the issue of when first level support should be offered is the importance of the business application. A telephone system that supports a revenue generating application is mission critical to the business. Whether an inbound call center or a telemarketing application, it is not enough to simply have the end-user call the telephone system vendor. The vendor may dispatch a qualified technician, but do they know where other sites may be that can accept rerouted calls? Also, the local vendor may not be aware of how the network is designed and consequently not understand what options may be open for rerouting calls. In such instances, the design and technology are too complex and important to place in the hands of the end-user or a local vendor.

The scope and complexity also dictate when first level support is appropriate. The WAN is a perfect example of this. If a WAN link fails or response time is unacceptable, business operations or revenue are at risk. WAN design and performance affect virtually every business unit. Moreover, the WAN is designed and implemented by the telecommunications department. There is no other resource within the company that will be qualified to provide support. Therefore, there is only one choice for first level support.

The visibility and perceived importance are also important factors in providing first level support. If a phone or PC is inoperable, it normally does not affect entire departments or divisions. That is why a local PBX administrator better serves these types of problems. Still, there are seemingly minor applications or simple technologies that would normally not warrant first level support, but are still mandated by senior management. I was once sent to Shanghai, China to oversee a small and simple PBX installation. The support was deemed necessary because my company was building a factory in a new and untested market. Because the startup costs were formidable and the risk high, executive management wanted to take every possible precaution to ensure that the new business venture would be successful. This mind-set also holds true for support issues. The telecommunications manager certainly cannot support everything, but there are times when the political climate will necessitate support.

8.3 The Help Desk

Smaller companies and telecommunications departments normally do not have help desks. The help desk usually evolves as an organization grows. Very often, the telecommunications manager realizes that there is no policy for servicing trouble calls. As calls randomly arrive at the department, he realizes that staff members are juggling trouble calls with project work. The problem is further exacerbated by the fact that many staff members are not always able to service the calls, because they have neither the technical tools, a thorough knowledge of the enterprise network, or problem-solving skills. Consider that a call may arrive for a data expert who is traveling on business. The call is then forwarded to a voice expert who has no idea how to even approach the problem. It is usually at this point that the telecommunications manager realizes that he must employ a more logical and well-defined system.

The help desk may grow from simple and rather humble beginnings. Because the function grows out of necessity, many telecommunications managers find themselves in a position whereby there is a problem but no funds allocated to solve the problem. The telecommunications manager may understand that a number of expensive issues will have to be addressed. For example, there may be a need for an ACD, various types of network monitoring software, education for help desk personnel, diagnostic equipment, office space, and computer equipment. The most expensive portion, however, will be the addition of unbudgeted personnel. Unless problems have reached astronomical proportions, it is doubtful that he will be able to justify the funds necessary to establish such an operation. Consequently, he may have to begin with simple designs and policies.

The basic premise behind the help desk is simple (see Figure 8.3). The end-user is given a number to call. This may be a standard 10-digit or toll-free number, depending on budgets and where the end-user resides (toll-free may be preferable for mobile employees). The end-user now calls a number for help, in lieu of calling an individual. As each call arrives at the help desk, it is assigned a trouble ticket number. In the early stages, the call may arrive at a bank of telephones set up in a simple hunting arrangement. The telecommunications manager will schedule staff so that the phones are manned during the most important business hours.

The help desk serves four basic functions or objectives. The first objective is to solve as many problems as possible solely with help desk personnel. In fact many calls will probably involve very simple problems. This may be as simple as having the end-user reboot a system or simply explaining how a technology works as opposed to how an end-user perceives how it should work. The second objective is for the help desk to assess the problem and contact or dispatch the appropriate person and expertise to resolve the issue. The fourth objective is to take ownership of the problem. A trouble ticket should be issued and the help desk (as an organization) will be responsible to make sure that the problem has closure. Finally, the help desk should be proactive in monitoring network activity so that problems are resolved before they become critical.

The ultimate goal of the telecommunications manager is to establish a staff of help desk specialists, perhaps using the group as a steppingstone to higher positions within the department. Prior to this, he or she must find ways to staff the help desk with existing resources, or find the funds to hire outside help. Staffing a help desk is a demanding undertaking. Callers will be angry, distraught, worried, and possibly insulting. Existing staff will probably look upon a move to the help desk as a demotion. Moreover, they will probably look upon the work as a nuisance, and seek employment elsewhere. The tele-

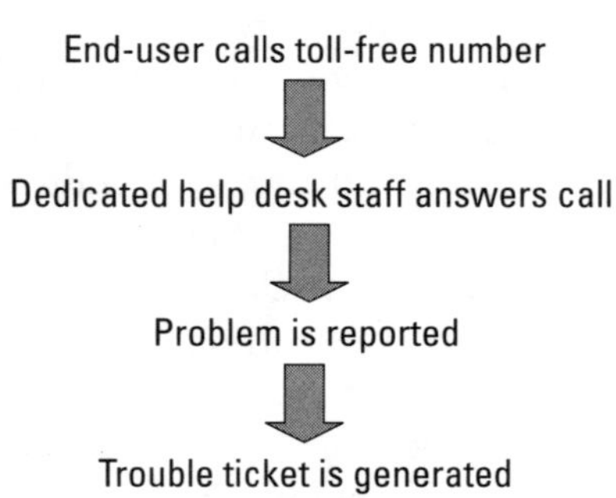

Figure 8.3 The basic premise behind the help desk.

communications manager needs to be careful how he staffs the help desk. However, in certain circumstances, it could be a career-enhancing move. In the case of an administrative support person who is in the early stages of learning telecommunications technologies, it could be an opportunity for a pay raise and the chance to learn more about the technology. Although the candidate may not view the help desk as a permanent career move, it can be viewed as a steppingstone to better positions such as junior telecommunications analyst.

The telecommunications manager may also consider using contract workers. Many recruiters also offer contract services, providing workers at an hourly rate. The advantage for the telecommunications manager is that there is no permanent commitment. While the hourly rate may be relatively high, it is usually not as expensive as a regular company employee collecting benefits. The manager also has the luxury of replacing workers that are not performing in an acceptable manner, with little or no repercussions. The downside of using contract workers is that noncompany employees almost never have the same commitment as regular employees do. However, using contract workers on a probationary basis, with the option to hire based on performance, can rectify this.

As the help desk function expands, so will the help desk budget. This will require demonstrating to senior management the importance of the help desk, and the monies that are required to support it. The most effective way to justify additional monies for the help desk is to relate all additions and modifications to corporate business objectives. If the requested monies do not ultimately protect or maintain revenue, it is doubtful that the monies will be approved. This concept must be maintained for all requests to executive management. For example, a network-monitoring tool will have little value for senior management unless the telecommunications manager conveys that there have been a number of WAN outages that caused lengthy downtime. The outages were at revenue-generating sites, which placed revenue at risk. If the network monitoring software were in place, alarms and error messages would have provided a warning to help desk personnel to take action before the problem became critical. Such an explanation brings significance to complex technology for senior management.

8.4 Policies for Problem Solving

The help desk is more than a group of people supported by various technologies. Once the call has been answered the help desk specialist must exercise advanced problem-solving skills. Help desk specialists will need to exercise these skills as a part of their day-to-day duties; however, the issue is not exclusive to the help desk. Other staff members will need to learn and exercise these skills. That is because complicated and highly technical problems will require that the more technically advanced staff members become involved in the problem-solving process. Consequently, all staff members must understand how to assess problems, apply their analytical skills, and take ownership in the problem, seeing it to resolution.

The concept of developing problem solving skills can be a nebulous concept. The telecommunications manager may take aggressive steps to counsel staff members or provide training, but there will still be varying capabilities within the staff. It takes a special type of person to keep a clear mind when problems occur and emotions run high. They must be able to ignore hostile attitudes and think clearly, working through complex technical issues. They must work with different personalities and attitudes, both end-user and vendor, keeping all parties focused until the problem is solved. It can be an exasperating process and a thankless endeavor. And just when one problem is solved, another one seems to surface. When all is said and done, regardless of how well the problem was solved, there may still be bitter feelings on the part of the end-user. While these types of situations may not be fair, they are not uncommon.

Problem solving is a combination of individual skills and departmental policy. Each staff member should be given training regarding basic problem-solving skills. Regardless of whether this is a formal class or an impromptu training, some form of remedial training should be mandatory for all staff members. The training, however, should only be viewed as a primer. The staff member, depending on knowledge, experience, and personality, may require continual training and reinforcement. In fact, it will be found that even the most seasoned of professionals will periodically require refresher courses.

Problem-solving seminars and training are often very theoretical and do not provide real-time experience. There is a big difference between discussing a problem in a classroom and actually having to deal with real-life situations. When a staff member fears for his or her job, or has to deal with a hostile end-user, even the best training may fail. Because of the visibility of telecommunications services, and the impact of failed services on the business, the development of problem-solving skills should be viewed as a continual process, and one that is critical.

The telecommunications manager needs to view the problem-solving process from two perspectives. First, each staff member should be individually trained and his or her skills continually refined. Second, there must be departmental guidelines defined so that staff members do not have to improvise every time a problem surfaces. In addition, the telecommunications manager must demonstrate that he or she is supportive. As previously stated, problems are endemic to the field of telecommunications. The staff must know that their efforts are appreciated and that, in the face of adversity, they have an ally in the telecommunications manager.

Once again, each staff member needs some form of training in problem solving. This will provide a good foundation for newer staff members, or serve as a refresher for experienced people. It is this author's opinion that the best training includes role-playing and invites opinions and participation from the students. The training then takes on more meaning for the staff members. Also, the telecommunications manager should strive for training that is more specific to the problems that the staff will face in real-life situations. I once attended a stress-management and problem-solving seminar that was a general course designed for all employees. The examples provided in the class and the recommended solutions were so unrealistic that I could not take the class seriously.

Each staff member will develop his own style of handling problems. Also, there will be staff members who will become very adept at solving problems and those who will forever struggle, not able to work effectively under pressure. In spite of their best efforts, telecommunications managers may have to move people from positions (such as the help desk) where problem solving is part of their daily duties. Still, telecommunications professionals are difficult to find and so every effort should be made to reinforce their problem-solving skills before any change is considered.

The problem-solving process begins with good interpersonal skills. As a call arrives at the help desk (or to an individual member of the telecommunications department), there may be a need to diffuse anger. Even if the caller is not angry, they may be distressed. The staff member needs to assess the emotional state of the caller and counsel him appropriately. The angry end-user will only become angrier if the staff member becomes defensive, is indifferent, or responds in anger. When addressing anger, staff members should be taught to disregard the anger and focus on the problem. But first they must try to diffuse the anger. There are a number of techniques that address this situation. One is to listen with a sympathetic ear, allowing them to vent their frustration. The staff member needs to understand that the anger is usually not vented toward the individual. Rather, the end-user is angry about his or her situation. Response time may be slow, the WAN link may be down, or customer calls cannot be processed. The bottom line is that business is being adversely affected or halted completely. Once the anger has been vented, the staff member must convey to the end-user that he or she will take control of the situation. It also often helps to provide the end-user with a name and trouble ticket, and assure him or her that they will be called with a status, good or bad. When the end-user understands that somebody is aggressively working on his behalf, he or she will more readily accept the situation.

Beyond the interpersonal issues, the staff member must then apply analytical skills to the problem. Before the staff member can apply his or her analytical skills, there are two factors to consider. First, interpersonal skills are often a necessary prelude to the analytical step. The more upset an end-user is, the more likely he or she will be to give inaccurate information. It is important to diffuse anger or even calm the end-user so that he or she can give the best possible description of the problem. Very often, experienced telecommunications professionals find that a problem is completely different from the original description, once they calm the end-user. The second factor is that the problem must be examined with the end-user on the line. The end-user is usually not well versed in the technology and may not be able to assess the problem accurately. Consequently, the trouble report cannot always be accepted at face value. It often helps to have the end-user duplicate the problem and describe it as it happens. The experienced staff member will automatically begin to eliminate various components of the design and zero in on the potential source of the problem.

Once the problem has been recorded, the staff member will analyze the problem and determine a course of direction. The trouble report should be logged in a database and assigned a control number with accompanying notes. As the help desk changes shifts, or if other staff members are asked to participate in the troubleshooting process, any staff member should be able to access the database to understand the problem and assess the status. When the help desk shift changes, there should be a turnover report for the next shift, so that they can continue with the troubleshooting process.

At this point, documentation bears special notice. Many problems encompass a number of services, various types of equipment, and multiple vendors. As the troubleshooting process progresses, there can be much trial and error. Many problems are actually solved through a process of elimination. Whenever a complicated problem is being addressed in this manner, it is necessary to thoroughly document the process, in order to avoid a duplication of effort. The documentation should be archived for future reference. It may serve as a teaching tool, or as a discussion document to affect changes in departmental policies. It may even point to design flaws in the network. Regardless of the benefit, it is important to document complicated and unusual problems.

When the problem is resolved, a report is generated. The telecommunications manager and help desk supervisor will systematically examine the data. They will want to evaluate the response and performance of individuals and the help desk in general. They will want to determine if any staff members need additional counseling or a refresher course. They will also want to evaluate whether the departmental policies are adequate.

The departmental policies for problem solving involve a number of factors, and one of the most critical is education. Problem-solving training has already been discussed, but quality problem-solving skills also require knowledge of the products and services that service the corporation. If a staff member is expected to support something, he should be given a thorough knowledge of the corporate telecommunications infrastructure. Moreover, he or she should be trained on individual system designs and how they are expected to operate in the current environment. This includes all aspects of the network: LECs, IXCs, CPE vendors, and so forth.

Beyond product knowledge, there are also the tools that are necessary to monitor, analyze, and troubleshoot problems. This will include written procedures, product manuals, reference materials, diagnostic tools, and the proper technical aids. Written procedures should be in place for the most common problems. For example, if the primary WAN link fails (e.g., a frame relay circuit), the written procedure should be to restore the service on a dial backup line (perhaps a BRI ISDN), then call the IXC and issue a trouble ticket. In addition to the basic procedure, there should also be escalation procedures written in the event that the problem is not resolved in a timely or satisfactory manner. Product manuals are also helpful when there is a question about a system capability. These will aid in the analytical process when there are discussions with other staff members or vendors. Reference materials should also be available, such as a telecommunications dictionary or technical reports.

The level of support offered by the telecommunications department to end-users varies from company to company. There are departments that will carry certified network engineers and technicians on their staffs and others that will rely completely on the technical support of vendors. Regardless of the skill levels that may be carried by the department, staff members need to be provided the proper tools to perform their jobs. This can range from simple hard-copy documentation to sophisticated and expensive diagnostic and monitoring equipment. The telecommunications manager must understand to what degree his people will be supporting the components of the enterprise network. With the input of various staff members, he must then provide the tools that will be necessary to troubleshoot the problems.

Problem-solving policies should be in a continual state of evolution. As each problem occurs, it should logged and then possibly examined in staff meetings to determine if a policy change is in order. For example, it may be determined that problems could be solved more quickly if the staff technicians had a sniffer (a device that can detect and analyze data protocols and errors). This is an extremely expensive piece of equipment but that concern may be relative when compared to the cost of an ineffective network. Regardless of the issue, the modus operandi should be for continuous improvement. Problems will never go away, but the department needs to managed as a team that continually strives for optimum performance.

8.5 Policies for Strategic Direction

The profession of telecommunications management is one that is filled with paradoxes. In many instances, the telecommunications department is viewed as overhead, much in the same manner as a utility. When this happens, the telecommunications budget is kept to a bare minimum, and the department is usually expected to provide technology only for existing business applications. However, this is not necessarily a steadfast rule. While the telecommunications budget is looked upon as being overhead, it does not mean that the department is immune to criticism if the services they provide fail to meet future demands. Regardless of how the department is viewed, it is folly to maintain a utility attitude and only provide what is absolutely necessary for the immediate application. Inevitably, the department will fail by not monitoring technological trends and preparing for the future. Even the most conservative of organizations must bow to technological trends.

Strategic planning encompasses three concepts. First, it is assessing the current technological capability and capacity of all products and services that currently serve the company. Second, it is the proactive effort to monitor technological trends for the future. Third, it is assessing the future needs of the company.

Assessing the current technological environment addresses short-term needs. For example, a PBX has three capacity issues that concern future growth. One is the current capacity that allows for additional stations or trunks via software changes. The "wired-for" capacity is the capacity of PBX to accept addition circuit cards with further equipment additions or modifications. The maximum capacity addresses the maximum number of ports (stations or lines) that the PBX will accommodate. If the maximum capacity is reached, the system will require replacement, known as a forklift. A staff member (usually an analyst) will prepare a facilities and demand (F&D) chart (see Table 8.2) that reflects trunk and station additions over a specified period of time. The F&D chart indicates when equipment should be purchased to accommodate future growth. It may also indicate if a system replacement is in order. Conversely, network monitoring tools will provide reports of bandwidth utilization. Staff members will examine reports that reflect WAN traffic trends and make

Table 8.2
A Facilities and Demand (F&D) Chart

	Installed	Capacity				
Station ports	224	352				
	January	February	March	April	May	June
Station additions	12	8	13	22	6	8
Remaining capacity	116	108	95	73	67	59

adjustments accordingly. It may be determined that the CIR on a frame relay link will have to be increased, or that a prioritization of data traffic is in order. This may necessitate an upgrade in the router software or an upgrade to the router itself.

Evaluating technology for the future is not an exact science. There are many technologies that receive much attention in the trade journals, but that may never become established in the business world. New technology carries a number of risks. For example, it is normally expensive, it may be proprietary (not adhering to international standards), and may even be unreliable because it has not been fully developed. Still, it is folly to arbitrarily dismiss any emerging technology. At the very minimum, emerging technologies should be tracked and understood. While new technologies may be expensive, the price always reduces over a period of time due to technological advancements and competition. The technology may be proprietary, but if it proves to be a valuable business application, international standards organizations will develop standards. If the technology is immature, like all telecommunications technologies, it will be refined.

As outlined in Chapter 10, there should be a logical and systematic approach to staff education. This includes monitoring technological trends. As each product or service comes to the end of its depreciation schedule or a contract is nearing the end of its term, decisions will have to be made. Systems will have to be replaced or contracts renegotiated. The decision might only affect a small portion of the company's business or it might have a corporate-wide impact. Regardless of the scope of the application, this is one of the times when strategic planning is in order. There are only two reasons to examine technology on a strategic basis. First, will the technology support the business application? Second, what is the most cost-effective manner of doing this? For as complex as the world of telecommunications can be, corporate telecommunications management always breaks down to these two factors.

Consider a corporate campus that supports a number of company disciplines. There are a number of types of data traffic transmitted throughout the corporate network, including bitmaps and video. Response time slows when large files or video is transmitted throughout the campus network and there is no set time when this happens. The routers and hubs have been purchased and have relatively little time left on the depreciation schedule. The problem has been deemed to be so critical that the remaining depreciation can be written off in order to solve the problem. Such an example invites a very pointed question: Why won't the current network support the data traffic requirements? If the telecommunications department knew of the requirements and failed to meet them, then this is certainly a black eye for the department. If the requirements were not known, then the question must also be asked: Why?

In order for the telecommunications department to effectively evaluate new and emerging technologies, it is necessary to understand the business applications to which the technology will be applied. Most businesses experience growth (or at least try to grow) and all businesses change over time. Of course there will be unexpected growth and changes, but every business has a plan that ranges for at least five years. Included in this plan will be growth that will include projections for new sites, augmentation to existing sites, or acquisitions. Included in this business plan will be head count and new business applications. The business plan will yield some very fundamental information that can be applied to strategic planning.

The number of devices that will need to be connected:

- The type of devices that will be used (e.g., telephones, PCs, wireless);
- The type of traffic that will be generated (e.g., IP, video, toll-free);
- The volume or bandwidth of the traffic;
- How robust devices and services will have to be in order to meet expected changes and new applications;
- Budgetary considerations.

Understanding the business plan will help staff members to understand two things. First, should existing equipment or services be modified or changed? Second, what will have to be provided for the future when the existing products and services have completed their life cycle? A high level understanding of future business plans offers guidance for research. In the case of the corporate campus, response time was affected by unpredictable surges in traffic. There are a number of factors that relate to this problem. Was there a system in place that allowed staff members to monitor WAN traffic and generate reports that reflected usage trends? There is an obvious bandwidth issue, but is the network robust enough to allow for segmentation, isolating high volume users to high capacity links? Is the physical layer sufficient to allow for upgrades to high capacity transmission systems (fiber vs. copper)? Finally, will the system allow for multiple types of traffic (voice, data, or video), and a means to prioritize and control that traffic? Assuming that many of these issues were present in the business plan, these are issues that should have been addressed before the system design and selection were finalized.

The telecommunications manager needs to stay informed of corporate business plans, present and future. Although there are a variety of methods of accomplishing this, the best method is to establish a relationship with executive management. While these people may have busy schedules, they may also be willing to share information through an assistant. The CEO will almost never have time to sit down with the telecommunications manager, but the chief financial officer or chief operating officer will have a vested interest in the financial and operational performance of telecommunications technologies. In many cases, however, this may not even be necessary. If the telecommunications manager reports to a chief information officer, this person will certainly be well versed in the present and future business objectives. It may also be advantageous to develop relationships with vice presidents of various divisions. Each division may have needs that require unique technical solutions.

8.6 Summary

A critical part of telecommunications management is to think strategically. This includes establishing standards for acquisitions, support, and setting strategic direction for the future. Standard processes should be in place for all aspects of these disciplines including RFPs, the help desk, a defined problem-solving policy, and policies for evaluating future technologies.

9

Human Resources Issues

9.1 Human Resources and Telecommunications Management

There is more to managing a group of telecommunications professionals than simply managing projects and technologies. There are unique human resources issues that need to be considered in order to provide morale, efficiency, and stability within the department. There are a myriad of books, articles, and seminars available on the subject of human resources and management techniques. With so much information available, it is not my intention to reinvent the wheel, nor to cover all possible human resources issues. Issues such as sexual harassment, equal opportunity employment, and disciplinary measures are covered by corporate human resources policies, and there is certainly no shortage of reference material on this subject. These are issues that are not exclusive to telecommunications management, and are generally applicable to all employees within a company. However, there are human resources issues that are more exclusive to telecommunications professionals that should be examined in detail. The reason for this is simple: Quality telecommunications professionals are often difficult to find and keep.

In addition to the shortage of quality telecommunications professionals, telecommunications people face many challenges and pressures that the average worker will probably never experience. There are long hours, highly visible and complex problems, complex technologies, and large, difficult projects. These people also face the daunting task of trying to keep their knowledge and technical skills current, while balancing this challenge with a demanding workload. When a company does not consider these issues, the department will inevitably become a revolving door and probably not perform up to its capabilities. The following sections will explore some of these issues and offer suggestions to help a manager and staff better cope with the demands of the profession. Staff members will be appreciative of efforts taken by the company and their manager to help them cope with their demanding environment. Companies and managers that take these steps will place themselves at an advantage because the demand for quality telecommunications professionals is increasing. A quality work environment will attract better employees and help to retain existing staff.

It should be noted that, in spite of the efforts a company or manager may take, there are no easy solutions nor any guarantees of improved staff morale or retention. Inevitably, every company and telecommunications department will face low morale and staff turnover in certain situations. The turnover will not always be the fault of the employer or the telecommunications manager. Economic conditions can force company policies that will render almost any effort by the manager as useless. For instance, how can a manager maintain high morale when he has been given the directive to reduce staff by 20 percent? Suppose that management has also told him (and the rest of the corporation) that more layoffs will be forthcoming if company performance does not improve. Regardless of how much a manager tries to maintain morale under such circumstances, staff members will probably perform poorly and look for another job because their workload will increase and their future is uncertain.

Nonetheless, the decision to leave should not be an easy one. If an employee has been treated fairly, there is always the chance that he will opt to stay, even during the bleak times that many companies face. Therefore, this chapter will offer suggestions and guidelines that relate to the nontechnical side of telecommunications management, the human resources issues that are unique to a complex field that is constantly changing. Whether a company is facing good or bad times, these issues do have a direct correlation to staff performance and stability.

9.2 The Telecommunications Manager: Coach and Mentor

Telecommunications managers should be coaches and mentors to the people who work for them. The manager is the person whose actions and policies most affect the personality and performance of the group. A quick glance at any department might indicate what types of managers are overseeing the department. Do they appear to be efficient and relaxed, or frustrated and uncertain? How do they react when there is a problem? Do they panic and point fingers or do they work together to solve the problem? Do the staff members openly exchange ideas and discuss new technologies, or do they simply go about their jobs, uninterested in discussing the technologies that are their lifeblood? Do they appear to be educated on current and emerging technologies, or do they appear to always be in a catch-up mode? While there are many reasons why a staff would demonstrate different traits, very often the manager is the person who has most affected the personality and performance of the group.

In my experience, some of the most counterproductive managers are those who ignore human resources issues, and simply dwell on the technical and business issues. These types of managers only focus their energies on department performance. Their criteria for evaluating employees are only based on quantifiable technical data. Do we resolve problems quickly and efficiently? Do we finish our projects on time and under budget? Is the network 99.99 percent reliable? Such issues are certainly important but they do not address the human side of managing a department. It is possible to run an efficient department and meet such objectives for a period of time without addressing human resources issues. Unfortunately, this often means pushing the staff to the limits of their capabilities, destroying morale, and inducing a high level of stress into the working environment. The end result of such managerial tactics is staff turnover. Prior to the turnover occurring, there will inevitably be dissension and a disintegration of the department's performance. Managing in such a manner requires that the department be viewed as a machine programmed to perform a function rather than as unique individuals, combining their collective talents, to achieve a common goal. When people are looked upon as head-count to achieve quantifiable goals, the process is dehumanizing. Moreover, no manager can expect quality performance from an employee who is not afforded respect as an individual.

As coaches, telecommunications managers should guide their staffs through projects, the general administration of the department, and through problems. They should offer guidance on complex projects, suggesting methodologies without getting mired in the details. They should keep the project teams focused and on track while still allowing their input and respecting their opinions. All of this is accomplished by using the resources that exist within the staff. That is, to use their skills and knowledge as a resource, rather then to view the staff as bodies hired to do a job, or bow to the manager's will. This does not mean that the manager's authority is undermined or that department performance is compromised. Just as coaches do not score the touchdown, they will ultimately decide what play would be used. The play, however, might have been added to the playbook by an assistant coach or player. The decision to use the play will ultimately rest with the coach, but the coach cannot do, or think of, everything.

Continuing with the concept, the coach is also responsible for maintaining morale and motivating the team (the department), through good and bad times. Winning is an easy concept for anyone to grasp and it is easy to coach when a team is winning. Losing is a different matter. This is when the coach must find the ability to keep the team going, motivating them through adverse conditions and finding positive things in a negative environment. The telecommunications department will inevitably face adverse conditions. There will be times when mistakes will be made, networks will fail, and projects will not be completed on time or on budget. The source of the problems could be the fault of individuals within the department, or they could have been caused by factors beyond anyone's control. Regardless, the manager (as coach) needs to stand by the staff, help solve existing problems, and then correct the source of the problem afterward. Solving the problem may mean counseling or educating staff, or implementing new policies. If this is addressed in the proper manner, the staff will feel that they are supported and that their manager is behind them.

As a mentor, the telecommunications manager should set an example and educate the staff with knowledge and actions. The manager should display good problem-solving skills and demonstrate quality analytical abilities. When a problem surfaces or a project is assigned, the manager should lead by example. The manager should openly discuss books he or she has read, or classes he or she has taken. The manager should teach the staff how to act when problems occur or how to prioritize when the workload becomes heavy and complex. When the staff understands that the manager has exceptional abilities and demonstrates them without imposing them, they often emulate the manager's style and incorporate it into the day-to-day activities of the department. It should also be noted that it is important for a manager to recognize when staff members possess knowledge superior to his or hers, or propose better ideas. Once again, the manager cannot do everything or think of everything. It is not a negative reflection of a manager's abilities when somebody in the department has superior knowledge or develops good ideas. Rather, if the manager recognizes and nurtures this talent, it is a sign of the manager's maturity and confidence to be willing to use available resources. In addition, it is to a manager's credit to have recognized, hired, and utilized quality talent.

When companies seek to hire a telecommunications manager, there is often a tendency to concentrate on the areas that are easily quantifiable, such as technical knowledge or project management skills. These are certainly important, but the position of the telecommunications manager needs to be a balance between technical and managerial skills. Managers who are only concerned with hard, cold statistics will inevitably lose people. The human side of management cannot be ignored. The problem is often acute in telecommunications professions because many managers are promoted for their technical skills and not their ability to manage people.

The problem of hiring telecommunications managers who lack human resources skills becomes exacerbated when companies are looking for a person to fix a myriad of problems. Consider some questions that might be posed to a managerial candidate. "We have tremendous problems. The telecommunications department has a terrible reputation and they have failed miserably on several high profile projects. We need somebody to come in here and take no prisoners. So do whatever you have to do, and don't worry about hurting anybody's feelings."

A manager who would take this directive literally would probably only solve a small number of the problems on a short-term basis. An appropriate response from the manager would be, "What are the true sources of the problems? And, is the staff really to blame?" The department may have been the victim of a poor manager or there may have been circumstances beyond their control. There are many reasons why a project can go awry, and a manager who would take the interviewer's comments at face value would only be displaying ignorance and inexperience.

Many of the subjects covered in this chapter will address human resources issues that are successful only if championed and implemented by the telecommunications manager. Once again, the manager is the leader of the department. The performance and personality of the department will often be indicative of his or her policies and managerial techniques.

9.3 Laying the Foundation: The Necessity for Good Recruiting Techniques

It is a very common theme in many of the trade magazines that good technical staff are difficult to find. The problem is not just finding a specific set of skills, education, and experience. It is also trying to find the right personality that will fit within the corporate culture and within the department. Just because a person has skills, knowledge, and experience, it does not mean they will be a good fit for the company or department. Because there are so many variables involved, it behooves the telecommunications manager to develop a logical plan for finding as many qualified candidates as possible. The manager may then have a choice of a number of quality candidates as opposed to offering the job to the first qualified person who comes along, simply because he was the only candidate available. When choices are limited, then the manager runs a higher risk of a "poor hire."

A poor hire can often have a devastating effect on the morale and performance of the department. A poor hire can happen in spite of a manager's best efforts, but poor recruiting techniques will almost invite it. Many job candidates become adept at writing good resumes and develop good interviewing skills. But this is certainly not indicative of what type of employee they will be.

Good recruiting techniques help a manager to find a greater number of quality candidates and understanding, through the interview process, what technical and business skills the candidate has to offer. Such efforts reduce the chances of a poor hire.

It should be understood what the impact of a poor hire can be. Inevitably, a poor hire will require more training. This is expensive and it slows the progress and performance of the department. Administrative costs are also increased by requiring more management intervention. Team efforts are often hampered because the poor hire will not pull his own weight. Inevitably, a poor hire will make more errors, and the result will be a negative image of the department to the rest of the corporation. A poor hire can destroy morale and possibly drive the quality performers away. Consider also that poor performers cannot be arbitrarily fired. It requires the intervention of the human resources department and strict documentation, which may require months to finalize. During this time, work is not being completed and morale continually erodes. If the bad hire is dismissed, it may be months before a new staff member is found and even longer before they become productive.

According to the Information Technology Association of America (ITAA), the current market for computing and telecommunications professionals stands at approximately 2 million positions, and 190,000 positions cannot be filled due to a shortage of qualified professionals. This statistic sends a clear message to telecommunications managers: Use quality methods to obtain good people and implement policies to keep them. Staff retention is affected by morale, recognition of achievement, compensation, chance for advancement, and a number of other issues that will be covered later in this chapter. It is also affected by recruiting quality people, a process that requires a concerted effort, something that should certainly not be perfunctory in nature.

9.4 Maintaining and Documenting Job Descriptions and Qualifications

Every telecommunications department should have a defined structure. Consider the organizational chart depicted in Figure 9.1. At the bottom of the chart are jobs that are more hands-on in nature. These may be technicians and administrators who work with equipment and software. Moving up the chart, the jobs begin to take on more analytical and managerial types of duties. These jobs may be titles such as network analyst, network designer, or project leader. Notice also that there is a clear delineation between voice and data. Each of these job titles will have a job description that will entail a specific set of skills,

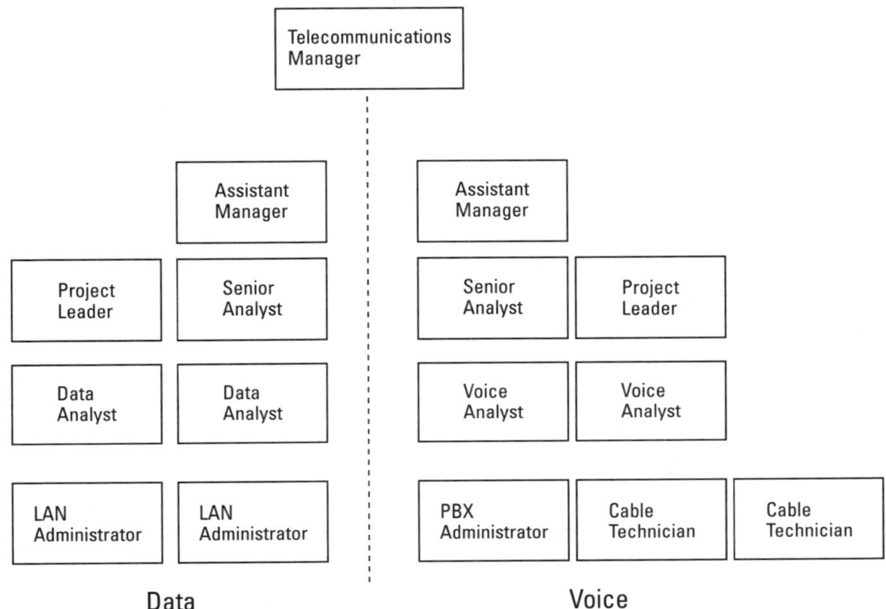

Figure 9.1 Organizational chart.

certifications, experience, and knowledge base. These job descriptions should be kept on file, both as hard-copy and in software.

Human resources departments mandate that job descriptions be defined and kept on file. However, there are not always clear rules for updating the job descriptions. For many job descriptions, this is not a problem because the skills do not change dramatically from year to year. A telecommunications manager will normally be able to find a job description on file for each position within his department. If the job descriptions are more than a few years old, they are probably obsolete. In the fast-changing world of telecommunications, the telecommunications manager needs to be cognizant of the skills necessary to support the modern communications environment.

The telecommunications manager will want to update all job descriptions that relate to his department. This does not have to be a major undertaking and can be done with the participation and cooperation of his staff. The job descriptions should be clearly defined in terms of duties, educational requirements and experience. Once these job descriptions have been completed, the manager will want to periodically update them as new technologies are introduced into the company. Once a year should be a sufficient time frame to keep department job descriptions up to date. If the company is in a state of rapid technological change, there might be a need for semi-annual evaluations.

Updating job descriptions yields a number of benefits. First, it ensures that the job descriptions are current so that employees other than the telecommunications manager can recruit, should some unforeseen event occur. Consider that the telecommunications manager may have a lengthy illness, resign, or unexpectedly be reassigned to another department. If several staff resignations occurred, the person overseeing the department could only rely on existing documentation, which may be obsolete. For example, a manager may see that a LAN administrator position requires that the person is a certified network engineer (CNE) because the company has been a traditional Novell shop. What the new person may not understand is that the company has recently made a decision to change the network operating systems (NOS) to NT, which is a completely different NOS.

A second advantage is that a job description review with a staff member is educational in addition to being a reality check. Very often, managers can get caught up in managerial duties to the point where they lose touch with what their employees are doing. A review with a staff member can be enlightening, showing the manager how the employee's job role is evolving. This is also a good time to review educational requirements and plan future classes. The manager may find that adjustments to staff duties are necessary, based on the outcome of these meetings.

A third advantage is that if a staff member resigns, there is accurate documentation of the employee's duties. This saves time and effort when writing want ads or communicating with recruiters. When an opening becomes available on the staff, the last thing the manager wants is surprises when a new hire begins work. The manager may suddenly find that the old employee did extra work that he was not aware of. Because the new employee is unfamiliar with the duties, the work is still not being done.

A fourth advantage is that the job description clearly defines the requirements of each position so that staff members can gauge themselves for future duties and perhaps promotions. For example, if an analyst expresses an interest in the position of project leader, the manager can simply print the job description and give it to the interested employee. This can provide clear goals for the employee in terms of experience and educational requirements.

A final advantage is that it helps the manager to clearly understand the department structure in terms of strengths and weaknesses. If the manager sees that certain employees hold unique skills that are not duplicated anywhere else in the department, it may be an indication that cross-training or a reworking of the job descriptions is in order.

9.5 Developing Staff: Starting Within

When a position becomes open in the department, the manager should not wonder how he or she will fill the position. There should be a recruitment file that provides a myriad of information and a logical plan of attack. However, prior to the manager looking outside the department, the question should be asked if anybody within the existing staff can fill the position. Would the position be a promotion for certain staff members? If these staff members are not eligible for the promotion, what is the reason?

When an opening offers a potential promotion to staff members, it sends a message that there is upward mobility within the department. This can be a strong motivational factor for the staff members who will always be seeking better compensation, more experience, and a better title. Promoting from within is often a win-win situation. It reduces potential turnover and minimizes the learning curve that is inherent to new employees. As an example, the manager may promote a technician to an analytical position. In this new role, the person will be writing RFPs, performing technical and financial analysis during the bid process, and providing general technical consulting services to the corporation. The person is weak in report writing and financial analysis, however he or she knows the company well and has a vast technical knowledge. He or she may be 80 percent of what the manager is looking for, but by promoting him or her, the manager has gained a number of advantages. First, the position has been filled immediately and therefore the workload is addressed immediately. Second, the person does not have to learn company policies. He or she already knows them, along with field locations, network design, CPE types and designs, and location contacts. A third advantage is that salary dollars may have been saved out of the budget. Consider that the technician was making $45,000 a year. The analyst who resigned was making $58,000 but was near the top end of his salary range ($50,000–$60,000). The technician is promoted and given a $5,000 raise (better than 10 percent). Immediately, $8,000 has been shaved from the salary budget. The manager may then opt to look for a technician who may command a salary of between $40,000–$50,000. Even if the technician is hired at the top of the salary range, there is still a potential savings in the salary budget.

Promoting from within has many advantages, however, its success is contingent upon the manager setting a promote-from-within policy. This means that staff members are cross-trained, encouraged to continue their education, and recruited on the basis that they may eventually be promoted. Of course, there will always be people who have no interest in advancing. These people may be content to occupy their current roles. This is not the manager's fault or problem if he has been proactive in encouraging people to improve themselves in anticipation of future promotions. Nor is it necessarily a negative reflection on the employee. There are employees who realize their limitations or who simply enjoy what they do. If the person is proficient and responsible, they can still be valuable employees.

A final advantage to maintaining the promote-from-within policy is that redundancy is established within the department. The promote-from within policy mandates that staff members are always training for the next position. Therefore, a smooth transition can be maintained, in spite of turnover. The worst possible situation is when a staff member resigns and there is a void within the department that cannot be filled by any existing staff member. Once this happens there is the inevitable recruitment period followed by a learning curve. This time frame equals lost time, money, and productivity.

9.6 The Search Process: Looking Outside the Company

Once it has been established that no internal candidates can fill the position, it is time to look outside the company. Many companies and managers view the recruiting process as a perfunctory exercise. They simply place an ad in the help wanted section of the local newspaper and hope for the best. A recruiting policy should be more than an ad that is offered to the general public. Specifically, the telecommunications manager will want to target sources that will provide the most exposure to telecommunications professionals. Recruiters often say that the want ads section of a newspaper represents only a small percentage of the available jobs in the general area. The same can be true of how many "qualified" candidates an ad in the local paper will attract. The manager can only hope that a quantity of quality candidates will see the ad in the short time that it will appear. Chances are that many will not.

Typically, the search will begin with local sources and then expand to a national level if necessary. A checklist will provide a plan that can be developed with corporate human resources personnel. For instance:

1. Local newspaper advertising;
2. The company's own Web site;
3. Local professional organizations; Telecommunications management associations (local end-users); Local user's groups; Local technology groups or councils that are not necessarily dedicated to telecommunications, but include it as part of their activities;
4. Local Internet Web sites;

5. Local recruiters;
6. Local universities that offer telecommunications courses and degrees;
7. National telecommunications newspapers and magazines;
8. National recruiters;
9. National universities that offer telecommunications courses and de-grees;
10. National Internet Web sites.

The telecommunications manager might set a time frame for each source on the list. If the first batch of resumes does not yield many quality candidates, the manager will want to move to the next step on the list. The plan will usually be to exhaust local sources before using the national resources. This is simply a matter of economics, because the company may not have to pay moving costs if the job is filled with a local candidate. It will be necessary to include the corporate human resources department in the process because control numbers are usually assigned to each source. The human resources department will also need to approve the ads to ensure that the information reflects corporate hiring policies and adheres to local and national hiring regulations. Human resources will also be involved because moving expenses or professional fees from recruiters may be incurred. In addition, human resources often keeps job descriptions on file in order to audit and track the hiring process so that all corporate and federal guidelines are followed.

The obvious first step with most companies (if the position cannot be filled internally) is to advertise in the want ads of local newspapers. This will provide a certain degree of coverage, but it is by no means all inclusive and I have seen where local advertising yields only a few qualified candidates. This is especially true of companies that are located in rural or semi-rural areas. Even when a company is situated near a major city, it still may be an hour or more for a candidate to drive from the city. With this in mind, consider that many local candidates do not want to move or are only marginally interested in changing jobs. The list of quality candidates then dwindles considerably. Local newspapers will usually provide a section in the want ads for computer professionals under which the telecommunications jobs are commonly advertised. Local job seekers will certainly reference this section, but not every candidate will have an opportunity to see the ad in the short time it will appear. The telecommunications manager needs to take steps that increase his or her chances of finding quality candidates within the local area.

The growth of the Internet is explosive and there are scores of companies that provide web sites. These web sites provide a wealth of information about the company, but they also provide career opportunities sections. While this is not necessarily advertising specifically to telecommunications professionals, it does provide one additional source of candidates. In addition, there is no cost to the telecommunications department since the web site has already been installed and paid for.

Locally-based professional telecommunications organizations may be ideal for recruiting purposes. These organizations usually meet on a monthly basis. In addition, there is usually a newsletter that is distributed before each meeting. The manager should call the president of the organization(s) before a meeting and ask if the job opening can be announced when the president addresses the membership. It might also be advantageous to provide the president with copies of the job description and have them available as handouts after the meeting. The second action would be to simply place an ad in the newsletter. This is usually very inexpensive if not free (a benefit of belonging to the organization). It may also be better coverage than the local newspaper, since it is distributed to most of the telecommunications professionals in the geographic area.

Telecommunications managers should proactively participate in local professional organizations. Not only do these organizations look good on a resume, they are also wonderful vehicles for networking. My own experience in such organizations has shown that many local candidates often look up such organizations when they are job hunting. The president usually receives many calls and may direct these candidates to a colleague if he knows of an opening.

The Internet is a powerful tool for attracting job candidates and the telecommunications manager will find a myriad of sites where a want ad can be placed. The local professional telecommunications organizations will probably have web sites. If they do not (because they are usually nonprofit and rely on volunteers), there will certainly be other sites that will advertise the opportunity for a nominal fee (or possibly no charge). The sites certainly do not have to be exclusive to telecommunications. For instance, in the Pittsburgh area there is an organization, The Pittsburgh High Technology Council, that is dedicated to business and educational issues that relate to advanced technology. While there are many technologies supported by this organization, telecommunications is certainly one category that is covered. There is a "Career Opportunities" section on the Pittsburgh High Technology Council web site where members can advertise.

Recruiters are often looked upon as a nuisance by many telecommunications managers. It is certainly true that some recruiters are aggressive, bothersome at times, and provide services of dubious quality. Quality recruiters, however, do exist and can be a valuable source of quality candidates. It is estimated that up to 80 percent of all telecommunications jobs are filled through a recruiter, but finding good recruiters can be a trying process. My experience has also shown that they vary greatly in quality. To some recruiters, the process of placing candidates is simply a numbers game. They feel that if they provide a large number of candidates, somebody within that group is bound to be hired. These types of antics only serve to waste time and money, and eventually frustrate potential employers. The telecommunications manager will want to weed out the less capable recruiters and concentrate more on the professional recruiters who understand the field of telecommunications.

There are recruiters who specialize in the field of computing and telecommunications. The good ones will understand the types of job titles, job descriptions, and have at least a journeyman's knowledge of the basic technologies. These types of recruiters often keep in touch with the top professionals in the area and retain good relationships with both candidates and employers. When recruiters provide this type of service, they save time and money. For instance, an ad in the local newspaper may yield hundreds of resumes. But of the hundreds of resumes, only a dozen have been submitted by qualified candidates. Any person who has ever advertised for a position in a newspaper will speak of the large numbers of resumes from people who are unemployed, unqualified, or both. Industry studies show that internal costs can be in excess of $200 per resume to screen each candidate. Company rules often mandate that each resume is logged and a legitimate reason for rejection documented. A good recruiter can avoid much of the time and cost associated with this process. In addition, telecommunications managers should be aware that less than two percent of all telecommunications jobs are filled via newspaper ads.

A good recruiter will ask many questions about the job title, the job description, the company, and the departmental structure:

1. What is the official title?
2. What duties will this job entail: voice, data, video, or all three?
3. What experience level are you looking for?
4. Is this a hands-on type of person or are you looking for somebody more high level?
5. What type of voice equipment are you currently using?
6. What type of data equipment are you using (e.g., routers, hubs, FRADs, DACS, modems)?
7. What types of LANs do you have in place?
8. Are you using any wireless (voice or data)?
9. What type of cable systems are you using?
10. Is there any upward mobility in the department?

11. What type of network will this person be supporting? Local, national, or global?
12. What type of educational requirements are you seeking?
13. Are the educational requirements firm or can they be balanced against experience?
14. Are you looking for any types of special certifications?
15. Will this job involve travel?
16. Do you require international experience?
17. What is the salary range for this position?
18. Are you willing to negotiate vacation time?
19. Do you have a 401K plan?
20. Describe your corporate culture to me. Formal (e.g., IBM, Big Six) or informal (Silicon Valley)?

There are many more questions, but the telecommunications manager should begin to find a comfort level when he sees that the recruiter not only understands the telecommunications business, but is also trying to gain as much information as possible so that the correct quality candidates are placed in front of the manager. Through experience, and trial and error, the manager will find a comfort level with the better recruiters. Even when there is not a job opening, the telecommunications manager will want to retain good relations with the better recruiters. This becomes an immediate resource when a vacancy occurs.

Essential criteria for a quality recruiter is that he or she acts in an unbiased manner, for both the candidate and the potential employee. A red flag is a recruiter who tries to sell the services of a candidate who has just been interviewed. A quality recruiter will understand that the candidate (his or her experience and qualifications) will sell the manager and nothing else. An additional litmus test is the consistent quality of the candidates. One unqualified candidate out of ten is simply a mistake or miscommunication. If nine of the ten candidates are inappropriate for the job, the recruiter is playing the numbers game, and has not listened to the qualifications set forth by the manager. There should also be few surprises during the interview. A quality recruiter will have reviewed a resume and already asked obvious questions of the candidate. "You changed jobs three times in two years. Why?" "You were making $45,000 two years ago and now you are making $60,000. It's highly unusual to realize such an increase in salary. Could you explain why?" The recruiter should convey this information to the manager prior to the interview. Recruiters are also often needed to act as mediators, clarifying issues and helping to negotiate human resources issues such as salary, bonuses, and vacation time. A good recruiter will be skilled at communicating and negotiating sensitive issues without offending either side. In addition, the better recruiters often offer a warranty upon placement of a candidate. If the candidate does not work out within a specified period of time (typically six months to one year), the recruiter will refund the placement fee.

Recruiters can be more than a source of candidates, they can also be an educational source. Many a manager has been surprised when a recruiter has said, "You'd better come up on the price a bit if you want a person with those skills and that much experience." Recruiters can let the manager know what the market will bear, and this could have a bearing on how he adjusts his annual budget. Good recruiters will even recognize inadequate job descriptions and make suggestions for refinements that will help the manager better focus on the correct skill set.

The ability to maintain confidentiality is an essential quality in any recruiter. This is more important for the candidate than for the potential employer who will probably be advertising the position. Managers should be sensitive to this issue because recruiters who do not exercise discretion are typically the ones of dubious ethical value.

Recruiters can work on either a contingency basis or on a retainer. A contingency basis means that the recruiter is paid a percentage of the employee's starting salary, and they are paid only when they place a candidate. Recruiters who work on a contingency basis typically offer varying degrees of quality. Recruiters who work on a retainer basis are paid an ongoing fee by the company. These recruiters tend to worry more about finding the right person for the job than simply filling a slot. Should a candidate be placed with a department, quality recruiters will follow up with the manager to determine if the candidate is doing well.

Beyond local recruiters, many universities are now offering individual classes and degrees in telecommunications. Universities and colleges are more than willing to post job openings on bulletin boards or make announcements in classes. Placing students in well paying jobs serves to boost the credibility of the telecommunications program and also helps the institution to recruit new students. The telecommunications manager should be aware of local telecommunications educational programs and contact the heads of the these departments when an opening occurs. The only caveat may be that experience could be a factor. It may be a concession on the part of the manager: lower pay and knowledge as a trade-off for experience.

Of special interest to the telecommunications manager may be internship programs. Many universities encourage these and the telecommunications manager may gain help for the department, without having to pay a full salary.

Once the internship is completed, there is always the option of hiring the intern who is well versed technically and has now gained valuable work experience.

Once local sources have been exhausted, it may be time for a national search. This would encompass the use of national telecommunications newspapers and magazines: in print, on the Internet, or both. Weekly newspapers such as Internet Week or Network World provide such capabilities. These periodicals are regularly scanned by many professionals along with the Career Opportunities sections of their web sites.

An additional source may be to use recruiters who work on a national basis. Once again, the telecommunications manager will want to carefully screen who is used, striving to find the professional who provides quality candidates in lieu of the less skilled recruiters who simply deal with a body count.

9.7 The Interview Process

When quality candidates are found, the interview process begins. The number of interviews depends on corporate and departmental policies, and the manager's individual preferences. Regardless of how many times a candidate is interviewed, the interview process should not be viewed as a perfunctory exercise. This should be a process where thought has been given to developing questions that will help identify the type of person who will best fit in the department and within the corporate culture. Of all the skills that I have observed in managers over the years, one of the weakest by far is that of interviewing. Many managers simply fall back on the same clichés. What are your strengths and weaknesses? Why did you leave your last job? How do you respond to criticism? What can you tell me about yourself? Conversely, many candidates become very adept at answering these questions to the point where the interview becomes a battle of clichés. This is counterproductive and tells the manager very little about a candidate. Many managers also have a tendency to dwell too much on history. How relevant is a person's work experience that occurred 10 or 20 years ago? Is it really necessary to dwell on why a person left each and every job over the past 15 or 20 years, especially if that person has had a relatively stable employment history over the past 10 years? In addition, high-tech skills change quickly. What a person did several years ago may be of no relevance today.

I have observed many techniques over the years that were counterproductive and never really yielded the best candidate. I knew of one manager who approached the interview process like a defense lawyer cross-examining a hostile witness. He would point to an item on a candidate's resume and say, "This wasn't a very intelligent decision. Why would you do something like that?" The manager's theory was that telecommunications is a tough field and that he wanted to weed out the faint-hearted quickly. The technique did weed out many candidates, many of whom were quality applicants. What the manager failed to realize is that nobody likes to be personally attacked, and the only message he was sending to the candidate was that he would be a bad manager to work for.

The interview process is an opportunity for a manager to truly understand what type of person he may be hiring. If the manager finds that he or she is constantly disappointed in the people that have been hired, but he or she has chosen people with seemingly good qualifications, the problem may be in the interview process itself. Telecommunications managers should take the time to identify the types of qualities that they would like to see in staff members, much in the same manner that a job description is developed. Consider some of the following qualities:

- The ability to remain calm in tense situations;
- The ability to think quickly and logically in tense situations;
- The ability to diffuse anger on the part of an end-user and then work with the person to help resolve the problem;
- The ability to evaluate and understand complex technologies and then convey those concepts in layman's terms to end-users;
- The ability to write technically accurate reports that are understandable to the layperson;
- The ability to work with a team, as either a member or a leader;
- The ability to work independently;
- The ability to be creative when designing networks or programming software;
- The ability to accept new ideas, even when the new idea is better than any that have been previously brought forth by the employee.

Notice that these are not necessarily technical skills, but rather interpersonal and business skills that often have a major impact on an employee's success. Managers should avoid the comfort zone they may have developed over the years with interviewing techniques, and try to understand how a person truly functions in a business environment. The resume will tell the manager about the candidate's technical knowledge and skills. The interview should tell him how he applies his knowledge and skills.

The interview should be a forum for exchanging information. The candidate wants to learn about the company where he is considering employment and the manager would like to understand the knowledge, skills, and abilities of his candidate. The interview should never be confrontational, but a forum for both parties to explore, understand, and sell each other.

The manager should develop an itinerary for the interview. The purpose of the interview should be to gather information on two levels. First, the manager wants to determine existing knowledge. This may be a simple matter of perusing the candidate's resume and asking detailed questions. The manager might want to start with general concepts: "Describe to me what you do in an average day." If the candidate provides a vague answer, the manager will want to further clarify it: "It says on your resume that you support several types of LANs. To what degree? Do you support the NOS, the cable, install the workstations, oversee the addressing? All of these or just some of these? Do you do hands-on work or design work?"

Once the manager has established general knowledge and responsibilities, more detailed questions may be in order: "I see that you implemented a campus backbone of ATM over fiber. What made you decide to use ATM? Was it a smooth implementation? I understand it's a complex technology to support. Did you find that to be the case? How much training did you need in order to support this? At what speed do you run ATM? Do you run it to the desktop? Do you run voice or video over the backbone in addition to data?" It may appear that these questions are being posed in a confrontational manner. That is not what is being suggested. These are only suggestions for questions and comments that might be used to keep the conversation flowing as lulls occur.

Beyond the general skills and knowledge, the manager will want to understand basic business and interpersonal skills. The importance of these skills cannot be overlooked because, while many people can be technically adept, they can also be social flops. This may not be the type of person that a manager would want in their department. A person who is moody, quickly blames other people for mistakes, or who shies away from high-risk projects will quickly destroy morale within the group. Consider some of the following questions and subsequent analysis.

Question: We like to do RFPs in this department. How would you go about preparing an RFP to replace all of the routers on the WAN?

Answer: I have extensive experience with routers. I've looked at them all and Cisco is the best. My experience has shown me that RFPs are a lot of busy work, and everybody knows what vendor will be picked from the very beginning. I'm the type of person who likes to get the technology installed and start helping the end-user.

Analysis: Assume that the manager is looking for a person who possesses a combination of business and technical skills. The answer given by this candidate indicates that he or she does not like the business aspect of telecommunications, and appears to be more of a hands-on technical person. This person will also probably shy away from reports that the manager may require. In addition, many companies require RFPs for any major acquisition, mandating a minimum of three bids. RFPs also address many issues beyond the technology. For instance, service, maintenance agreements, technological upgrade clauses, and so forth.

Question: You receive an angry call from an end-user. You have just installed new hubs and a router at their location. They scream at you, using a great deal of profanity, and go on to tell you they are losing time and money. It's your fault and, if the problem is not fixed immediately, they will certainly start escalating throughout the company, telling everybody and their brother how incompetent you are. How would you address the confrontation with the end-user and the problem?

Answer: The first thing I would do is diffuse the anger. Otherwise you'll end up arguing with the individual and nothing productive will happen. The first thing I tell an angry end-user is that I understand how serious the situation is and that I am also concerned about the outage. If the anger persists I tell them that as long as they yell, we're not working toward resolving their problem. I then tell them what I am going to do, and promise to call them within a specified time frame. I always make sure that I call them back with a status. This way, they understand that somebody is working on their problem, and they are not nearly as frustrated.

As far as how I would actually solve the problem, you haven't given me much information. I assume there would be documented procedures and escalation processes that have been defined for the department. Is there a toll-free number that I would call? Is there a defined response time in our service agreement for major outages? If the vendor hasn't responded in the specified time frame, what are the escalation procedures? But you also said that both the router and the hub were down. This makes me think that the electricity has gone out and the UPS capacity has been expended. Before I came to any conclusions, I would ask more questions.

Analysis: The candidate obviously has had extensive experience with trouble reporting procedures and has had training in interpersonal skills and conflict resolution. He or she knew how to deal with conflict and was not distracted from his or her real job, which was to resolve the problem. In addition, this candidate examined what information was given to him or her, concluded that it was inadequate, and decided that more information would be required in order to determine the true source of the problem. The manager learned two things from his or her response (1) the candidate has good interpersonal skills and (2) the candidate is methodical when approaching a problem.

Question: We just purchased a company and we need to design and implement a WAN for them. They have 10 locations throughout the midwest, all of them currently using analog, private lines. Here is a network map—how you would design this network?

Answer: The candidate recommends frame relay service for all locations, all of which will be routed back to the new corporate headquarters. The candidate contends that money will be saved by going to frame relay and response time will improve because the company can now take advantage of the bursting feature that is inherent to frame relay.

Analysis: The candidate did not ask any further questions, and simply drew a quick and easy solution on the board. It is true that the candidate's solution would probably save money and improve response time. What the candidate failed to do was ask for any further detail. What type of traffic would be transmitted over the WAN—IP? SNA? Do we know how much traffic is being sent over the WAN? The access circuit and the CIR should be determined based on bandwidth requirements, not a best guess. What type of CPE is currently connecting the analog private lines? Point-to-point modems or routers? Would the CPE have to be replaced and would that incur a termination liability to an existing contract? What about backup facilities, such as analog dial-up, switched 56 Kbps, or ISDN?

Question: Tell me about what courses you have taken, what magazines or newspapers you like to read, and what you try to do to keep your knowledge current?

Answer: I really don't have time to read. I take courses as I need them or as management recommends.

Analysis: The candidate probably does not have time to read, but the answer might also be indicative of a person who only learns what he has to learn. These types of people can still be valuable in a department (albeit at the hands-on support level), but the manager may also want to weigh this against the answers given by other candidates. Ideally, the manager wants people who aggressively pursue education and who bring knowledge and ideas to the department when new projects surface. Technological change is rapid, therefore technical skills will rapidly change. When employees are self-motivated and aggressively pursue their educational needs, the operation of the department flows more naturally. Employees who are not self-motivated are often in a catch up mode. This only hurts the department.

These are just a sampling of questions that could be asked by the telecommunications manager. The manager is only limited by his or her creativity.

Notice that each question is devised to provide answers that help the manager evaluate the employee's business and interpersonal skills, not necessarily just technical knowledge. There will often be different positions in the department and questions should be developed that are exclusive to those positions. The manager should also list issues and problems that need to be corrected within the department and present these as questions or problems to be solved. Throughout this process, the manager will gain a better appreciation of the candidate's ability and chances will be better that his or her decision will yield a quality employee who will fit well into the department.

9.8 Compensation: The Basics

There is an old cliché in the business world that dictates the five motivational factors for employees: fear, greed, guilt, need for approval, and exclusivity. Whatever motivates an individual, money is the most obvious and probably one of the strongest of these factors. The telecommunications department should be structured so that compensation is reasonable and customary, with similar job descriptions, in similar industry segments, and in accordance with pay scales that are reasonable within the geographic location. An open invitation to poor morale and high turnover is to offer substandard pay. It does not matter how well an employee is treated or how good the working environment is, if employees see significant financial advancement elsewhere, they will probably leave. If the pay scale is within reasonable proximity, then other factors will affect department stability.

Telecommunications managers should be cognizant of salary levels on a local and national basis, and how those salaries relate to duties and job titles. This will enable them to retain staff, recruit better talent, or to solicit an increase in the department budget should the salary levels begin to cause excessive staff turnover or destroy morale. They should also be cognizant of performance bonus plans, which can often be used to enhance productivity, boost morale, and aid in the recruiting process.

The most basic and common form of compensation is salary, benefits, vacation, and an annual merit increase. This is the platform on which much of the staff morale and stability will be built. There are certainly other methods of augmenting compensation, such as performance bonuses, spot bonuses, or employee stock option plans. The foundation, however, will be the employee's annual salary, because all other methods of compensation will be dependent upon company, departmental, or individual performance. Seasoned telecommunications professionals will realize that there are always factors that will be out of their control, which may impact whether a bonus will be forthcoming.

Therefore they will often opt for the "sure" money and consider bonus money to be just that: a bonus.

Compensation is a combination of salary and benefits, of which benefits equal an additional 25–35 percent of the base salary. Department salaries often comprise 30–35 percent of the total telecommunications department budget. Different companies will structure their departments in distinct ways, but most departments should reflect a pyramidal structure (see Figure 9.2). At the bottom of this pyramid are the lower paying jobs such as LAN Administrator, PBX system administrator, or staff technician. Gradually, as jobs become less hands-on and more managerial or analytical in nature, the salaries will increase, the highest being the telecommunications manager. The structure of the pyramid will often remain stable, however, there will also be times when the balance will be upset. For instance, a 1995 salary survey conducted by Datapro revealed that people with Internet skills were commanding higher salaries than those in the more traditional fields of voice or data communications. This remained consistent in a 1997 survey conducted by Network World. However, as more people become skilled with newer technologies, the price for such people will certainly diminish, falling more in line with other salaries. In the interim,

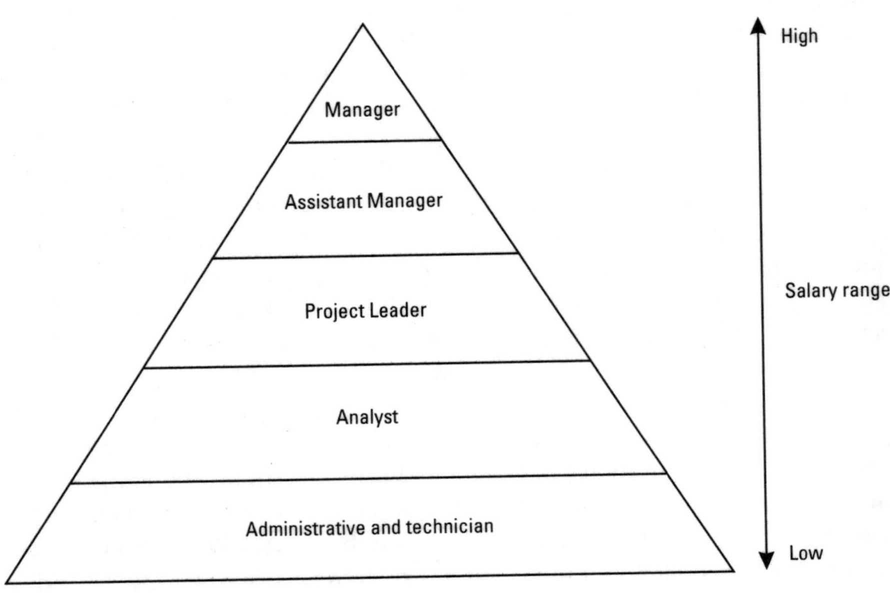

Figure 9.2 A pyramidal structure of employee compensation based on position in the company.

managers will have to realize that skilled people in emerging technologies that are in demand will command a premium price.

Part of a manager's duties is to prepare a budget for the department, forecasting what costs will be incurred for the upcoming fiscal year. A portion of this budget will, of course, be salary. The salary part of the forecast should be based on a balance of historical information and future projections. On the historical side, the manager needs to examine the existing salary structure and determine if it has had an effect on department performance, morale, or stability. This will also be compared to average salaries in the industry, average annual increases, and the job market in the area (low unemployment and an expanding economy gives people the opportunity to move quickly).

As part of an annual or semi-annual task, the telecommunications manager should compare department titles and duties to studies that are published in industry publications. This can be accomplished by creating a simple spreadsheet (see Table 9.1). In this example, it can be seen what the average salary of telecommunications managers are in various regions of the country. Similar charts can be found for telecommunications analysts, LAN administrators, and other job categories. Below the salaries and average raises are listed. A quick glance at these statistics indicates that network-based help is commanding a higher raise than the average employee in the business world.

Department performance can be tied to morale, but there are also incidents where a manager may be trying to support technology without the proper expertise. For example, the telecommunications department may have been asked to provide Internet connectivity with a firewall. Assume that there was no such experience within the department, but the project was still mandated by the company. The result was a large learning curve on the part of the staff and a project that was deemed to be of dubious success by senior management. Because the manager had no open requisitions for new hires, he or she had to support the project with his existing personnel. The manager tried to make a

Table 9.1
Network Manager Average Salary by Geographic Territory

Region	Pacific	Mountain	South Central	North Central	Mid-Atlantic	New England	South Atlantic
Average Salary	$83,360	$72,900	$64,026	$65,540	$81,649	$82,993	$77,979
Average Raise	4.9%	4.7%	6.8%	6.6%	8.6%	9.1%	7.1%

case that the company should consider hiring an experienced person, but there was no room in the budget for a person who might command a $70,000 salary (the going rate in the area for a person with strong Internet experience). In this instance, the manager has determined that he or she needs Internet expertise in order to support this expanding application and has budgeted for an additional staff member in the coming fiscal year.

Compensation has a strong effect on morale and stability in two ways. First, the existing staff members may feel that they are overworked, underpaid, or both. Second, there may have been a series of resignations related to the issue of inadequate pay. This tends to have a snowball effect, because resignations will increase the workload of existing staff.

Overworked employees can only have their problem resolved by hiring new people. This is not a situation that is easily rectified. Simply stating that his employees are overworked will not be justification for expanding the budget to hire new employees. There needs to be concrete information in the forecast that specifies how department performance is being affected by inadequate staff. For instance, if the department provides a help desk function, the manager will want to use ACD reports to show how support has diminished over a period of time due to increased workload, lack of experience, and reduced staff as a result of resignations (see Table 9.2).

Charts and statistics such as this offer concrete proof that department performance is impaired because of inadequate staff. Additional statistics may be meantime to resolution or meantime to installation. For example, the telecommunications department may have been installing new telephones for new employees within three days of the request in March (see Table 9.3).

Due to a series of resignations, by September meantime to installation has deteriorated to 10 working days. But this statistic still may not be enough to justify salary increases and additional staff. The manager will want to supply documentation in the form of letters or e-mails that lists complaints from the end-users. This will give senior management an indication of the severity of the problem and the cause. The formula then becomes very simple: It will take X

Table 9.2
Help Desk Service Levels

	March	April	May	June	July	August	September
Service level	92%	91%	85%	79%	62%	65%	63%
Abandoned calls	52	49	56	72	125	132	127

Table 9.3
Telephone Meantime to Installation

	March	April	May	June	July	August	September
Working days	3	3	5	7	8	10	10
Complaints	0	0	0	5	10	18	21

amount of additional operating dollars to resolve this problem. This is presented in the following ways:

1. Adjustment to existing salaries, based on average salaries in the region. This is a move that is aimed to retain the existing staff and necessary because salary levels are below average;
2. Adjusted salaries for new hires to replace departed employees;
3. Additional hires to support new technologies or reduce existing workload.

Managers may try to incorporate as many statistics as possible in their forecast, but many seasoned managers will often complain that the process is as much art as science. The reasons for this are the extraneous influences that can affect department budget, such as company financial performance, the state of the economy, or internal company politics (to name but a few). Regardless, managers need to understand the market for telecommunications professionals and base their budgets on a balance of their own company's requirements and the demands of the business world in general. For instance, consider a manager who has projected 3–5 percent raises for his or her staff for the forthcoming fiscal year. The manager knows that current staff salaries are only average compared to similar telecommunications positions located in the region. The telecommunications manager is told by his superiors that the company is expecting poor market conditions and that the amount allocated for salary increases will be 1–3 percent. The manager, who has researched the issue of salary, knows that average raises for telecommunications professionals in the region are ranging from 5–7 percent. He or she knows that 3–5 percent will not be received well, and 1–3 percent may cause a mutiny. The telecommunications manager also knows that he or she has not been privy to proprietary information and has simply referenced nationally published newspapers and magazines, of which the staff has had the same access to.

Such scenarios are common in the business world and many a manager has engaged in vicious battles over such budgetary issues. Ultimately, the manager can only budget what is allowed by senior management and there may only be a limited amount of money for each department. However, he or she should be prepared to offer facts and figures that can be substantiated, along with existing problems and potential results of offering the staff substandard salary and raises. In addition, there should be contingency plans for the use of outsourcing, consultants, or contract workers to either relieve some of the current workload or take up the duties of staff members who have resigned. In many companies, it is much easier to allocate monies for such services because the financial liability is limited. Although outsourcing organizations, contract workers, and consultants can command a high price, a company does not have to pay benefits and the termination of such services is relatively easy and risk free.

In addition to considering outside workers, the manager may also have options of utilizing the salary budget in creative ways. For instance, replacing a worker can often be less than the industry average for a given geographical average. In fact, statistics citing average salaries can be skewed because they reflect many people who have a great deal of tenure. Moreover, salary surveys are not all encompassing. The publishers of such surveys are at the mercy of the public's willingness to participate in them. Consequently, the "average" may not always be applicable. Some surveys actually publish replacement worker costs, which are often lower than the average cost. For example, a 1997 salary survey in Network World cited a network manager position as being $78,022 and a replacement cost as $55,170. Such statistics argue that all turnover is not necessarily bad. In fact, a small degree of turnover can actually help the manager to make better use of his budget. The critical difference is whether people leave because they are being compensated fairly, or if it is just a natural evolution of individual career advancement. Managers should not be lured into thinking that a lower replacement cost is necessarily a good thing. Turnover costs money because the workload is not addressed and new employees have an inherent learning curve.

Promoting from within can also offer advantages in terms of managing the budget. A department should always be organized so that staff members are cross-trained to move laterally or upward. This methodology ensures that a resignation (by any staff member) does not cripple the group. A telecommunications manager can either create a new position (such as an assistant manager) or promote to an existing vacancy. Assuming that a person is qualified, the raise that is consistent with such a promotion will still usually place that person at the bottom or middle of the new position's salary range. This would allow the manager to hire a new person at a lower level, which may not be as difficult to fill.

Another option, which carries some large risks, is to not hire when a vacancy occurs and use the additional money to boost raises. The obvious problem is that the work load, if severe, will remain so and this may cause severe turnover, in spite of generous raises.

Regardless of how much research a manager does, or how much ingenuity is applied to the salary budget, the manager will still fall victim to market conditions and company policies. In the latter part of the twentieth century, the American worker has abandoned the concept of lifetime employment. Companies are downsized, right-sized, sold, and reengineered. This results in lost jobs, redefined jobs, or a change of venue for many workers. When major changes are made, there is often a reluctance on the part of many companies to rehire staff that will increase operating costs. Consequently, many departments learn to do a lot with a little. Because of such conditions, managers need to be aggressive when they budget for salary because while workers may be laid off, the company seldom reduces the workload to compensate for the lost staff.

Considering the much publicized shortage of telecommunications professionals, some organizations have employed "preemptive hiring" in order to prepare for future projects and technologies. Preemptive hiring is simply overhiring from a shrinking labor pool to be assured of adequate resources as growth continues. This policy is not without its caveats. Budgets are often scrutinized carefully and preemptive hiring for potential work will be difficult to justify to senior management. Regardless, the fact that preemptive hiring has been employed (albeit by more liberal and forward thinking companies) is an indication of how difficult it can be to recruit telecommunications talent.

A final option that relates to both hiring practices and budget management is the use of contract people. This policy offers a number of advantages. First, it provides an opportunity to "audition" employees and allows the manager to determine if an employee will actually fit in with the staff. This is an added bit of insurance because, however much the telecommunications manager tries to employ good recruiting and interviewing techniques, the interview process still has its limitations.

A second advantage is that the employee is not a true hire, and therefore does not incur costs for benefits. This is a small savings, but more importantly, the company is only financially liable for the time frame specified in the contract. This means that services can be terminated without financial or legal liability at the expiration of the contract. Contract workers can also be hired on a contingency basis, which means that their services are brought in on an as-needed basis.

Contract workers can be used either as potential future hires or remain as permanent contract workers. A method of making the best use of salary budget is to assign general administrative duties to contract workers, freeing valuable time for the rest of the staff to perform more important duties.

9.9 Vacation: New Policies

Vacation has never traditionally been a bargaining tool or something that a manager could use as a bonus or award. Employees were simply given two weeks after a year of service. As an employee accumulated years of service, additional vacation was awarded via strict company policy. If an employee changed companies, it was assumed that the additional earned vacation would be lost, and the employee would wait the traditional year before vacation could be taken. Several factors have changed this situation. First, employees no longer think of life-long commitments to their employer. Companies have downsized, reengineered, merged, and declared bankruptcy. Consequently, employees no longer hire on with a firm, thinking it will be a final career move. The second reason is that the demand for telecommunications professionals is continually increasing to the point were it is an employee's market in certain areas of the country. As a result of this, recruiting for telecommunications professionals is aggressive, and tempting offers are often being placed in front of staff members.

In the field of telecommunications, job candidates have begun to use vacation as a negotiating tool. Understanding that a company needs their services, they will demand an equal amount of vacation time that they are receiving from their current employer. For the sake of one or two weeks of vacation time, many employers are willing to award this in addition to waiving the traditional one year waiting period before vacation time can be taken. In order to grant additional vacation time, a manager will have to clear the policy with the human resources department. This may not be an option in many companies, however, there are companies that have lost quality candidates for the sake of a single week of vacation time. Moreover, the business world is changing regarding the subject of vacation and many companies are relaxing their policies. Before the recruiting process begins, the telecommunications manager will want to determine if he has any leverage with vacation time. Many recruiters who specialize in the field of telecommunications recruiting insist that three weeks has become standard for experienced candidates.

A third reason why vacation has become an issue are the liberal human resources policies that have been implemented by many high-tech and emerging companies. These companies, in a effort to attract top technical talent, have implemented policies, such as signing bonuses, longevity bonuses, extra vacation, and liberal comp time. The end-user community often competes with these firms for technical talent and have had to respond with similar policies. When a manager is in the process of establishing qualifications and experience levels during the recruiting process, he or she should be aware that certain experience levels will mean that many candidates will have accumulated vacation. A manager may think that an attractive salary may be a substitute for the lost vacation. Unfortunately this does not always work. Telecommunications professionals work long, hard hours. What may be more important to them is the time off that they know will be necessary for a periodic sanity check. Increased salary is not always a primary reason why telecommunications professionals change jobs. Many candidates know that long hours will be required, but they also want relief so that they can balance their personal lives. Employees who spend long periods of time away from home often experience problems with their families. Managers should also be aware that personal problems often have a way of filtering into the office environment. Extra vacation time is a tool that helps to combat this problem.

In an environment where long hours and worked weekends are the norm, managers should also take care to understand how the overtime is balancing against vacation time. Comp time is not always a company policy and, when it is awarded, it often does not equal the total number of overtime hours an employee has worked. When this happens, the staff member begins to feel as though he is not actually getting a vacation. I knew of one colleague who spent a tremendous amount of time working long days and weekends. He once told me that when he considered the weekends he had worked and weighed that against his vacation, the overtime canceled the vacation.

An additional problem that sometimes surfaces is that vacation is lost because employees simply did not have time to take it due to heavy work schedules. Many companies do not allow employees to carry over unused vacation for a number of reasons. Primary among them is that vacation time carries a dollar value and if an employee resigns, a company is obligated to pay the value of unused vacation time to the employee. The policy of not carrying over vacation also assures the company of a guaranteed amount of work time from the employee. Accumulated vacation time can have an impact on the structure and morale of the department. If an employee is allowed to accumulate large amounts of vacation time, and it is taken in one large sum, it can reduce the labor pool required to perform the workload. The manager must understand the workload of each employee, balance that against vacation time and make a judgment about the impact on overall department performance.

9.10 Compensation: Bonuses and Perks

Bonuses and perks can be included as part of the employee's compensation plan. These monetary incentives can also be used as motivating factors for performance, staff retention, or added insurance that a project or tasks are completed on time and under budget. Bonuses and perks can be offered in a variety of ways:

1. Sign-on bonus;
2. Bonus as part of salary;
3. Spot bonuses;
4. Project specific bonuses;
5. Educational bonuses;
6. Departmental bonuses;
7. Longevity bonuses;
8. More frequent merit increases;
9. Nonmonetary bonuses used as incentives or spot bonuses.

Given the shortage in some areas of quality telecommunications professionals, companies that compete for such talent are often forced to develop innovative ways to recruit and keep talent. A method that has been gaining some popularity in recent years is the sign-on bonus. This is simply a flat amount paid to a new employee for hiring on with a company. Sign-on bonuses are common in areas such as Silicon Valley where many high-tech companies compete for talent. This is especially true for skills that are in demand such as candidates with Internet expertise. Bonuses of $2,000–$3,000 are common and, depending on skill level and the geographical area, bonuses have been reported as high as $10,000.

When a bonus is assigned as part of an employee's salary, it is normally contingent upon specific and tangible goals, which are often financial in nature. Telecommunications managers often receive bonuses based on balancing their budgets, and perhaps added incentives for coming under budget. Bonuses can also be assigned to other staff members (such as assistant managers or analysts) who may oversee a part of the budget (e.g., toll-free numbers, outbound calling, router maintenance contracts). Bonuses as part of salary are not often connected with project management, because there are often too many variables associated with the successful completion of a project. Inevitably, if employees are denied their bonuses because of factors that are outside their control, this will have a detrimental effect on morale.

When a portion of an employee's salary is contingent upon bonuses, the company (or telecommunications manager) should consider two factors. First, what percentage of total compensation equals the bonus? Second, is the goal achievable? These two factors will have an impact on recruiting (many candidates will look for the "sure" money) and staff retention (if the goal is not achievable). If, for instance, the bonus is 10 percent, the base salary is probably manageable for the employee during the course of the year. If the bonus is awarded at the end of the fiscal year, and the employee does not meet the required objectives, 10 percent should not leave him destitute. Conversely, the bonus cannot be so low that there is no incentive for the employee to meet it. As a general rule of thumb, 10 percent is a good working figure for bonuses that neither scare away good candidates nor place them on the brink of financial ruin, should they not meet their objectives.

It should be noted that there is a negative side to a bonus and salary plan. As explained in previous chapters, telecommunications management entails a number of responsibilities, such as financial analysis, technical analysis, strategic planning, project management, technical support, and so forth. If a portion of the manager's compensation is dependent upon balancing the budget, there may be a tendency to concentrate exclusively on financial matters and ignore other issues. For instance, in researching this book I was told the story of one manager whose compensation was a mix of 10 percent bonus and 90 percent salary. If the manager balanced the budget, he was awarded a 10 percent bonus. However, there were additional bonuses set as milestones for percentage below budget. If the manager was five percent below budget, there was an additional two percent. Ten percent below budget was another two percent bonus, and so on. There was a maximum to the manager's allowable bonus which could not exceed twenty percent. The advantages to such a plan for a company are obvious. Consider a five million dollar telecommunications budget. If the manager comes in at 95 percent of plan, this is a savings of $250,000 in operating expenses. Consider a manager who may be making $90,000 per year. Two percent of $90,000 is $1,800. This is in addition to the ten percent bonus already awarded, which was included in the original $5 million budget ($9,000). For $250,000 in operating expense savings, the company paid only $1,800.

The problem with such a system is that the manager often concentrates solely on financial matters to the detriment of other critical areas. In the example cited above, the manager spent an inordinate amount of time switching long distance carriers and denying education and travel authorizations to his or her staff. The results of his or her actions were that problems and critical business applications were ignored while the staff was directed to implement the "long distance carrier of the month." The staff also fell behind on technical skills because they were denied education, and critical problems in the field were not addressed because staff members were not allowed to travel. The problem was further exacerbated when a staff member resigned. At the time, the workload was heavy and the staff could ill afford to lose a member. The manager, in an effort to shave dollars off his budget for the fiscal year and make his bonus, delayed hiring a replacement. This meant more work for the existing staff, which resulted in low morale and resentment. Financially, this manager was successful, but the complaints piled up from the end-users and a common complaint from the staff members was, "All he cares about is his budget."

If the bonus is not awarded the second factor needs to be considered: Is the goal achievable? As an example, think of a company that has provided the same bonus and salary plan cited in the example above. Now consider that the company may be in the midst of an explosive growth mode. This growth entails the purchasing of many new companies which require additional frame relay circuits, routers, and hubs. The company has not provided a special account for unanticipated growth and so the new services and products are charged to the existing telecommunications budget. The manager is allowed a 10 percent variance on the budget, but there is no allowance that relates to the bonus. If such occurrences happen every year, this will obviously have an effect on the manager's morale (or any staff members who may be under a similar plan). The purpose of the bonus is to act as an incentive and such circumstances will only render the bonus ineffective.

Spot bonuses are awarded by managers as a reward for excellent employee performance, normally at the successful conclusion of a complex and highly visible project. The bonuses are awarded at the discretion of the manager, which can often involve much subjectivity. A portion of the budget is allocated for miscellaneous expenses, of which a portion can be used for spot bonuses. There has to be an internal procedure for awarding spot bonuses, which includes approval processes through layers of management and specific forms that are processed through the accounting and payroll departments. The key difference between a spot bonus and a salary bonus is that spot bonuses are normally associated with specific projects while salary bonuses are normally tied to the budget.

These types of bonuses can have a positive effect on the morale of individual employees. A spot bonus will often be interpreted in a more favorable light than a salary increase. For instance, if an employee is making $60,000 and receives a six percent raise (two percent above the average for excellent performance), this would equal $3,600 per year or $150 per pay period. After taxes, the employee will see approximately one hundred dollars extra pay. While this will certainly help the employee, it will not have the impact of $2,000 awarded for a job well done. Certainly less than $3,600 but $2,000 in hand will have more meaning to the employee. The message this sends to the employee, and to the rest of the staff, is that hard work is recognized and rewarded handsomely. This may motivate the employee and the rest of the staff to perform well in the future.

Unfortunately, there is a downside to dolling out spot bonuses. Consider that a spot bonus was awarded to an employee who finished a large project on time and under budget. Were the other staff members given the opportunity to head, or participate, in the project? If not, are there other projects that provide them the opportunity to earn a spot bonus? In addition, under what circumstances is the spot bonus awarded? For highly visible projects or for any project that may be assigned to the telecommunications department? If spot bonuses are only awarded for project-related work, then how is this perceived by staff members who are highly involved with demanding (and necessary) administrative work?

Spot bonuses can be an effective motivational tool, but the telecommunications manager needs to have a plan for awarding such bonuses and be cognizant of the perceptions of his staff members. Staff members should understand how bonuses are awarded, when they are awarded, who will be eligible, and under what circumstances. Cash awards cannot be arbitrarily awarded by a manager, even if he might have legitimate reasons for recognizing an employee. Any award needs to be a managed perk and distributed on a fair and reasonable basis. However, the manager should be aware that no matter how fairly the spot bonuses are distributed, there will probably be staff members who will be dissatisfied with the results. There is no right or wrong way to approach the concept of spot bonuses. If spot bonuses are awarded privately, there is still the risk of word leaking to the rest of the staff. If awarded openly, there is always the risk that resentment and jealousy will surface within the ranks. While there can be great morale boosting benefits, there are also many caveats. Managers need to be cautious and exercise discretion when using this system.

Project specific bonuses differ from spot bonuses in that the potential bonus is known up front and tied to a specific project. The staff involved in the project understands the dollar amount and the criteria for earning the bonus. This includes a project budget and time frame for completion and perhaps certain performance criteria for the technology. One of the advantages of the project specific bonus is that the manager can work with the staff up front to determine reasonable time frames and budgets. Project specific bonuses are more effective when they are tied to a project team for the simple reason that few projects are accomplished by a single individual. Many projects also have a team leader, which could merit more money. However, the bonus money should be distributed as equitably as possible in order to avoid negative repercussions. Once again, the manager also needs to be cognizant of how the projects are assigned so that all staff members have an opportunity to earn bonus money.

Educational bonuses are implemented to encourage the continuing education of staff members. This may be in the form of a one-time bonus or it could mean placing an employee in a higher salary range. Continuing education improves the overall quality and performance of the department. This can be a double-edged sword because advanced education also makes an employee more attractive to other potential employers, especially if the employee has been recently certified for a technology that is in demand. That is why some companies set dollar values to employees who have certain degrees or have completed specific certification processes. Under the same category, there are some companies that also award bonuses for publishing in industry specific journals or for making a presentation at a major seminar.

Departmental bonuses can be awarded as a flat amount at the end of the calendar or fiscal year. These bonuses are normally tied to overall department performance. My experience with colleagues who have had departmental bonuses as part of their overall compensation is that they always felt that their base salary was sub-par. In addition, they also felt that the annual bonus did not compensate for the low salary. Once again, candidates who are considering employment at a company will probably be more concerned with the sure money in lieu of a bonus.

Many high-tech companies have found a certain degree of success with longevity bonuses which often are offered in the form of stock options. The premise is that a candidate for a job is offered a stock option equal to a specific dollar amount. For instance, the option may be for $50,000. However, the employee is awarded 25 percent each year he remains employed ($12,500 per year). There are usually guarantees that the stock will equal the dollar amount promised to the employee at the time employment begins. This type of bonus plan can be effective in areas where unemployment is low and the demand for telecommunications professionals is very high.

When a manager is not given much leverage regarding bonuses, there still may the option of increasing the frequency of merit increases. While many companies may not offer official bonuses, a bit of creativity may be applied to provide an increase on a semi-annual (or even quarterly) basis, in lieu of the traditional annual raise. If the manager opts for more frequent increases, the amount should be meaningful enough so that the employee will feel that it is a true bonus. If the manager only offers a pittance, perhaps one or two percent, then the raise will have little meaning. If the raise is more in line with what is normally awarded, the raise will have more value as a bonus. Managers should be aware that increasing the frequency of raises can fuel the inflationary factor of their salary budget. It can also consume the monies that have been allocated for all salary increases before the fiscal year is complete.

Bonuses do not always have to be offered in the form of money. Many telecommunications departments are finding that recognition can be offered in a variety of forms such as parties, get-away weekends, dinners, and so forth. These bonuses can be awarded to individuals or to the entire group. Parties, luncheons, or dinners that are held for the entire group can be extremely effective motivational tools. Get-away weekends for the department are also effective morale building tools. These awards can take the form of white water rafting trips, camping weekends, paint-ball competitions, and the like. Activities planned for get-away weekends can be effective ways for staff members to bond. The camaraderie and bonding that happen during these activities often overflows into the office on Monday morning. This can be a very effective tool for getting people to work together as a team.

Every company has a different corporate culture. While one company may give managers much freedom with bonus money, other organizations may be more restrictive or conservative. Telecommunications managers should understand that bonuses do not necessarily have to be large. As a morale building tool, the gesture can be significant to employees, however small the award. Even if the manager pays for a lunch out of his own pocket, it can have a very positive effect on the department. It can give the group a chance to get away from the office and perhaps the manager can give some small, inexpensive awards to recognize the efforts of certain individuals or the entire department. Very often, the size of the bonus is not as important as the fact that excellent efforts and results are recognized.

Bonuses can be an effective tool for recognizing achievement and boosting morale. However, they must be awarded with discretion and on a reasonable basis so as not to undermine their purpose. Also, bonuses are not a replacement for a fair and reasonable salary. Rather, the purpose of a bonus is either as an incentive or a morale builder. Once again, there is no right or wrong way to administer a bonus, and each manager must weigh the positive and negative aspects of using them. It should also be realized that while money is a strong motivational tool, it is not the complete answer to providing good morale, teamwork, and stability within the workplace. In fact, many employees who resign and are offered more money to stay very often decline the offer. Surprisingly, the two most common reasons for employees resigning are to develop new skills and increase their responsibilities. That is why managers need to be cognizant of the fact that money is only a piece of the puzzle and there are other factors that affect the department.

9.11 The Issue of Morale and Factors that Affect It

The concept of morale is one that is intangible. How can a manager effectively judge the morale of each individual or the collective morale of his or her department? Even if you could gauge such a nebulous concept, how could you quantify it? Would you assign a dollar value to it or perhaps monitor individual or department performance with statistics? And where would those statistics come from? Unfortunately, there is no precise way to evaluate or control morale. It is a concept that is open to interpretation and affected by a number of uncontrollable variables, one of which is human emotion. A staff is made up of unique individuals. Each individual within the department will have different motivational requirements and unique personalities that will affect the morale of the group. There will also be company-based policies that will affect morale of which the telecommunications manager will have little or no control. Considering all these factors, it is obvious that the concept is complex, vague, and probably in a constant state of flux. But this does not mean the issue should (or can) be ignored. As nebulous as the concept may seem, ignoring the issue of morale often has severe detrimental effects on the stability and performance of the department. Low morale results in poor productivity and staff turnover, which will ultimately reflect on the manager. If turnover becomes an epidemic, senior management will want to know why. That is why every company and department should be concerned with maintaining and boosting morale.

A morale boosting policy actually takes little effort, time, or money. The rewards, however, can be very handsome. In all of my years as a telecommunications professional, the most common reason cited to me by colleagues for changing jobs was dissatisfaction with their previous position. While money certainly has a hand in people changing jobs, the most common complaint I hear is related to morale. And while there may be no exact way to gauge morale, a proactive policy will result in lower turnover, higher productivity, and higher quality work.

Morale should be an issue that telecommunications managers address on an ongoing basis. The following items are suggestions for improving and maintaining higher morale within a telecommunications department. It is by no means an all-inclusive list, and telecommunications managers should use their creative abilities to augment what is outlined below and continually evolve their morale-building policy. Many of these suggestions will strike the reader as being simple common sense, and in many ways they are. But many employees, including management and staff members fall into the drone of day-to-day activity, forgetting the obstacles that their fellow workers face. That is why it is often necessary to shake the cobwebs in our minds by continually taking courses, attending seminars, reading books and articles, and integrating good morale-boosting policies into our work environments. This applies to both management and staff members.

I have often thought of the telecommunications manager as the coach for the department. Since this person is the leader, he or she will have an immense impact on the morale of the department, and much of the responsibility for boosting morale will fall on his shoulders. This, however, does not mean that staff members will be absolved from participation. Staff members can indeed have a major impact on morale. Common problems can be chronic complainers, angry outbursts in tense situations, or domineering employees who will not listen to other ideas or recognize a different approach to working a project or solving a problem. When the telecommunications manager sees these types of problems occur within the group, it is necessary to become proactive and counsel employees. How can a manager understand the morale of his staff? While there is no exacting formula, there are behavioral patterns and alarms that can be identified.

One of the first signs of low morale is behavioral change. An employee who was once content now openly and loudly complains about the job and the company. A staff member who was once open and cheerful is now withdrawn and often broods at his or her desk. These are classic examples of employees who are dissatisfied and, perhaps unknowingly, are sending a signal.

A second sign of low morale is declining productivity. A manager begins to receive complaints about an employee, or perhaps a certain group of employees within the department, that work is sub-par. The manager knows that historically this has not been the case. Employees who are not happy will not be enthusiastic about doing their job. Deliverables will be slow in coming and ideas will not be forthcoming.

A third sign of low morale is when a staff member openly talks about how good the working conditions are at another company. This is a clear signal that something is bothering the employee. Another possible sign is when an employee expresses that he or she is not being fulfilled. An employee openly states that he or she is bored or does not feel challenged by the work. Perhaps he or she states that they watch the clock during the day and cannot wait until quitting time.

A final, and very telling, sign of low morale is turnover. Employees who leave, and are very vocal when they leave, are perhaps a sign that overall staff morale is low. Managers should examine all resignations and complaints seriously. There will naturally be individuals who are chronic complainers and who will never be happy, regardless of where they work. The loss of a good employee who was once a positive influence on the staff is another matter. This may be an indication that one defection is only the tip of the iceberg. How does staff morale deteriorate? It is a number of factors that begin with the day-to-day administration of the telecommunications department.

Very often, managers will ignore the accomplishments of their employees on a day-to-day basis and fall into the habit of taking them for granted. The result is that the employee often begins to feel neglected and abused. This is a problem in almost any job, but it becomes more magnified in a high-pressure job such as telecommunications. Staff members will often spend long hours away from home. During this time they might be working on difficult technical problems. The trouble-solving process in telecommunications can be exasperating. The source of the problem is not always obvious and staff members need to engage in a great deal of brainstorming and trial and error before a solution is found. During the course of this process, they often need to coordinate the efforts of various vendors, while also dealing with angry and frustrated end-users. The stress level is enhanced more when staff members receive pressure from upper management. Even when problems are not occurring, staff members are designing, evaluating, modifying, and supporting technologies that support the corporate telecommunications infrastructure. The daily routine can be both chaotic and mind numbing. Insensitivity to this environment or the application of poor interpersonal skills quickly exacerbates frustration levels. Headhunters feed off of this type of environment, and quickly add names to their list of highly qualified candidates who inevitably will feel that the grass is indeed greener on the other side of the fence.

Concerning the day-to-day aspects of telecommunications department management, the telecommunications manager will need to be concerned with three concepts that have a direct impact on morale: exemplary behavior, good interpersonal skills, and recognition of achievement.

One of the most important aspects of management, and one that is often ignored, is to lead by example. When staff members see a manager throwing a tantrum, belittling or scolding an employee, or being arrogant to staff, there will be no benefit to promoting good interpersonal skills. The staff will also not accept any counseling that the manager may offer. Such actions only destroy morale and cause the manager to lose credibility. If the manager does not practice it, the employees will not learn it, nor will they practice it. There cannot be a double standard and all staff members, both management and staff, must be treated with respect. If this is not applied on a daily basis, the foundation for morale is destroyed.

Consider the example of one manager who was probably the calmest person I had ever observed under stress. When a problem struck the department, he never panicked or looked for somebody to blame. He remained calm, gathered the appropriate personnel, and set a course of resolution. After staff members were assigned tasks for the problem-solving process, he would periodically check in for a status report, but at no time did he make accusations of incompetence. He would often ask if he could help or sometimes offer suggestions. His modus operandi was to guide the staff through the problem-solving process, not create further problems. Once the problem was solved, the manager would hold a meeting to see if there was any way to prevent the problem from happening in the future. This behavior promoted teamwork on the part of the staff and allowed them to work effectively under pressure.

Conversely, I remember another manager who had an explosive temper when problems occurred. He was known to insult and belittle employees, was intolerant when problems happened, and demanded immediate resolution. The resentment within the ranks swelled and morale was nonexistent. It was curious that this manager had a pleasant demeanor at official company functions and at quarterly meetings. He was polite and often recognized the accomplishments of various individuals within his department. Unfortunately, the praise rang hollow and only fomented further resentment. After receiving the recognition, the employees would often grumble about unpleasant incidents that had happened in the past. "I'd rather be treated like a human being on a day-to-day basis than to receive this false praise once a year," was a fairly typical statement by employees.

Both of these examples demonstrate that leading by example has a dramatic effect on the performance of the department. In comparing the two departments, staff members who worked for the manager in the first example developed good problem-solving skills. They learned that to become emotional only clouds your judgment. They were also not concerned with being blamed. Without the fear of a reprimand or the possibility of a public humiliation, the staff members remained focused and worked hard to resolve the problem. The most obvious patterns that evolved with the manager in the second example was dissension, finger-pointing, crankiness, and general paranoia. Not all employees will react this way, but this type of behavior will certainly become more common under such circumstances, and it destroys all notions of teamwork.

Beyond setting an example, managers need to employ a variety of interpersonal skills, which starts with the most basic: respect. This ranges from a simple "please and thank you" to complimenting and recognizing the actual work or task, however mundane. While these are simple suggestions and concepts, my experience has shown me they are the most commonly ignored or overlooked in a day-to-day environment. There is an old cliché in the business world that states, "Treat people as you would like to be treated." While I shudder to fall back on clichés, they often survive because they offer solid and time-tested advice. Too often, I have observed people who become self-absorbed in the daily activities and become oblivious to the feelings of their co-workers. That is why managers need to periodically exercise a degree of soul searching to see if they have fallen into bad habits. They should practice good interpersonal skills and periodically remind their staff that such actions will only improve their work environment.

Telecommunications managers should always be concerned with staff morale and take concrete steps to both improve or maintain it. The foundation of morale begins with the day-to-day management of the staff. This is where the staff will understand whether the management-staff relationship will be cooperative or adversarial. A small litmus test is for the manager to ask himself if a major portion of the day is spent in confrontation with staff members. If it is, there is certainly cause for concern and probably room for improvement. While many seasoned managers may scoff at applying efforts to such skills, it is never too late to adopt a back-to-basics approach that will inevitably boost morale. The following is a list of recommendations for improving interpersonal skills with employees:

1. Avoid using a confrontational style. I once knew of a manager who used to speak with his staff members (about general day-to-day issues) as if he was a lawyer cross-examining a hostile witness. Such tactics are counterproductive. It will only place the staff member on the defensive and make him nervous. If the employee feels he or she is being attacked, they will not be able to think clearly and not provide good, accurate answers.
2. Do not demand things. Ask staff members to do things politely, using the time-tested and proven methodology of please and thank you. Demanding things (especially if the manager mandates a due date for the task) will only foster resentment.
3. Allow employees to express themselves and listen to them. I once worked for a manager who made many decisions independent of his staff members. He would hand assignments to people and, if there was an objection or alternate suggestion, the person was aggressively silenced. Staff members stopped offering suggestions and ideas because they always fell on deaf ears. The telecommunications staff is a great resource for innovation and new ideas. Allowing the staff members to express themselves encourages them to bring forth their ideas. This only benefits the overall performance of the department.
4. Avoid using e-mail for corrective measures or sensitive issues. It is human nature for people to be more aggressive when they use e-mail. In addition, the use of e-mail comes across as being cold and

impersonal. The same manager who suppressed the ideas of his staff was also very adept at responding to e-mails with strong reprimands or corrective actions. This impersonal approach only serves to alienate and anger the staff.

5. Do not address corrective measures or sensitive issues in front of other staff members. This is common sense. Nobody likes to be reprimanded in front of their peers and such actions only foster resentment.
6. Be sincere. If you do not mean it, do not say it. There are numerous management classes offered to the business world in general, often offering faddish management concepts of dubious practical value. Such courses often produce catch-phrases or behavioral techniques. Cloning these concepts from a class will not ring sincere with employees. Do not fall back on clichés, simply be sincere and be yourself.

Many of these ideas are common sense and many readers may ask why the obvious is being stated. The reason for this is twofold. First, anybody who has even a modicum of experience in the business world knows that the application of these ideas is usually the exception and not the rule. Second, we are all human and have a tendency to fall into ruts and bad habits. It does not matter how good a manager is, chances are there will be days when the workload is taxing, problems seem to be ubiquitous, and no progress seems to be made on anything. With these two concepts in mind, all managers need to give themselves a wake-up call and ask if they employ the most fundamental but effective interpersonal skills.

As outlined in the previous section, bonuses serve as a means of recognition, a building block for boosting morale, productivity, and quality work. Beyond the monetary advantages of a bonus, the simple act of recognition goes a long way toward motivating employees and maintaining morale. Above the day-to-day issues, recognition should always be given for major accomplishments, such as the successful completion of a highly visible project. In addition, recognizing milestones within the project will also enhance and maintain the morale of the department.

There are any number of ways that achievement can be recognized. At a very basic level, simple verbal recognition at a meeting will invoke a positive response. Once again, if staff members are ignored at all other times, the recognition will ring hollow. Also, the recognition should be continuous and sincere. Managers can ensure this by placing "recognition of achievements" as an action item on meeting agendas. Have we accomplished anything in the time frame since the last meeting? If the answer is yes, recognition is in order. In addition to recognizing the achievement, the manager should also verbalize the scope of the work that was accomplished along with his appreciation. “I know you all have put in a lot of overtime on this. It’s tough to be away from your families, and I appreciate what you did.” It will normally only take a few minutes out of the meeting to draw attention to an employee’s accomplishments. The good will that is generated by such gestures is immeasurable.

These types of gestures should be done, even if the scope of the work is within the employee’s job description. In fact, one of the worst things that a manager can do is to state that very fact. I knew of a manager who, when employees were involved in critical problems that involved excessive overtime, only asked the employees if the problem was solved. When an employee once complained of the overtime, and the fact that he had to spend a holiday in the office away from his family, the curt reply was, “That’s in the job description.” Needless to say, there were a number of resumes always circulating from that department.

Letters and e-mails are a good way to recognize accomplishments. When they are directed to high level managers, citing the work of the employee, and copying the employee, the positive reinforcement goes a long way toward boosting morale. These memorandums can also be copied by the employee to be hung on walls or filed for future reference. If the letter is hung on the wall, it will serve notice to other employees that good work is both recognized and rewarded. Recognition of achievement will be noted by each staff member and good news will travel via word-of-mouth. Managers should not dismiss the power of word-of-mouth communications within the department. As any manager will know, the rumor mill is strong within any given department, and the rumors are usually of a negative nature. Positive stories also circulate, and so morale can be boosted via the same vehicle that normally disseminates negativity.

Publishing achievements on a companywide basis also helps tremendously in boosting individual and department morale. There are company newsletters and quarterly reports that might provide a forum for recognizing individual or department achievement. In addition to promoting morale, publishing achievements also serves to sell the services of the department to the corporation in general.

Beyond written or verbal recognition there are some companies that are finding success with awards above and beyond the standard compensation plan. As stated in the section related to bonuses, these awards can take the form of parties or luncheons to celebrate the completion of a project or the close of a particularly good month or quarter. But managers should not necessarily limit themselves to the completion of an event or project. There may be times when the pressure is very high and the stress factor is palpable. It is at these times when a party or luncheon can act as a great source of temporary relief. One of the best ideas I have encountered was a manager who distributed an invitation to all staff members inviting them to a picnic. The invitation stated the time and date of the picnic. It instructed staff members to bring bats, gloves, balls, Frisbees, badminton sets, and so on. It went on to say that beer, soft-drinks, hamburgers, and hot-dogs would be served and the only restriction was that staff members were not allowed to discuss work. The picnic was a huge success and the manager told me that each staff member showed up for work on Monday morning refreshed and energetic.

Awards can be made even more attractive if the budget will allow it. There are some companies that are finding success with awards that range from monetary gifts to get-away weekends, gift certificates, or stock to name but a few. Normally, such awards are provided at the completion of large, highly visible, and expensive projects. The awards are compensation for finishing the project on time and under budget.

Good interpersonal skills certainly have a positive effect on department morale, but there are still other factors that need to be considered. Compensation has been discussed and substandard compensation will certainly yield low morale. But in spite of fair compensation, many telecommunications professionals still change jobs. One of the primary reasons is personal fulfillment and job satisfaction. This can relate to the work environment, more challenging work, the chance for advancement, and the opportunity to work with different, new, or more challenging technology. This brings up one of the most critical factors to understanding staff morale: communication.

There are two ways that the manager communicates with the department: as a group and as individuals. Each methodology will offer staff members the opportunity to voice their concerns, frustrations, ideas, and goals. Group meetings will afford staff members the opportunity to voice opinions that affect the group. One-on-one meetings will offer the opportunity for individual employees to offer opinions regarding their individual careers, opportunities, ideas, and concerns. Both methods of communication are essential for a number of reasons.

The first benefit of proactive communication is that the manager will gain an understanding of staff morale. By giving his or her staff a forum in meetings, he or she will allow staff members to express their opinions, criticisms, and frustrations. While this may be an unpleasant undertaking, it is important to feel the pulse of the group. Otherwise, the low morale will fester through the rumor mill and problems will never be addressed. Managers who, as a matter of policy, allow for open exchanges with employees will probably not be surprised too often. Managers who implement the policy for the first time may receive some very unpleasant surprises followed by some very long sessions. Allowing for communication can be simply opening a meeting to the floor for a short period of time, or calling special meetings for the distinct purpose of allowing employees to voice their concerns. For instance, the department may be in the midst of a very complex and demanding project. By calling a special meeting, the manager may be able to gain an understanding of project progress, and take corrective action for problems that are beginning to surface. Many managers who do not encourage open exchanges with their employees are surprised when an employee resigns, leaving a wake of bitter criticisms about the company and the manager. A very common response is, "I did not know." This is only a valid excuse if the manager offered the staff members opportunities to express themselves.

Once issues have been placed on the table, they can be addressed by the manager. It is essential that managers listen to their employees with empathy and act, when they can, on the concerns of their staff members. This is a second advantage to open communications simply because employees will be less frustrated, even if all of their concerns are not addressed. Any reasonable person will understand that there are factors outside the control of the manager. The simple fact that an employee's voice has been heard will raise morale because, very often, a staff member simply needs to vent his frustration. Staff members will be understanding and appreciative when a manager listens and acts when he can. This is a policy that is fair and reasonable, and that is what employees expect from their manager. A common source of low morale is when staff members feel as though they are not being treated fairly.

9.12 Burnout and Stress

Telecommunications environments can be stressful and when the stress level reaches certain proportions, all other factors become secondary. I once had a colleague who was under a great deal of stress. The pressures of the job were severe but he thought he was coping with the demands of the job. However, one day he began to experience chest pain and was rushed to the hospital. Numerous tests revealed no heart problem and, after asking this person a number of questions, the doctor made a diagnosis similar to, "I don't care what you do, or how much money you are making, you have to quit this job and find something else to do." Several years later, I saw this person at a professional meeting. He was making less money but was much happier with his job. When this person was facing the severe pressures that caused his body to revolt, his manager and company had failed to take into consideration the human side of management. They simply piled on the work and expected it to be accomplished on time and accurately, not considering the demands it was placing on their employees. They had failed to consider that the working environment needs to be reasonable and that employee performance and department stability is improved with good working conditions and a concerted effort to maintain good morale.

I would like to think that my friend's plight was unusual, but it is more common than one might think. Modern business has become reliant upon telecommunications as a mission critical part of the business infrastructure. When communication links fail, it means that orders are not entered, product is not shipped, and customers are not serviced. In many instances, it is not just a matter of lost productivity, but also lost revenue. Consequently, failed communications services will inevitably attract unfavorable attention from high ranking management. This results in a downward pressure for the telecommunications manager and department to resolve problems as quickly as possible. Unfortunately, telecommunications problems can be complex and time consuming to resolve. There is also the inevitable application of Murphy's Law, because no matter how well a project is planned or something is designed, strange things do happen. That is why it is necessary for the telecommunications manager and department members to be sensitive to the pressures of the job and apply interpersonal skills when dealing with vendors, end-users, and co-workers in the problem resolution process. This facilitates a faster meantime to resolution, and promotes better morale once the problem has been solved. What a department does not need is ubiquitous resentment once a problem has been solved.

Beyond major problems, there is also the day-to-day administration of telecommunications products and services. Telephones, fax machines, PCs, and printers need to be physically installed. Cable needs to be installed or cross-connected in order to connect the devices. Software changes need to be made in the PBX, the ACD, a router, or a hub. Bills need to be paid, RFPs written, and documentation needs to be updated. In most corporate environments, this heavy work load is accomplished with minimal staff. The net result of this formula is a high level of stress and burnout.

Once again, the telecommunications manager is the person who shapes the personality of the telecommunications department. As stated before, he or she is the coach and mentor. The policies set forth by this person affect the performance and morale of the department and its reputation within the company. The telecommunications manager also has a dramatic affect on the stress level within the department and the burnout factor. A manager who is not sensitive to the pressures that his or her staff faces usually experiences low morale, quick burnout, and high turnover within the department. How can a telecommunications manager effectively address the problem of stress?

Stress management is a learned skill. Both the manager and the staff members need to develop stress management skills and there needs to be a policy in effect that provides for fine tuning these skills. A first step in this direction is to establish a stress management policy. This can be approached from two levels: the manager and the staff.

The manager needs to address stress from two levels. First, is the manager a source of stress? In most telecommunications environments, stress will be an inherent part of the job. There is no need for the manager to aggravate this situation unnecessarily. If the manager is a source of stress, then the manager needs to perform some self-appraisal and take steps to help him or her better manage the staff. Once the manager has examined his or her own role in the department, he or she needs to set an example and counsel employees in order to reduce stress within the working environment.

Stress can come from a variety of sources. In a telecommunications environment three primary sources are heavy workloads, highly visible and complicated problems, or interpersonal conflicts with management or fellow staff members. There are certainly other sources of stress (e.g., problems at home), but there is nothing that a manager can do to prevent these in the office. Human resources departments will have policies regarding stress and how to deal with stress induced from outside sources. Although these types of circumstances can certainly have an affect on an employee's performance, the purpose of this section is to identify sources of stress that are often inherent in the field of telecommunications and to make recommendations to help both the manager and employees reduce and cope with stress.

Regardless of the sources of stress, there are tell-tale signs that an employee is beginning to burn out or experience stress. Some of the most common symptoms are as follows:

- Negative emotions, including dissatisfaction, frustration, anger, and depression, usually accompanied by anxiety;
- Interpersonal conflicts, sometimes accompanied by emotional outbursts or withdrawal;
- Health complaints (colds, headaches, insomnia, backaches) and signs of fatigue;
- Impaired performance accompanied by boredom, lack of enthusiasm, or impaired concentration;
- Abuse of substances such as alcohol, drugs, coffee, and cigarettes, accompanied by reduced or increased food consumption;
- A "So what?" or "Why bother?" attitude reflecting feelings of meaninglessness or cynicism, where once the person had been enthusiastic and dedicated.

One of the first steps in addressing the problems of burnout and stress is to understand the workload and the employee's ability to accomplish it. Consider an employee who is overburdened on a day-to-day basis. He or she is caught in projects that seem to never end. These projects are also offering unpleasant surprises (both financial and technological). The employee has trouble prioritizing tasks because of the sheer volume of work. When he or she leaves at the end of the day, the employee feels that he or she has accomplished nothing and fears coming into work the next day to start the same agonizing process over again. He can be seen whispering to other employees, and gives sarcastic answers, or becomes defensive, when asked about the status of a project or task.

This is a classic example of burnout and an employee who will probably be looking for another job. This is also a classic example of an overburdened employee. A manager who will discipline an employee under such circumstances will only make matters worse. This would be ironic since the true cause of the problem might be the manager's inability to properly disseminate the workload. In fact, an employee will become more resentful if he or she sees a fellow staff member who has a manageable workload.

A perceptive manager can intervene before the problem becomes acute. There are a number of ways that the problem can be addressed, but the first (and most important) step is to proactively reduce the workload. A second method is to offer gestures that will act as pressure valves to release the stress. A third method is to counsel the employee by providing advice for more efficient work habits.

Workload can be reduced in a number of ways. The manager may want to review the workload and prioritize the daily, weekly, and monthly tasks: "Put this one on the back burner. If it doesn't get done, it's no big deal. If anybody complains, send them to me!" Workload may also be disseminated among available staff members who are not as overburdened: "John's in the middle of a major project and we need your help. Once this is over with, we'll get back to normal."

If the manager perceives that the workload is reasonable, the problem may be that the employee needs to learn, or improve, his project management skills. This may entail a temporary relief of the workload while the manager counsels the employee.

Consultants or contract workers are another way that workload can be reduced. There may not be the proper expertise or manpower within the department to offer any relief for the project in question. Consequently, the only other option may be to hire a consultant. This is often an effective way to address busy work that is a nuisance and hindrance to a person who is overseeing large complicated projects. I was once working on a project that was simply overwhelming in terms of size and complexity. Part of the project entailed a great deal of interaction with an LEC to order hundreds of adds, moves, and changes on a Centrex system. My manager suggested hiring a contract worker who was an ex-employee of the LEC and the move proved to be a determining factor in the success of the project. The consultant was not only a excellent worker, she also knew how to effectively deal with the LEC bureaucracy so that orders were placed on time and accurately.

A second method of relieving stress is to provide relief. This can come in the form of extra time off. While some managers may shudder at this suggestion (for it may break the formal personnel rules of the company), it has always been my experience that telecommunications professionals work long and unusual hours. For salaried employees, there is usually no compensation for these extra hours. It has been my experience that a company never loses when these extra hours are awarded. In fact the company is usually ahead of the game, because the awarded hours never equal the number of hours worked. An afternoon off here and there will go a long way toward helping reduce stress levels and the employee will feel as though he is working for a fair and reasonable manager and company.

An additional method of relieving stress is to offer educational diversions. For instance, there may be local vendors that offer presentations, a professional organization that meets monthly, or conferences that are held nationwide. A manager might want to suggest that employees attend some of these events, so that they can concentrate on something work-related but not face the day-to-day demands and stress of their jobs.

In spite of all these suggestions to reduce workload or provide relief, the telecommunications manager must also perform some self-examination and perhaps take blame in the excessive workload. There are many managers who, in their eagerness to provide service to the company, never learn to say no. A manager who blindly accepts all requests from the field will inevitably find himself "drowning in opportunities." This is an especially dangerous policy considering that most departments are run with minimal staff.

By not exercising discretion, the manager will overburden his or her staff members. Consider an employee who is overwhelmed and continually sees tasks coming across his or her desk or through e-mail. When the employee complains about the never-ending stream of new tasks and projects, the manager explains that, "We are here to service the end-users. Without the end-users we would not have jobs." Such a statement would do little to assuage the frustration of the staff member. The employees will begin to feel like Sisyphus, the character from Greek mythology who was condemned to push a large boulder up a never-ending hill for eternity.

What is the solution to this dilemma? The manager cannot simply refuse to do work for the end-users. This tactic would be folly resulting in a deteriorating reputation for the telecommunications department. The manager needs to employ some diplomatic skills: "We are inundated right now and, based on the current workload, I don't think we could do this project justice. However, in three or four months, we should be able to allocate some resources. I would like to meet so that we can discuss the scope of the project, and when we could do it. In addition, we may want to discuss the use of consultants or contract workers if this project is a priority and needs to be done now."

The manager can bring staff members into this process, and they can help to outline the scope of the project to the end-user. This would entail costs, time frames, resources, logistical issues, and so forth. When the telecommunications manager proactively works with the end-user, the end-user will be more understanding of the difficulties in assigning resources to projects. In addition, the staff will learn better corporate relations skills and feel that the manager is working on their behalf. It should be noted that these suggestions will not significantly reduce the workload. That can only come from increased staff. On the contrary, these suggestions are aimed at helping the manager reduce stress and avoid burnout in the midst of a heavy workload.

Stress does not only come from a heavy workload. It cannot be ignored that the telecommunications manager may be one of the sources of burnout and stress, a result of his managerial style. Managers who are confrontational, belligerent, quick-tempered, unforgiving, and suspicious will elevate stress levels and contribute to burnout. Quality interpersonal skills are essential when addressing or preventing stress.

I once saw a manager (whom I did not report to) call a number of employees who directly reported to him a bunch of "dummies" in a meeting. He went into a tirade, asking questions such as, "What did I do to deserve this bunch of idiots? They really don't know what the %*#@& they are doing! That's why we are having this problem."

Stress management begins with respect. All telecommunications professionals should understand that problems and heavy workloads are the nature of the beast. Problems are not solved by attacking people. In fact, my own adage has always been, "Attack problems, not people!" Whenever the manager addresses issues with a staff member, there should always be an effort to maintain self-esteem. When a manager does not respect the employee, the employee will put up a wall. This is counterproductive when a problem needs to be solved. Once again, the cliché bears repeating, "Treat people as you would like to be treated."

Communication and teamwork are also essential factors in dealing with stress. A dictatorial manager will inevitably find himsel for herself in constant confrontational situations. Stress will be high on both sides of the fence and the dictatorial style breaks a fundamental rule of stress management: respect. Dictatorial managers distribute the workload without asking who within the staff is best qualified and who has the least heavy workload. Dictatorial managers are also unforgiving when projects bog down or when problems occur.

Communication and teamwork, on the other hand, bring the staff members into the department management process. Their input gives the manager guidance, and he or she is better able to understand the workload and what staff resources are available to meet it. When there is teamwork, the manager can make more reasonable decisions, based on what his or her staff tells him. During interactive meetings, the staff will communicate project status and even offer suggestions that will improve department productivity or solve problems.

Beyond communications and teamwork, managers should strive to empower their employees. Empowerment means that employees should make their own decisions and be allowed to make their own mistakes. This philosophy stimulates new ideas and productivity. If employees are not allowed to set their own goals and make mistakes, they will not try any more. When employees are interactive with the management of the telecommunications department, it helps to reduce burnout and stress because they understand why the workload is heavy and where it comes from. They are less prone to be negative because their voice is being heard and their opinion is an integral part of the department's operation.

Stress and burnout are also caused by a manager not standing up for his or her people. If the manager knows that his or her people are giving an honest effort, and that they are skilled and competent, he or she must stand up for them when criticism strikes the department. There are often extraneous factors that affect the success (or perceived failure) of a project. There are also times when things simply do not go well or when employees have made mistakes. These are critical times, and the manager must exercise sound judgment, or the burnout and stress factors will accelerate.

Consider a project to replace existing dedicated lines in the WAN with frame relay service. The justification for the project is to save money; however, the telecommunications department feels that response time problems might be solved due to increased bandwidth and the bursting feature inherent to frame relay. But response time problems are not solved in several locations and these locations were originally experiencing the worst problems. The end-users are furious and send angry e-mails, copying high-ranking management. The telecommunications manager, in turn, begins receiving angry e-mails and phone calls from these high-ranking managers.

This is a very common occurrence in the field of telecommunications, and one that managers must be prepared for on an on-going basis. In this example, let us assume that the response time was not guaranteed to be fixed, but that frame relay was to be the first step in providing a more dynamic and flexible WAN service. It is then the manager's job to reconfirm the original objective of the project to the end-users, and to reconfirm that the project was successfully completed according to the original objectives. The manager also needs to convey the quality of the work his or her people have done, and the extra effort that was given during the course of the project. This is to be done, of course, in the most diplomatic manner as possible. For example, "I understand that you are frustrated with the poor response time. We are always concerned when an end-user is having difficulties with the products and services that we provide. We will make it a priority to resolve this issue. But I would also like to say that my staff completed a large and complex project, faced many obstacles, and still succeeded in completing the job to my satisfaction. As you may remember, the original scope of the project was to provide a more dynamic and versatile WAN capability. Now that we have this new capability, it will help us to resolve this issue in a more timely manner than was ever available before."

Under no circumstances should the manager copy the staff members on the response whether it is e-mail or interoffice correspondence. If the staff members read angry remarks in the original e-mail or correspondence it will only invoke angry and defensive responses. In these instances the manager is a buffer between the end-users and his or her staff and acts as a filter to keep unnecessary information away from them.

The example just cited does not address when mistakes are made. Mistakes do happen and they are committed by both staff and vendor, both important participants of a project. Once again, the manager must be diplomatic with the end-users, but he or she must also stand by his people. The manager must also be honest with the end-user. If his nor her department committed an error, he or she must own up to it, apologize for the inconveniences, and explain that policies will be implemented so that it does not happen again.

Once the manager has addressed the issue with the end-user, he or she must address the issue with the staff. Mistakes should be addressed via one-on-one conferences. The issue should be addressed dispassionately and professionally: What was the reason for the mistake? Was it something that was simply overlooked? Was it inexperience? Was it a lack of education? Was it simple human error or was the workload so much that details were overlooked? Regardless, the manager should allow the employee to explain the problem and to take corrective actions. "In the future, I think we need to set up a checklist for these types of items."

During meetings, the manager can address complaints or criticisms that have been directed to the group. Once again, a confrontational style will only build walls and result in angry responses. The manager should explain what the complaint was (but not necessarily the origin) and how he or she responded to the problem. This is in accordance with the manager's role of coach and mentor. When the staff members see how the problem was handled, many will learn from the manager's actions and adopt similar professional behavior. The manager should ask for suggestions from the group about how to resolve the problem, or what the correct behavior would be. He or she should also strive to maintain and enhance the self-esteem of the group at all times.

When the staff members understand that the manager is behind them, the staff begins to work as a group. They will aggressively bring forth new ideas, and work hard without fear of reprimand. This is a critical part of a telecommunications manager's job. When I have spoken with colleagues in the past, and we have discussed different companies and managers, one of the questions asked most often is, "How does he handle problems?"

With so many things seemingly stacked against the telecommunications manager and department, is there really anything that can be done to help people to better cope with the stress? While there is no panacea, there certainly are measures that can be taken to alleviate some of the pressure. It is not the intention of this book to be a complete resource for stress management. There are countless numbers of books, courses, seminars, and videos available in the business world that claim to address this issue. For the most part, my own experience with media relating to stress management is that these tutorials do not address the real world. However, they do sometimes serve a useful purpose by providing a wake up call to the telecommunications manager and the department. By providing stress management materials to the department, it makes managers and staff stop and think about how they are doing in their working environment.

It is naïve to think that every person will be able to manage stress in the same manner. In any given group of people, there will be those who can remain calm in times of duress and those who will literally fall to pieces. There are even those who lash out in rage, which usually results in resentment among their co-workers. Regardless of a person's given nature, companies and telecommunications managers should counsel employees and apply programs that enable telecommunications professionals to manage the stress that is an inherent part of the job. Problem-solving skills can be learned; people can be taught to manage their time better and to be more productive; and departments can be structured so that employees work together rather than as adversaries. It should be noted that there is no absolute remedy to aid people in dealing with a stressful work environment. In fact, in many instances, the solution to the problem is not teaching employees how to deal with stress, but hiring more employees to relieve an unreasonable workload.

9.13 Empowering Employees

The concept of employee empowerment is one that has been given much attention in recent years. It is a term that has been bandied about so much, it has been reduced to a cliché. However, the concept continues to gain acceptance because it does, when properly implemented, provide positive results that equate to higher morale, decreased turnover, and higher productivity. Additional benefits are that employees buy into and contribute to departmental and corporate objectives.

Like many of the issues covered in this chapter, empowerment is not a concept that can be accurately measured. In addition, there are many managers who do not believe in the concept because it is interpreted as relinquishing power and control. Quite the contrary, empowerment actually makes the manager more powerful because he is using the full capabilities and knowledge base of his department. It is an especially rewarding concept in the field of telecommunications where there are so many technologies that are changing so rapidly. A single telecommunications manager cannot thoroughly understand or manage all of these concepts. That is why it is advantageous to poll the collective talents of the group and empower them to participate in the design, implementation, and management of these technologies.

The dictatorial method of managing a department can be effective, but it is a limited concept. There are many companies where managers have traditionally made all of the decisions, and employees are not encouraged to act independently. While this system can still work, it continually becomes more ineffective as technology and business advance rapidly. The first reason for this is the sheer scope of the technologies that are supported by a telecommunications department. No single person can be an expert on everything. Therefore, telecommunications departments often develop experts in various technologies out of sheer necessity. Also, the telecommunications manager cannot conduct detailed research when he or she is inundated with the day-to-day management of the department. In these instances, the manager needs to use the resources within his department. It is a natural progression of events that a manager becomes less technical once he or she assumes the manager's position. There is simply not enough time to learn every intricate detail of a technology when one is concerned with budgets, personnel issues, and the like. Unfortunately those details can become very important when selecting a technology. The best knowledge on a subject may be within the group, either through independent research or experience. To not utilize this knowledge is to perform a disservice for the group.

The dictatorial method can also cause dissension within the department. Because the manager is not as knowledgeable as his or her staff members, there will be resistance and resentment if a system or technology is forced upon them. Managers who are proponents of this top-down method normally do not take criticism from their staff very well and the situation tends to fester, causing morale to deteriorate. This methodology tends to facilitate staff turnover. Two of the primary reasons cited by employees for changing jobs is the chance to work with new technologies and a more challenging environment. The dictatorial method of management usually allows for neither of these opportunities. Staff members become increasingly frustrated as the manager dictates what technologies will be used, how they will be designed, and how they will be supported.

The concept of empowerment, on the other hand, introduces the concepts of teamwork, an open culture, and the opportunity for staff members to make decisions and take on more responsibility. Implementing the concept of empowerment is not always easy in a department that may have a history of the manager making all decisions. Employees who have worked in such an environment may also be hesitant to take on responsibilities along with the manager who may be reluctant to give them. The transition may be difficult for all concerned parties but it is one that results in a win-win situation for both management and staff. The first step is to adopt the policy of empowerment and then enable a systematic method for change.

According to the Gartner Group, there are five mantras for effecting change:

- Open culture is key;
- Hire necessary people or things won't get done;
- Don't forget learning curves;
- Let people manage their managers;
- Think "people using technology" not just "technology implemented."

The open culture promotes the free and open exchange of ideas. This means that all staff members are allowed to voice their opinions regarding issues such as technology, vendors, project management, and so forth. Of course, not all ideas are good or will be accepted, but the proliferation of ideas ensures that better ideas are incorporated into each aspect of the department's operation. It is important that the manager is diplomatic and professional when ideas are rejected. He or she should convey to the staff member that while all ideas are welcome, not all ideas are accepted, including his own. The manager should also offer advice for positive change so that future ideas have more merit.

When the culture is open, it also allows interested staff members to become involved in new projects. This is a tremendous morale builder because it is frustrating when a staff member wants to branch out, but is continually given the same type of projects. The staff member soon finds himself or herself in a rut, feels that his or her career is at a dead-end, and that the company has painted him or her into a corner. In addition, when people are given more freedom to choose their own projects, they have a stake in the project and its success.

Hiring the necessary people is essential. Empowerment will not work if the people are overworked and on the verge of burnout. This can only be avoided if management understands the workload and provides appropriate staff. Otherwise the employees will refuse to participate in the open process. When employees are overworked, they become negative and cynical. Empowerment requires that employees are upbeat and eager to participate in the team concept. It encourages them to bring forth ideas. But if they are miserable they will not volunteer ideas. In such situations, if the manager is lucky enough to get them to participate, they will usually tear the fabric of the meetings apart with their cynicism.

Learning curves will be a natural byproduct of an open system where employees are participating in projects which involve new and unfamiliar technology. Empowerment means that employees are being trusted to enter new territories with which they will be unfamiliar. This means that mistakes will be made. This should be expected and accepted because education and experience are never achieved without painful moments or mistakes. This is where the telecommunications manager can exercise his role as coach and mentor. By guiding the projects and assuring employees that mistakes are allowed and that the group will learn from them, the staff members will be positive about their work and continue to apply their efforts.

Allowing people to manage their managers is probably one of the most difficult aspects of the empowerment process. This is commonly interpreted by many managers as a loss of power, or perhaps letting the inmates run the asylum. The concept should not be taken literally. Employees who are empowered also understand the organizational chart. Ultimately, the manager will have to make decisions, perhaps nullifying ideas or setting a project back on a logical course. If this is done discreetly and with sound interpersonal skills, the staff members will usually accept it. Remember, they have been given more power and input than before, therefore they are still ahead of the game.

The concept of "people using technology" not "technology implemented" is often a philosophical difference that may require retraining on the part of the staff. I have always been an advocate of two fundamental principles when it comes to choosing, designing, and implementing technology. First, understand the needs and desires of the end-user. If the telecommunications department works with other departments and end-users, they will be more understanding when problems occur. Inevitably, problems will occur, and if the end-users feel that something has been forced upon them, they will react in a hostile manner. This makes for unpleasant interactions and will also erode the reputation of the department. Consider that an end-user might be talking with a fellow employee: "How did your implementation go?" "Not bad. A few bumps and bruises, but the telecommunications department is working closely with us, and we're working things out. They told us to expect a bump or two, but they would hold our hands during the process." Now consider the same question being asked and the reply being: "A disaster! We implemented this new LAN and the PCs freeze every 20 minutes. We have no idea why we did this; the old LAN worked just fine. Those 'techies' keep telling me this is a superior system, but what did it buy us? It doesn't work! And you can't get those guys to return a phone call. When they do, they're smug and arrogant, baffling you with all these technical terms that nobody understands!" These are prime examples of a department that worked with the end-user and a department that simply implemented a technology with little or no interaction with the end-user.

The second principle is to walk in the shoes of the end-user. This means, before selecting or implementing a technology, understanding the end-user's business requirements and environment. Telecommunications professionals often have a tendency to get lost in acronyms and technical talk. They forget that the average person does not understand such jargon or concepts, but also needs to use these tools. Staff members should think about how to talk with end-users, avoiding arcane acronyms and jargon and expressing themselves in simpler terms. They should also spend time observing the work environment of the end-user and what the impact of the new technology will be. If the technology fails or is not operating properly, how would that affect them? If the technology does fail, does the end-user know what to do? Are there redundancy or backup capabilities provided with the technology? Do the end-users thoroughly understand the concept and have they been properly trained? These types of questions make the telecommunications staff begin to think, "people using technology," not "technology implemented."

Empowerment is not an easy concept for employees who have never been entrusted with responsibility. In fact, there will always be employees who will resist the concept, no matter how hard management tries to implement the concept. "Just tell me what to do and let me do it!" is not an uncommon attitude. However, the modus operandi of any company or manager should be to push forward, implementing new ideas for positive change.

9.14 Conducting Reviews

The performance appraisal process is not one that many employees always look forward to. One technical writer likened them to dental appointments: get the unpleasant task over with as quickly as possible. Many employees state that performance reviews are simply perfunctory gestures and little effort goes into them.

Performance reviews are normally conducted on an annual basis and their purpose is to gauge productivity, document progress, suggest improvement, or even lay the groundwork for termination. When taken seriously, the performance review can be a productive vehicle for quality staff development. Unfortunately, many managers only view it as a periodic task. It is seldom that a manager gauges the quality of his reviews or sees it as an ongoing process that should be continually developed. In fact, many managers often cite the major blunder that occurred six months ago or only remember the past six weeks, the time when they began to think about the pending performance review. An employee may have had a relatively good year except for a few negative incidents. In the field of telecommunications, this is not unusual, given the complexities of the technologies. It is therefore unfair to make the employee feel that he is being evaluated on these isolated incidents if the overall performance has been good.

Common employee complaints demonstrate how ineffective the process can be:

- Manager hasn't gathered adequate performance information;
- Standards by which employees are judged are unclear;
- Manager rushes through the review or doesn't take it seriously;
- Manager isn't honest;
- Employees don't receive rewards for excellent performance;
- Not enough attention is devoted to employee development;
- Employee is surprised by criticisms of performance.

If it is so common that performance reviews are not effectively administered or well received, what are the reasons? Unfortunately, many technical managers are not promoted for their abilities to manage people. They have reached their position because of their core technical competencies. This is a fundamental reason why many telecommunications managers are ill-prepared to deal with human resources issues.

There are two policies that telecommunications managers can adopt to make the performance review process more effective and minimize its negative connotations. The first policy is to adopt an ongoing interactive process with employees during the course of the year. This can entail the use of mini-appraisals and addressing unique problems and issues as they surface during the course of the year. Applying such policies has a positive affect on morale, allows them to grow professionally, and has a positive impact on productivity.

Mini-reviews do not necessarily have to be long, nor do they have to be often. New employees might merit reviews on a more frequent basis until they are more acclimated to the department and the corporate culture. Experienced employees might merit less frequent reviews. Perhaps quarterly reviews or semi-annual reviews.

The manager should choose a simple set of critical issues so that the meeting is not long or cumbersome. This is necessary because the manager does not want to be overburdened with many reviews that take up a great deal of time. A time frame of 30 minutes to an hour should be sufficient. He or she should choose issues that he considers to be important such as:

- Professional development (education);
- Project progress;
- Workload;
- Recognition of recent achievements;
- Corrective issues;
- Counseling, recommendations, and advice.

The manager should strive for the review to be positive, even if corrective action needs to be taken. The employee will receive a status report and should receive tangible goals for the remainder of the year. Rest assured, the rumor mill will also be at work. Staff members will inevitably complain to their co-workers about a negative review. This induces paranoia within the group and generally yields overall negative morale. That is why it is better to hold mini-reviews, so that employees understand their status, and have a reasonable opportunity to make corrections. Moreover, employees should feel that the process is being conducted for their benefit, so that they can develop professionally.

There may be times when a special review is in order. Perhaps there is a project that is not going well and the staff member is the project leader. The employee will know that things are not going well and will be concerned. A mini-review can address and solve many issues. For instance, why isn't the project going well? The manager needs this information in order to report it to his manager. But that is only one reason why the meeting is being held. What are the true sources of the problems? Were they a result of employee inexperience, incompetence, neglect, or were there extraneous factors that were beyond everybody's control? The manager should work with the employee to take corrective measures. Does the employee need additional resources? Does the employee need additional training for future projects? Does the employee need guidance for the remainder of the project? In this situation, the manager will want to correct as much as possible in order to salvage the project. Employees should feel, after their reviews, that the manager is working in their best interests, and that if corrective actions are taken, their performance will be viewed favorably.

If managers take a proactive approach to reviews, then the employee will not be as apprehensive when the annual review arrives. Rules that can be considered for an effective review are:

- Viewing the appraisal as a process, not an event;
- Continuously documenting employee performance;
- Being careful to appraise behavior, not personality;
- Making sure that nothing comes as a surprise to the employee;
- Outlining a plan for employee development or improvement;
- Maintaining the employee's self-esteem;
- Allowing the employee to express himself, and respecting his opinion;
- Performing a self-appraisal and always thinking about how the process can be improved.

How can a manager know that his reviews have been effective? First, are most of the reviews confrontational? If so, this is a good indication that something is wrong. A single confrontational review may only mean that the employee is difficult and unreasonable or there are difficult circumstances. If most of the reviews are this way, it usually means that the process is flawed. Second, are employees surprised? Regardless of whether the review is negative or positive, the employee should not be surprised. Third, do the employees feel that the reviews are fair and reasonable? Once again, the rumor mill is always in effect in every business. Employees will be asked, "How did it go?" Most employees, if they are given a structure to follow and are treated fairly, will report that the process was fair and positive.

9.15 Summary

Quality telecommunications professionals are difficult to find and often difficult to keep. Consequently, telecommunications managers should develop human resources skills and policies that address the specific demands of the telecommunications environment. Since many telecommunications managers are promoted or hired because of their technical skills, it is often a difficult transition for them to deal with the human resources issues that can have such a dramatic affect on department performance and stability. These are policies that need to be examined and continually refined so that the staff can perform optimally under reasonable policies and conditions.

10

Staff Training and Education

In the modern business world, there are many professions that require little educational enhancement throughout the course of a career. Perhaps the employees of a particular department might attend a short class every few years, adding small bits of information to their knowledge base. In such professions, the employees often feel that their schools days ended when they left high school or college. They assume that most job knowledge will come from experience and on-the-job training. Conversely, there are also many professions that involve complex subject matter, where change is constant, and continuous education an absolute necessity. Telecommunications is such a profession.

There are many telecommunications managers who place little emphasis on educational policies. Their modus operundi is to hire the most qualified candidate they can find when there is an opening. Otherwise, they often leave educational development up to the individual staff members. A telecommunications manager once told me, "Some staff members are industrious, and some are not. An educational policy won't change that situation. Education is up to the individual." These words probably sum up the philosophies of many telecommunications managers. The problem with this philosophy is that it is contrary to the manager's role of coach and mentor. It also ignores education as a continuing goal and objective of all telecommunications professionals.

10.1 Establishing an Educational Policy

The aforementioned quote, while a flawed philosophy, is also a true statement. I have often classified telecommunications professionals as "have-to-learn" and "want-to-learn" people. A have-to-learn person is the type who only takes initiative when presented with a challenge. These people learn about a technology when they are assigned a project, or charged with supporting a specific technology. As strange as it may sound, these people are still very intelligent, and can be a valuable part of the telecommunications staff. They simply lack intellectual curiosity or are not self-motivated. Surprisingly, I have seen many people rise to the occasion when issued a challenge, learning complex technologies very quickly.

The want-to-learn people are staff members who seek out knowledge, regardless of whether they have an immediate need for it. These people regularly subscribe to telecommunications magazines, proactively read books, and actively seek out classes and seminars. These people also regularly attend meetings held by professional telecommunications organizations and actively exchange information with colleagues. Within a telecommunications department, the want-to-learn people are more likely to move up within the organization or be recruited away. The reason for this is very simple. Staff members who sit in the middle to upper portion of the triangle are people who can think strategically. This means being able to project what technologies will be applicable for the future needs of the company, or any new applications that may require unfamiliar technology.

Regardless of what mix of people reside on the telecommunications staff, the telecommunications manager is responsible for setting many of the departmental policies. One of these policies certainly should be to establish an educational policy. Leaving education to each individual staff member does not address the overall needs of the department. While there may be staff members who actively seek to further their knowledge, they may go in directions that have little value for the department. Moreover, others will simply ignore education, only undertaking this endeavor when the need arises.

One of the first steps in establishing an educational policy is to make education a part of each employee's performance review. When each employee is hired, they are informed of the educational policy, which should be presented in writing. For example:

- Staff members are expected to attend a minimum of one telecommunications course each year. This does not include vendor specific training such as a router configuration course, or a PBX adds, moves, and changes class. More specifically, the course should educate the staff member about fundamental technology related to the field of telecommunications.

- XYZ Company encourages active participation in professional organizations. When seminars are available, and the project load will allow it, staff members are encouraged to attend such meetings and to openly communicate with colleagues in the telecommunications field.
- XYZ Company maintains a corporate telecommunications library, which contains books, periodicals, videotapes, audiocassettes, and computer-based training (CBT). It is open to all staff members who are encouraged to actively use the library. Staff members are also encouraged to actively maintain the library, which includes recommending books, periodicals, or research services that may benefit the entire department.
- Staff members are often assigned technologies for which they will become experts. While they may not be involved in a current project related to this technology, they are expected to maintain a core competency in this technology. For example IP, Internet, or ACD.
- Staff members are expected to educate other staff members when necessary. This includes formal classes or the informal exchange of information via a variety of media.

When staff members understand that education is heavily emphasized, and part of the performance review process, the entire departmental culture is affected. All staff members will more aggressively pursue education when they realize it is a mandatory goal and objective. This holds true for the "need-to-know" types of employees. These people normally need some form of motivation in order to pursue education. Education, as part of the performance appraisal process, provides that motivation.

There are many types of education that need to be addressed in the field of telecommunications, and the needs of various staff members are not strictly limited to technology. Included in the mix will be basic business skills such as accounting and managerial skills. There will also be project management, department management, basic supervisory skills, and team building skills. The most common types, however, will be telecommunications technologies. The telecommunications manager will want to develop a list of essential skills. This should include a master list for all positions, and an individual list specific to each skill. The lists will be divided into two basic categories: business skills and telecommunications skills. These skills and knowledge requirements will then be listed in the department job descriptions. As people are hired into the department, promoted, or reassigned, the manager will need to understand what educational deficiencies need to be addressed. For example, a PBX administrator may be promoted to the position of analyst. However, this person may have never performed financial analysis or been a project leader, although he holds very strong technical skills. In this instance, the telecommunications manager may want to augment on-the-job-training with a basic finance course and a project management course. This becomes a critical part of making the pyramid work (as defined in Chapter 3), because upward mobility is not only available, there is also a systematic method for providing the proper skill sets for staff members who have been promoted or assigned new duties. When staff members are promoted, there will always be a high degree of anxiety and apprehension about their new role. One of the worst things that can happen is to throw the newly promoted person to the wolves by not providing any guidance. Documenting a recommended list of educational resources that relate to the position's core competencies alleviates part of this dilemma. Newly promoted people can then read books and articles, or use CBT and videos during business hours or during leisure time. When the work schedule and budget allows, they will also be able to attend various courses.

The list of educational resources, cross-referenced by job description, also affords the opportunity for aspiring staff members to cross train for future jobs. Consider, once again, the example of the PBX technician. This person might have prepared for the position by proactively reading and using CBT or videos from the corporate telecommunications library. This was accomplished with the approval of the technician's supervisor, understanding that the employee has career objectives, and is taking positive steps to move up within the organization. At some future time, the same person may go to his supervisor to discuss moving from the voice-related position to one that is data oriented.

10.2 Education as Part of Everyday Culture

Education is more than attending classes a few times a year. It should be a process that is entrenched in the everyday existence of the telecommunications department. The staff members should understand, from the day they are hired, that education is an integral part of an employee's responsibilities. They will be expected to pursue knowledge and skills enhancement, both officially and unofficially. There should also be proactive actions by all staff members with supervisory responsibilities to maintain this policy.

Educational plans begin with very simple, but effective policies. For example, when every staff member is hired, they should be given a list of magazines and newspapers to subscribe to. Consider that subscriptions to most telecommunications periodicals are free. Each staff member should be encouraged to subscribe to as many free periodicals as they can. In addition, the department should have a file where they keep the free subscription forms for new employees. Corporate mail departments generally abhor the telecommunications department. Each staff member is often bombarded with stacks of magazines, newspapers, and advertisements on a daily basis. There is nothing that can be done about this because it is simply the nature of the beast. In fact, if staff members are not receiving scores of reading material, it should be cause for concern.

Any person who has visited the offices of a typical telecommunications department would probably be struck by the chaos of the environment. Part of this perception would be due to the stacks of magazines and newspapers that seem to flourish on every desk. Most of the periodicals would probably appear to be untouched, and this perception would undoubtedly be true. This is because few telecommunications professionals have the time to read every bit of information that is published in the field. Moreover, few professionals even have the time to tap a small percentage of what is published. If this is true of most telecommunications departments, what is the point of having staff members subscribe to so many periodicals? Isn't this policy self-defeating? Actually, there are a number of advantages to such a policy.

First, the staff members have copious educational material at their fingertips. Whenever there is a break in their schedule, they have a wealth of material to reference. Second, the simple act of cleaning off one's desk can be educational. I once handed a stack of subscription forms to one of my employees. He asked, somewhat cynically, if I really expected him to read all of the information. I told him that I readily recognized that he would only read a very small percentage of the periodicals. I also went on to say that if he, at a minimum, only read the headlines before he threw them away, he would at least understand what the trends were in the industry. If he read one or two articles a year, I had provided some education (however small) to this employee and had not spent a dime. A third advantage is that staff members have their own copies of educational material that they can use as their schedule allows. A fourth advantage is that there is safety in numbers. If everybody in the department receives the same magazine, the telecommunications manager has a greater chance of somebody reading an important article that may provide a timely answer in a staff meeting.

There are many telecommunications managers who will order department copies of magazines and place them on a universally accessible magazine rack. This certainly has its advantages, but experience shows that the periodicals seldom get read or they are often out of date, in addition to mysteriously disappearing.

Beyond free subscriptions is the issue of paid subscriptions. Certainly the budget has its limitations and, given the fact that there is so much free reading material on the market, paid subscriptions should be addressed more cautiously. Not every staff member will, or should, have paid subscriptions to all available magazines. Staff members should be asked to evaluate what magazines they feel are essential and supervisors should evaluate and approve each request on a case-by-case basis. While it is not a major issue that staff members do not read free periodicals, paid subscriptions that are unread are a waste of precious operating dollars. On the other hand, dismissing paid subscriptions completely may deprive the department of valuable information. As a rule of thumb, the telecommunications manager may want to provide one copy of each paid subscription per group (e.g., voice or data), and then give supervisors the authority to approve individual subscriptions on a case-by-case basis.

The issue of keeping up with reading is problematic with all telecommunications professionals. With so much material being published today, it is simply impossible for any telecommunications professional to read it all. If one even tried to attempt this, he or she would do little else. Working a full-time job that demands extra hours exasperates the situation. What then is the solution? Unfortunately, there is no solution for the entire problem. The modern telecommunications professional can only pick and choose, carefully selecting the material that he reads.

The telecommunications manager can address this problem in a number of ways. When staff meetings are held, members should be encouraged to offer ideas and suggestions. How do they manage their time in order to keep up with their reading? What periodicals do they find to be most valuable? Have you found a new periodical that nobody has seen before? Have you recently read an article that you would like to share with the staff? One staff member may suggest that he or she leafs through his magazines at the end of the day, tearing out the most relevant articles and placing them in a folder. At the end of the week, the folder is taken home and he may read some articles on a Saturday morning or perhaps a Sunday afternoon. Another staff member may recommend that he or she gains the most value from one particular magazine. There can be many more ideas, but the interactive sharing of ideas provides input for continuous improvement. Staff members find that while they cannot read everything, they can make time to choose the best and most relevant material to enhance their knowledge and skill sets.

As the previously mentioned examples illustrate, a dilemma that the modern telecommunications professional faces is trying to balance education against a demanding workload. With so much time consumed by critical projects, when can a person find time for education? One partial solution is to hold "Lunch and Learns." The lunch and learn is a concept that has gained a great deal of popularity in recent years and they can be approached in a number of ways. The concept is simple. The department gathers in a conference room at lunchtime for an hour of education, lunch is brought in, and a speaker makes a presentation.

The telecommunications manager may assign a staff member to deliver a talk about a new technology at a lunch and learn. The technology may be something that will be used in a forthcoming project, or an enhancement to a technology already used by the department. The talk may even be about an emerging technology that may have no current application within the department, but appears to be a significant trend in the industry. For example, voice over IP may not be deemed to be a high priority within the department. But this technology is receiving much publicity, and the cost effectiveness cannot be ignored. Consequently, the telecommunications manager wants to educate his staff and monitor the progress of the technology, which currently has technical limitations.

A second and more common method is to have a vendor make a presentation and also provide lunch. Typically, the vendor is a strategic business partner of the company. For example, the provider of all routers for the company will make a presentation about a new networking concept, and the new product line that will support it. The situation is win-win. The staff learns about the new technology, they receive a free lunch, and they have not been taken away from their normal duties. From the vendor's perspective, they have strengthened the knowledge base of their customer, and possibly their relationship with the telecommunications department, for a little time and the price of a nominal lunch. One cautionary note is that the telecommunications manager should refrain from having lunch and learns with vendors that are not business partners of the company. If the department has not purchased equipment or services, and is not sure if they will in the future, it is unethical to ask them to buy lunch for the staff, however nominal the price may be. Generally, lunch and learn sessions are well received by staff members. It makes learning fun and is a good way to promote group morale in a relaxed atmosphere. Regardless of the benefits, there are a few more cautionary notes. First, vendors should not feel as though they are being forced to provide lunch for the staff. It should be an activity that they view as a positive move to strengthen their relationship with their customer. The telecommunications manager should always ask if vendors would like to participate, but never order them to do so. The telecommunications manager should also strive to make the lunch inexpensive for the vendor. Pizza or sandwiches are usually adequate. Lastly, it is common sense that all staff members should be invited when feasible, even those who are not directly involved with the technology. Of course, this is all within reason. If the entire department is comprised of 75 people, supporting many disparate technologies, it is unreasonable and unethical to expect a vendor to provide lunch for the entire group.

The telecommunications manager should make time in staff meetings to discuss education. He or she certainly does not have to address this issue in every meeting, but the subject should be proactively addressed. For example, a staff member may have recently attended a course or seminar. The telecommunications manager may ask the staff member to provide a brief overview of the course, perhaps providing a handout with a brief outline. The telecommunications manager may also periodically ask staff members to provide copies of articles they have read. Throughout the course of the year, the forum in meetings should be open, encouraging ideas and exchanging information. When this type of atmosphere is promoted, it tends to have a snowball effect, as staff members proactively bring materials to meetings and exchange information. The atmosphere should not be limited to meetings. Staff members should be encouraged to copy articles and distribute them throughout the department.

10.3 The Corporate Telecommunications Library

The concept of a library conjures images of infinite rows of bookracks, magazines, and card racks that reference the Dewey Decimal system. A small contingent of employees keeps the racks full and orderly, headed by a librarian who must categorize the large volume of materials. This is certainly not the case of a corporate telecommunications library. In most cases, the library will only take up a few bookcases or file cabinets and will only require a minimum of administrative support.

The corporate telecommunications library should be in a central location, and all staff members should know where it is and how to use it. Each work should be cataloged and kept in a simple database located on a server that can be accessed by all staff members. A simple policy should be put into effect so that the staff has maximum use of the materials. For example, staff members may take any book out of the library for use during the day, but must sign out the book if they are taking it home. This can be done with a simple sign-out sheet that is taped to the bookshelf. The same procedure holds true for all other materials.

A corporate telecommunications library will be comprised of the following items:

- Textbooks;
- Telecommunications dictionary;

- Newspapers;
- Magazines;
- Technical reports;
- White papers;
- CBT;
- Video courses;
- System descriptions;
- Marketing materials;
- Catalogs.

Within the library there may be 30 or 40 books. This is dependent upon the size of the department and the array of technologies that are supported. The books can be textbooks related to basic technologies or specific to vendor products and services. Staff members should be encouraged to read the books and to offer their opinions about the quality of the work and whether it would benefit other staff members. They should also be encouraged to peruse catalogs to determine what works are available for potential future purchases.

Every telecommunications department should have a telecommunications dictionary. In fact, depending on the size of the department, there may be a need for three or four. A good telecommunications dictionary should be all encompassing with easy-to-understand definitions. If staff members are constantly frustrated because they cannot find information in the dictionary, the telecommunications manager may want to reevaluate the type of dictionary or consider buying a more updated version. Moreover, if the staff members cannot easily grasp the concept after reading the definition, this may also be cause for change. Regardless of how learned the staff members are, there will always be new terms and acronyms surfacing with which they will not be familiar. Consequently, the telecommunications manager should strive to provide as much reference material as possible for the staff. A good telecommunications dictionary is a solid beginning and becomes worth its weight in gold to the staff.

Newspaper and magazine subscriptions are generally free. As previously mentioned, each staff member should get their own copy, however, there should be a department copy that is archived for a period of one year. Since most staff members will probably throw their own copies away within weeks of receiving them, a department copy should be available if an issue or question surfaces regarding an article. At the end of each calendar year, an administrative support person should discard the old issues and start archiving for the new year. It should be noted that it is not necessary to enter each issue or individual articles in the computer database. Staff members will know that the newspapers and magazines are archived and can rummage through the stack when they have a need. It is important to keep a library of reference material, but it is not necessary to make the administration of the library a full-time job. Keeping the periodicals in neat stacks by title should be sufficient for most departments.

Technical reports can be provided via Internet services, CD-ROM, or hard copy. The department may subscribe to a service such as Faulkner Technical Reports or Data Pro. There may be a myriad of reports available via the service. One of the advantages of subscribing to a service that is provided via a computer medium is the convenience of storing vast quantities of information on a CD or dialing into the Internet. CDs should be stored in a centrally located server so that all staff members have access to the information. Typically, a subscription includes an updated CD each month. Administration of this service would simply be to replace the CD stored on the server each month. White papers, system descriptions, and marketing materials need to be categorized and placed in the database. When there is a question regarding a technology, staff members should be able to pull up the simple database, reference the available materials, and where it is stored. Marketing materials and catalogs do not necessarily need to be logged in the database, but should be categorized by subject and company for easy access.

An administrative person can be assigned to manage the library, or staff members can rotate this duty on a semi-annual or annual basis. The procedures should be documented so that the responsible person understands and adheres to the schedule and the library is kept up to date. The list of responsibilities can be simple. For instance:

- Catalog each magazine and newspaper in the proper stack on a weekly or monthly basis;
- Dispose of all archived magazines and newspapers at the end of the calendar year;
- Receive new books, videos, and CBT; assign corporate ID number, and enter into library inventory file;
- Replace CDs once a month in server;
- Store catalogs and vendor specific information in the appropriate files.

The corporate telecommunications library can be a tremendous resource for new staff, new projects, cross training, and knowledge enhancement. With a minimum of effort, the department can have a powerful tool that will make staff members more knowledgeable and the department more efficient.

10.4 Courses, Seminars, and Degrees

Regardless of how much a telecommunications manager tries to integrate education into the daily routine of the telecommunications department, the demands of the job will take precedence. Staff members can certainly take their own initiatives to learn at home, but this too has its limitations. As explained in Chapter 9, telecommunications professionals need to balance the demands of their profession with their personal lives. If a staff member is spending 9 to 10 hours a day in the office, and perhaps putting in a weekend or two, he or she will be hard pressed to find the extra time to take a CBT or video course in the office or at home. Moreover, he or she will probably not want to read a book or magazine that relates to work. Even if he or she wanted to, the demands of his family life will probably take precedence. There is often no substitute for getting away from the office and taking a telecommunications course where the staff member travels away from the office and is able to concentrate solely on learning.

There are numerous telecommunications courses offered today. They run the gamut from independent organizations that are not credited, to credited university programs. At this point the telecommunications manager needs to understand the difference, and make a distinction, between a course, a seminar, and accredited courses.

Either vendors or independent organizations can hold a seminar. It can address almost any aspect of telecommunications and are generally of short duration. A telecommunications course usually lasts between two days to one week, sometimes longer. A seminar only provides limited information about a particular aspect of telecommunications. The time ranges from a few hours to a day. For example, a staff member may attend a seminar about xDSL. The seminar may only approach the subject from a high level: what it is, how it works, what are the applications, and how much it costs. The staff member will walk away from this one-or two-hour seminar knowing only the basics of the technology. Conversely, if the staff member were to attend a four-day course, he would be expected to have a more in-depth knowledge.

There are a number of organizations that offer telecommunications courses. These tend to be rather expensive and so the telecommunications manager needs to take concrete steps in order to gain the most value from his precious educational budget. In fact, when bad economic times strike the company, one of the first expenses to be cut may be the educational budget. Therefore, the telecommunications manager must make the most of the budget when the monies are available. This includes trying to use the best educational organizations and the best of the courses offered from each organization. Catalogs are listed as being an essential component of the telecommunications library. Included in the mix should be catalogs of telecommunications courses. As new projects surface, and as staff members are promoted or need to be cross-trained, these catalogs can be a valuable source of information for the telecommunications manager. These courses should also be an essential part of employee developmental programs.

The basic telecommunications course is offered by an independent organization that specializes in telecommunications education. Once again, these courses can range from a few days to several weeks. Normally, upon completion of the course, the attendee is provided a certificate of completion. It should be noted that these certificates hold little value other than confirming that the employee attended the course. It does not mean that the material was actually learned or that it was even a very good course. The pace of the average telecommunications course is also normally very fast. Sometimes they are so fast-paced that many students are lost after the first day. While there is nothing that the telecommunications manager can do to change this situation, there are a number of steps that can be taken to make sure that the maximum advantage is gained from these courses.

First, the telecommunications manager should always understand the knowledge level of the person who is being sent to the course. If a person has no prior data communications experience, it will be self-defeating to send him or her to an advanced course about TCP/IP networking. Sending the staff member to such a course without the proper prerequisites will also frustrate the staff member. I once had a manager who handed me a book and told me to program a UNIX-based program that would be used for call detail recording (CDR). What he failed to understand was that I had had no prior UNIX training, so I stumbled through the program until I finally asked for UNIX training. Once I had this, the use of the program became much easier. Preparing a staff member for a course can be every bit as important as the actual attendance of the course. Perhaps the prerequisites can be satisfied by simply pulling material from the corporate library. The telecommunications manager might even have to send the staff member to a completely different course as a prerequisite. Regardless, the telecommunications manager should understand whether the staff member is prepared to take the course.

A second step is to require feedback from the staff after they attend a course. All staff members should be required to write a report about the course that they have just attended. The report should include a short synopsis, an outline of the course, and their evaluation of the course. Some standard and basic questions should be asked for all courses. For example:

- Was the course designed in a logical manner?
- Was the instructor knowledgeable?
- Did the course provide additional textbook material for independent reading, above the classroom material?
- Would you recommend that the department use this organization again?
- Would you recommend that the department use this specific course again?

The telecommunications manager may also want to have the staff member rate various aspects of the course on a scale, perhaps 1-5. The results of the reports should be kept on file in the library for future review whenever a staff member requests a course.

Beyond basic telecommunications courses, there are courses for certification and degree programs at many universities and colleges. Certification curriculums differ from basic telecommunications courses in that they require the student to pass a test, or a series of tests, before the certification is awarded. Upon completion, the student is often entitled to use the title next to his or her name. For example, CNE stands for certified network engineer. Certification courses can be quite involved and, in some instances, they last as a long as several years. Knowledge and proficiency are proven by the certificate which was awarded because the student passed a series of demanding tests.

There is a large difference between a person who has attended a basic telecommunications course and a person who is certified. The certified person has been given classroom instruction in addition to practical hands-on training. Also, part of the certification process often involves intensive problem-solving labs that must be passed before certification is awarded. Consequently, certified staff members are already acclimated to many of the problems that affect network design, operation, and problem solving. Therefore there is less of a learning curve when they are hired or challenged with new problems or applications. Certified people are not inexpensive and they have great value in the job marketplace.

In addition to certification programs, many colleges and universities are beginning to offer degree programs in telecommunications. These degrees range from undergraduate programs (2- or 4-year) to advanced degrees such as a masters program or a Ph.D. Telecommunications degrees are becoming more common, but telecommunications professionals who possess one of these degrees are still relatively rare. One of the great values of a university level program is that the student receives a well-rounded education. These programs cover a range of skills and technologies that are often only stumbled upon via on-the-job training during the course of a career. For example, a university level curriculum will cover computer languages, basic programming, telecommunications law, regulatory issues, voice communications, data communications, the public network, wireless communications, telecommunications department management, and emerging technologies to name but a few. When a person has such a well-rounded education, there is less need to learn a technology from scratch when a project surfaces or a question is asked. While the staff member may not have current knowledge of the technology in question, he or she will have a conceptual view, and also probably know how to approach the project. Other staff members may have to start from ground zero.

A staff member who embarks on a 4-year degree or a certification process takes on a momentous task, and sacrifices much of his personal life. For a person with a family, this means long hours away from home because of work and classes. When the staff member actually is at home, he or she is often inundated with homework and projects. The point is that these types of programs differ significantly from the basic telecommunications courses. The telecommunications manager should understand that staff members who participate in such programs are making sacrifices to further their career opportunities. Consequently, they cannot assume that this person will be content to stay in their present position or at their current salary level once the educational process is completed. Moreover, the educational institution will probably have a placement department that offers salary ranges and opportunities for people with similar degrees. Inevitably, this will be greater than what their present position offers. Upon completion of the degree or certification process, the telecommunications manager should reevaluate the role of the staff member and what their future is in the department. Of course, if there is no position available, there is nothing that the telecommunications manager can do. However, every effort should be made to bring the staff member in to a position and salary level that is commensurate with their educational level. If this is not done, there will be a morale problem that will certainly lead to turnover. I once had a colleague who made great sacrifices to obtain his master's degree. While the telecommunications manager knew that he was working toward a master's degree, he did nothing to use the employee's newly developed skills. Upon completion of his degree, it was not even acknowledged that the employee had earned an advanced degree. The employee felt angry and unappreciated. Needless to say, he found a new job within weeks.

10.5 Summary

A well-designed educational policy will work towards continually developing the knowledge base and professional skills of the staff. This includes planning educational goals and objectives for each staff member on an annual basis, establishing a corporate telecommunications library, and proactively encouraging education in every facet of department operations.

Glossary

Address Resolution Protocol (ARP) An Internet protocol used in IPv4 to map an IP address to a MAC address.

Asymmetric Digital Subscriber Line (ADSL) A copper-based access technology that allows cable TV, video, telephony, and other multimedia services to be carried over a single twisted-pair cable that can carry from 1.5M to 8M bps downstream, and between 16K and 450K bps upstream.

Asynchronous Transfer Mode (ATM) A cell-switching technique using the cell relay method of transmission using 53-byte fixed-size cells to provide high-speed (150 Mbps and higher) local and enterprise-wide WAN transport.

Automatic Call Distribution (ACD) A method of distributing calls to groups of employees known as agents. An ACD can be either a stand-alone system or an adjunct capability of a PBX. ACDs also supply reports of a real-time and historical nature.

Automatic Number Identification (ANI) A method by which the caller's telephone number (known as the CLID) is transmitted to the called destination. The CLID can be used to either identify the caller or can be stored in a database for future analysis.

Basic Rate Interface (BRI) A version of ISDN also known as 2B+D. There are two Bearer channels rated at 64 Kbps and the data channel which is rated at

16 Kbps. BRI is typically used for smaller applications that require either high bandwidth for data or simultaneous voice and data transmission.

Bearer Channel The B channel of an ISDN service equal to 64 Kbps which can be used to transmit voice, data, or video.

Call Detail Recording (CDR) A method by which telephone data is sent from a telephone system (either PBX or key system) to a computer system (typically a PC). The data is formatted via special CDR software into reports that monitor costs and employee telephone activity. CDR reports also provide telephone traffic information that is used to engineer trunk lines.

Caller ID (CLID) The telephone number of a calling party that is transmitted through the PSTN to the dialed destination.

Carrier Sense Multiple Access/Collision Detection (CSMA/CD) An LAN access technique in which multiple stations connected to the same channel are capable of sensing activity on a channel and deferring transmission while the channel is active.

Cellular Digital Packet Data (CDPD) A method of using existing cellular networks to offer remote and mobile computing. Based on OSI routing at 19.2 Kbps.

Central Office (CO) The local telephone company building that houses a telephone switch that serves all local customers for access to the PSTN.

Central Processing Unit (CPU) The main component of any computer or communications system that performs all processing functions.

Centrex A method of providing telephone system functionality through the CO. Telephone sets are connected to copper that is directly connected to the CO, which provides features such as transfer or conference.

Certified Network Engineer (CNE) A person that has undergone a certified educational program to design, maintain, and troubleshoot local area networks.

Class A/Dual Attached Stations (DAS) A device that is attached to both the primary and secondary rings of an FDDI.

CODEC A device that samples an analog sine wave and translates the signal into a digital format or vice versa.

Committed Information Rate (CIR) The transmission rate that is guaranteed by a carrier for frame relay service. If a customer subscribes to a 16 Kbps CIR, the transmission speed should go no lower than 16 Kbps, even if the carrier network becomes congested and frames are discarded.

Competitive Access Provider (CAP) A provider of local access that competes with the LEC. CAPs only provide dedicated services such as T-1, not dial tone or telephone numbers.

Competitive Local Exchange Carrier (CLEC) An LEC other than the incumbent local carrier. CLECs are relatively new and have limited access as compared to the LECs established by the old Bell System. CLECs offer dial tone and telephone numbers.

Computer Telephony Integration (CTI) A method of integrating a telephone system with a computer in order to achieve specific functions. For example the CLID may be captured by an ACD and cross-referenced through a database. The call and customer information can be offered to an agent simultaneously. This is also known as a screen pop.

Copper Distributed Data Interface (CDDI) See Fiber Distributed Data Interface (FDDI).

Customer Premise Equipment (CPE) Any communications device or switch owned or leased by the customer that connects to the PSTN.

Cyclic Redundancy Check (CRC) An error detection technique which generates a series of two 8-bit block check characters that represent the entire block of data. These block check characters are incorporated into the transmission frame and checked at the receiving end.

D Channel See Data Channel.

Data Channel The D Channel of ISDN. This channel is used for out-of-band signaling which allows for faster call setup time and the delivery of information such as the CLID. The data channel is 16 Kbps on ISDN BRI and 64 Kbps on ISDN PRI.

Data Link Connection Identifier (DLCI) The circuit number of a frame relay link. Each site on a frame relay network has a DLCI and network design and operation requires the inclusion of the DLCI into the routing tables in data communications devices.

Datalink Switching (DLSw) A method of transporting SNA and NetBios traffic reliably across a multiprotocol backbone.

Dialed Number Identification Service (DNIS) A method of identifying different toll-free numbers or different types of calls from individual toll-free numbers. For example, a customer may have two toll-free numbers that are directed to the same trunk group. Unique digits are forwarded from each toll-free number to the ACD that allow the system to differentiate the different types of calls.

Digital Subscriber Line (DSL) Generic name of a technology that provides high-speed access over existing voice lines. There are five variations: ADSL, HDSL, RDSL, SDSL, and VDSL. The whole group is known as xDSL.

Direct Distance Dialing (DDD) A method of dialing long distance directly through the PSTN, without using any form of special access. Typically, DDD rates are more expensive than dedicated access.

Direct Inward Dialing (DID) A method of dialing directly to a PBX station without the assistance of an operator. The end-user subscribes to DID service by purchasing a block of telephone numbers and special DID trunk lines. There must also be a DID trunk card in the PBX to accept the digits from the LEC.

Divestiture A term commonly used to describe the breakup of AT&T on January 1, 1984. AT&T retained Bell Laboratories, Western Electric, and its long distance network. The RBOCs were divested from AT&T and established as independent companies.

Domain Name System (DNS) A system used in the Internet for translation of names of network nodes into IP addresses.

Electronic Data Interchange (EDI) The electronic exchange of trading documents, such as invoices and orders.

Electronic Tandem Network (ETN) A private network of voice grade tie lines that interconnects a company's telephone systems. A customer could save money by placing most calls over tie lines to reduce DDD rates. An additional feature was an independent dialing system that allowed for four or seven digit long distance calling to on-net sites.

Facilities & Demand (F&D) A chart used by telecommunications professionals to chart the capacity of telecommunications systems against the current demand.

Federal Communications Commission (FCC) Regulatory agency, located in Washington, DC, established by the Communications Act of 1934, charged with regulating all electrical and radio communications in the U.S.A.

Fiber Distributed Data Interface (FDDI) A LAN standard specifying a LAN-to-LAN backbone for transmitting data at 100 Mbps speeds over fiber optic (or copper) media.

Field Engineer (FE) A technician employed by telecommunications vendors, specifically trained in the hardware and software of CPE. An FE is dispatched for installations, upgrades, and trouble calls.

Foreign Exchange (FX) A type of dedicated private line that connects a telephone system with a distant CO. The customer is charged for dedicated circuit, however, a call made to the city of the distant CO will only be charged at local rates.

Frame Relay Access Device (FRAD) A data communications device used specifically for connectivity to frame relay circuits.

High Bit Rate Digital Subscriber Line (HDSL) Part of the family of DSL technologies, HDSL gives network or Internet access over existing voice lines at 1.5 Mbps on T1 lines with 2 pairs combined.

Incumbent Local Exchange Carrier (ILEC) The legacy LEC that was established through the old Bell System.

Independent Telephone Company (ITC) A local telephone company that was not owned by the old Bell system, but was still completely interconnected to the PSTN. ITCs did not compete on a local basis with the old Bell system the way that a CLEC does today.

Institute of Electrical and Electronics Engineers (IEEE) An organization of engineering professionals that defines many standards within the telecommunications industry such the 802 series of standards for LANs.

Integrated Services Digital Network (ISDN) A family of transmission technologies which includes BRI, PRI, and Broadband. All ISDN service is structured to have Bearer and Data channels. Bearer channels can carry voice, data, or video. Data channels are used for out-of-band signaling or data transmission.

Interactive Voice Response (IVR) A method of using telecommunications technologies to provide information to an end-user. For example, a customer may call a toll-free number provided by a bank. The caller is prompted to enter account information via a touch tone phone (or verbally). Account balances are then provided to the caller.

Interexchange Carrier (IXC) A long distance carrier that provides services via their own network. AT&T, MCI and Sprint are all examples of IXCs.

Intermediate Distribution Frame (IDF) A smaller distribution frame, perhaps a small communications closet, that is used as a distribution point for riser or horizontal cable.

International Telecommunications Union (ITU) The international standards organization that defines telecommunications standards for interconnecting services and devices on an international basis.

Internet Packet Exchange (IPX) A widely used routing protocol, based on Xerox's XNS, developed by Novell.

Internet Protocol (IP) The Network Layer protocol of the TCP/IP suite including the ICMP control and error message protocol.

Internet Service Provider (ISP) A vendor that specializes in providing Internet services including access and DNS.

Inverse Multiplexer (IMUX) A device that bonds data channels for higher bandwidth capacity. For example, an IMUX will bond the two B channels of a BRI for an aggregate rate of 128 Kbps for videoconferencing applications.

Key Telephone System (KTS) A smaller, less sophisticated telephone system than a PBX. Typically, the end-user will choose a button to seize an outside line. Key systems have been becoming more sophisticated in recent years to the point where they often rival PBXs.

Least Cost Routing (LCR) The capability of a telephone system to route calls to a number of paths, choosing the most cost effective first, and then overflowing the call the next most cost effective path.

Local Access and Transport Area (LATA) The geographic area that is defined as a service area for the LEC. Once call crosses an LATA boundary, it is considered to be a long distance call.

Local Area Network (LAN) A high capacity data network contained within a small geographic area that relies on a specific transmission method. LANs have unique topologies and methods for error detection and flow control.

Local Exchange Carrier (LEC) A local telephone company that provides local services and various methods of access.

Logical Link Control (LLC) A protocol developed by the IEEE 802 for data link-level transmission control.

Main Distribution Frame (MDF) The wiring frame that is the central point of distribution for cable in either a building or campus. The MDF is typically where entrance cable provided by the LEC is terminated. The MDF is usually located in the main telephone closet.

Management Information Systems (MIS) The reporting system of an ACD. The MIS offers historical reports for a number of factors, agents, ACD groups, and trunk lines.

Media Access Control (MAC) The method that enables network stations to access network media and transmit information. This corresponds to the second layer of the OSI model.

Metropolitan Area Network (MAN) A network, usually data, which interconnects sites located within a metropolitan area.

Modem A device used by computers to transmit over analog telephone lines. Modems have become very common with the advent of the Internet and portable PCs.

Multipoint Control Unit (MCU) A device that allows for multiple (three or more) connections in a videoconference. MCUs have dial-out or dial-in capabilities and can perform protocol conversions.

Multistation Access Unit (MAU) A hub that used to connect Token Ring stations.

Music On Hold (MOH) The concept whereby a port on a telephone system is connected to a music source (such as a CD player) to provide music when a calling party is placed on hold.

Network Information Center (NIC) A central group responsible for assigning network IDs for the Internet. There are three classes of IP licenses; A, B, and C. The classes of licenses relate to the size of the network where the addresses will be used. A network ID is assigned to a specific class. The user is then responsible for assigning the host ID within the network ID.

Network Interface Card (NIC) The circuit card that plugs into a PC and allows dedicated LAN access. NIC cards are specific to various types of LANs such as 10BASET.

Network Operating System (NOS) Software that usually resides on a server within an LAN. The NOS provides a number of capabilities for managing the PCs, which includes security, permission levels, and management reports.

North American Numbering Plan (NANP) The 10-digit sequence of digits used in North America for dialing locally (seven digits) or long distance (ten digits). The NANP standard consists of an NPA, an NXX, and a four digit subscriber code.

Number Portability (NP) The ability to "port" a number from one carrier to another. Local and toll-free numbers can be ported. With portablilty, an end-user can change carriers, but not have to change telephone numbers.

Numbering Plan Area (NPA) Also known as an area code, this is the first three digits of the ten-digit standard of the NANP. An NPA covers a large number of subscribers. The NNX narrows the focus to a specific CO, and the remaining four digit subscriber code relates to individual trunk or line.

Off Premise Extension (OPX) A dedicated private line that connects a PBX with a telephone located off premise. The telephone connected to the OPX line would have full functionality of the host PBX.

Open Systems Interconnection (OSI) The ISO Reference Model, a seven-layer network architecture used for the definition of network protocol standards. This enables all OSI-compliant computers or devices to communicate with each other.

Other Equipment Manufacturer (OEM) The process whereby a vendor contracts another vendor to manufacture equipment but brands it as their own. OEM equipment is very common in the telecommunications industry.

Packet Assembler/Dissasembler (PAD) A protocol conversion device and/or program that permits end-users to access a packet switched network.

Permanent Virtual Circuit (PVC) A circuit that is defined in software, but does not have a dedicated physical connection. The concept is used in frame relay where one site will direct all transmissions exclusively to a second, even though there is no physical connection. The connection is established via network addresses.

Personal Communications Service (PCS) A form of wireless digital communication in the 900 MHz range. PCS networks are still relatively new and not as ubiquitous as analog cellular.

Plain Old Telephone Service (POTS) An analog line provided through the LEC that includes dial tone and a telephone number.

Point of Presence (POP) The long distance switch of an IXC. When a long distance call is made, the call is first routed through the LEC, which will have a number of trunks dedicated to each long distance carrier's POP.

Primary Interexchange Carrier (PIC) The method of selecting a long distance carrier at the LEC CO. When a customer "PICs" a carrier, all long distance calls are routed to the chosen IXC POP.

Primary Rate Interface (PRI) A form of ISDN that includes 23 B channels and one D channel. Each B channel is 64 Kbps. The D channel on PRI is 64 Kbps.

Private Automatic Branch Exchange (PABX) Another name for a PBX.

Private Branch Exchange (PBX) A telephone system that consists of telephones, trunks, lines, a CPU, a switching matrix, and controlling software. PBXs are feature rich, offering capabilities to enhance employee productivity or to help manage network costs.

Public Switched Telephone Network (PSTN) The PSTN is the summation of all carriers and switches that are interconnected and communicate via standard addressing and communications protocols.

Public Utility Commission (PUC) A regulatory organization at the state level that oversees issues relating to utilities. Included in this mix are local telephone companies. The PUC has authority over rates and service issues.

Regional Bell Operating Company (RBOC) A term that refers to the original seven Regional Bell companies that were part of the old Bell system.

Regional Holding Company (RHC) A term that refers to the original seven Regional Bell companies that were part of the old Bell system.

Remote Network Monitoring Specification (RMON) A standard for network remote monitoring, used by telecommunications professionals to monitor network activity and troubles.

Request for Information (RFI) A document used by telecommunications professionals to gather information about telecommunications products and services. The RFI is often used as a precursor to the RFP process.

Request for Proposal (RFP) A document used by telecommunications professionals to evaluate vendors and telecommunications products and services. The RFP has specific sections for technical information, pricing, and vendor information. It is submitted to potential vendors who are asked to respond to the questions, sections, and specific format of the RFP. Decision matrices are developed to bring structure to the decision-making process.

Request for Quote (RFQ) A document used by telecommunications professionals to gather information about the specific pricing of a telecommunications product or service.

Responsible Organization (**RespOrg**) Assigns toll-free numbers to a customer or client, and this assignment is often registered with national directories. A RespOrg can actually be any type of organization. They are registered with the services management system 800 (SMS/800) and pay a fee for the right to access the national database. Typically, a RespOrg will be an IXC.

Routing Information Protocol (**RIP**) A TCP/IP link-state protocol, which supports the exchange of information between hosts and gateways.

Service Provider (**SP**) Any carrier that provides a form of service (whether voice or data) from the PSTN.

Shielded Twisted Pair (**STP**) Communications cable that is encased in a protective sheath to protect transmissions from outside electromagnetic interference.

Simple Network Management Protocol (**SNMP**) A protocol used for managing TCP/IP networks, internetworked LANs, and packet switched networks.

Single Pair Symmetrical Services Line (**SDSL**) Part of the DSL family, SDSL provides access over existing voice lines at 1.5 Mbps asymmetric or 768 Kbps, both ways over 8,000 feet distance.

Small Office Home Office (**SOHO**) The concept of small or personal businesses that require small but robust telecommunications services. ISDN BRI (2B+D) is a service that is ideal for SOHO applications because of its ability to carry simultaneous voice and data transmissions.

Spanning Tree Algorithm (**STA**) IEEE 802.1D committee standard for bridging LANs.

Switched Multi-Megabit Data Service (**SMDS**) A packet-switching data service which provides wide area data transport speeds up to 45 Mbps. SMDS offers group addressing, connectionless transport, and strict packet delay objectives.

Synchronous Optical NETwork (**SONET**) A high-speed fiber optic transport network with transmission rates ranging from 51.84 Mbps to 2.5 Gbps.

Time Division Multiplexing (TDM) A method of obtaining a number of channels over a single path by dividing the path into a number of time slots and assigning each channel its own repeated time slot.

Transport Control Protocol/Internet Protocol (TCP/IP) A protocol suite for networking and internetworking that occupies the middle layers (3 and 4) of the OSI Reference Model.

Uninterruptable Power Supply (UPS) A system that employs batteries or a diesel generator to provide continuous power in the event of a failure.

Unshielded Twisted Pair (UTP) Communications cable that is not encased in a protective sheath to protect transmissions from outside electromagnetic interference.

Value Added Network (VAN) A packet switching network capable of providing protocol conversions.

Virtual LAN (VLAN) An LAN network that is connected logically (i.e., software) rather than physically.

Virtual Private Network (VPN) A network that does not have any physical connections. Connectivity is achieved via software programming and addressing.

Wide Area Network (WAN) A corporate data network comprised of various data communications services and a CPE that connects geographically dispersed sites to one or more computers or servers.

Wide Area Telecommunications Service (WATS) A bulk discount, usage sensitive, long distance service developed and marketed by the old Bell system. Many IXCs also marketed WATS after divestiture, but the name is not used very often today.

Wireless Private Branch Exchange (WPBX) A PBX that uses wireless transmission technology in the 900 MHz range to connect telephones in lieu of copper wire.

List of Acronyms and Abbreviations

ACD automatic call distribution

ADSL asymmetric digital subscriber line

AOL America Online

ANI automatic number identification

ARP address resolution protocol

ATM asynchronous transfer mode

AWG American Wire Gauge

B Channel bearer channel

BRI basic rate interface

CAD/CAM computer-aided design/computer-aided manufacturing

CAP competitive access provider

CBX computerized branch exchange

CCIA Computer Communications Industry Association

CDR call detail recording

CDPD cellular digital packet data

CDDI copper distributed data interface

Centrex central exchange

CIR committed information rate

CLEC competitive local exchange carrier

CLID caller ID

CNE certified network engineer

CODEC coder/decoder

CO central office

CPE customer premise equipment

CPU central processing unit

CRC cyclic redundancy check

CSMA/CD carrier sense multiple access/collision detection

CTI computer telephony integration

DAS class A/dual attached stations

DBX digital branch exchange

D Channel data channel

DDD direct distance dialing

DID direct inward dialing

DLCI data link connection identifier

DLSw datalink switching

DNIS dialed number identification service

DNS domain name system

DSL digital subscriber line

DS digital signal

EDI electronic data interchange

EPSCS enhanced private switched communications services

EIA/TIA electronics industries association/telecommunications industry association

ETN electronic tandem network

FAQ frequently asked question

FCC Federal Communications Commission

F&D facilities & demand

FDDI fiber distributed data interface

FE field engineer

FHSS frequency hopping spread spectrum

FPS frames per second

FRAD frame relay access device

FX foreign exchange

Gbps giga bits per second

GHz giga hertz

HDSL high bit rate digital subscriber line

IDDD international direct distance dialing

IDF intermediate distribution frame

IEEE Institute of Electrical and Electronics Engineers

ILEC incumbent local exchange carrier

IMUX inverse multiplexer

IP internet protocol

IPX internet packet exchange

ISDN integrated services digital network

ISP internet service provider

ITAA Information Technology Association of America

ITC independent telephone company

ITU International Telecommunications Union

IVR interactive voice response

IXC interexchange carrier

Kbps kilo bits per second

KSU key service unit

KTS key telephone system

LAN local area network

LATA local access and transport area

LCRt least cost routing

LEC local exchange carrier

LED light emitting diode

LEOS low earth orbiting satellite

LLC logical link control

MAC media access control

MAN metropolitan area network

MAU multistation access unit

Mbps mega bits per second

MCU multipoint control unit

MDF main distribution frame

MEOS medium earth orbiting satellite

MFJ modified final judgment

MIS management information systems

Modem modulator/demodulator

MOH music on hold

NANP North American Numbering Plan

NIC network interface card

NIC network information center

NOS network operating system

NPA numbering plan area

NP number portability

NSI Network Solutions, Inc.

OC optical carrier

OCC other common carrier

OEM other equipment manufacturer

OPX off premise extension

OSI open systems interconnection

PABX private automatic branch exchange

PAD packet assembler/dissasembler

PBX private branch exchange

PCS personal communications service

PHY physical protocol

PIC primary interexchange carrier

PMD physical medium dependent

POP point of presence

POTS plain old telephone service

PRI primary rate interface

PSTN public switched telephone network

PUC public utility commission

PVC permanent virtual circuit

RBOC regional bell operating company

RespOrg responsible organization

RFI request for information

RFP request for proposal

RFQ request for quote

RHC regional holding company

RIP routing information protocol

RMON remote network monitoring specification

SAS class B/single attached stations

SCC specialized common carrier

SDN software defined network

SDSL single pair symmetrical services line

SMDS switched multi-megabit sata service

SMS/800 Services Management System 800

SNA systems network architecture

SNMP simple network management protocol

SOHO small office home office

SONET synchronous optical network

SP service provider

STA spanning tree algorithm

STP shielded twisted pair

TCP/IP transport control protocol/internet protocol

TDM time division multiplexing

UPS utninterruptable power supply

UTP unshielded twisted pair

VAN value added network

VLAN virtual LAN

VPN virtual private network

WAN wide area network

WATS wide area telecommunications service

WPBX wireless private branch exchange

XNS Xerox Networking Services

Selected Bibliography

Alexander, S., "1998 Outlook: IT Salaries Will Continue to Soar," *Infoworld*, November 17, 1997.

Alexander, S., "Managing IT Turnover," *Infoworld*, July 6, 1998.

Alexander, S., "Where Will the Opportunities Be?" *Infoworld*, November 24, 1997.

"AT&T Frame Relay Services," *Datapro Reports—Communications Networks Services*, April, 1998.

"Communications Managers Survey," *Teleconnect*, October, 1994.

"The Computer in the 21st Century," *Scientific American*, 1995.

Bates, B. J., and D. Gregory, *Voice and Data Communications Handbook*, New York: McGraw-Hill, 1998.

Black, U., *Emerging Communications Technologies*, Upper Saddle River, NJ: Prentice Hall, 1997.

Bonvissuto, S., "Justifying Integrated Network Management," *Networking Management*, January, 1993.

Boothroyd, D., "Wireless Communications Technologies and Standards," *Faulkner Information Services*, April, 1998.

Brickates, E. V., "Tapping into Training," *Network World*, February 16, 1998.

Bruno, C., "Hire Learning," *Network World*, August 18, 1997.

Bruno, C., "Party On! Net shops are hosting luncheons and hitting the rapids to reward team players and high performers," *Network World*, November 3, 1997.

Caldwell. B., "Waste is Expensive," *Information Week*, September 22, 1997.

Caruso, B., "Soft Skills Can Be Hard For Tech Managers," *Information Week*, May 11, 1998.

Chan, V. W. S., "All-Optical Networks," *Scientific American*, September, 1995.

Chiaramonte, J. and Budway, J. N., "Finding and Keeping Good People," *Telecommunications*, February, 1991.

Cone, E., "Staffing: Short Supply," *Information Week*, November 3, 1997.

Corcoran, C. T., "How to Find an IT Management Job," *Infoworld*, December 1, 1997.

Costello, R., "Central Office Switches: Overview," *Datapro Reports—Managing Voice Networks*, April, 1998.

Csenger, M., "Getting Tough About Time Off," *Network World*, October 27, 1997.

Daggatt, V., "Satellites for a Developing World," *Scientific American*, September, 1995.

Dertouzos, M. L., "Communications, Computers and Networks," *Scientific American: The Computer in the 21st Century*, 1995.

Devoe, D., "Interns Can Bring Relief to IT Labor Shortage," *Infoworld*, December 15, 1997.

Dickenson, S., "Videoconferencing: Hard Sell, Soft Dollars," *Data Communications*, May, 1996.

Dix, J., "Unsolicited Management Advice," *Network World*, November 3, 1997.

Dooley, B. J., "Internet Infrastructure," *Faulkner Information Services*, April, 1998.

Doster, S., "Career Shock: How to Avoid It," *Teleconnect*, September, 1992.

Duffy, T., "Hangin' at the High Tech Playground," *Network World*, March 16, 1998.

Duffy, T., "Write Your Ticket," *Network World*, September 1, 1997.

Dyer, F., et al., "Networking as a Career Option," *Management 899, Telecommunications and Networking*, March 11, 1998.

Emerging Broadband Technologies, Business Communications Review, August 7–9, 1995.

Fickel, L., "Watching the Clock," *Network World*, June 1, 1998.

Frank, H., "Empowering Employees Through Networking," *Networking Management*, October, 1992.

Frank, H., "Moving to Customer-Driven Network Management," *Networking Management*, June, 1993.

Gable, R. A., *Inbound Call Centers: Design, Implementation, and Management.* Norwood, MA: Artech House, Inc., 1993.

Gable, R. A., *Toll Free Services: A Complete Guide to Design, Implementation, and Management.* Norwood, MA: Artech House, Inc., 1995.

Gallant, J., "The Seven Habits of Highly Effective Net Managers," *Network World,* February 23, 1998.

Gantz, J., "Network Managers Move Into The Field," *Networking Management,* October, 1992.

Gantz, J., "Surviving the Re-engineering Revolution," *Networking Management,* January, 1993.

Gascoyne, K. L., and C. A. Skvarla, "Toll-Free Services: Overview," *Datapro Reports —Communications Networking Services,* April 1998.

Gates, B., *The Road Ahead,* New York: Viking Penguin, 1995.

"Globalization—The Converging World," *New World—The Siemens Magazine,* January, 1998.

Gore, A., "Infrastructure for the Global Village," *The Computer in the 21st Century, Scientific American,* 1995.

Green, J. H., *The Irwin Handbook of Telecommunications,* Chicago: Irwin Professional Publishing, 1997.

Green, J. H., *The Irwin Handbook of Telecommunications Management,* Chicago: Irwin Professional Publishing, 1996.

Green, J. H., *Telecommunications Projects Made Easy,* New York: Flatiron Publishing, Inc., 1997.

Harler, C., "Fishing in the Talent Pool," *Business Communications Review,* May, 1997.

Harler, C., "How Does Your Staff Size Up?" *Business Communications Review,* December, 1997.

Herick, C. N., and C. L. McKim, *Telecommunications Wiring,* Englewood Cliffs, NJ: Prentice Hall, 1992.

Herman, J., "Network Management Is the Age of the Information Superhighway," *Business Communications Review,* August, 1994.

Horak, R. and M. A. Miller, *Communications Systems & Networks,* New York: M&T Books, 1997.

Hulme, G., "How to Conduct A Review," *Network World,* August 11, 1997.

"Installing and Maintaining a Premise-Based Voice Wiring System," *Datapro,* June, 1995.

Isaacs, N., "Use Job Interviews to Evaluate 'Soft' Skills," *Infoworld*, April 6, 1998.

Jacobs, P., "IT: Trade or Profession?" *Infoworld*, June 29, 1998.

Johnner, M., "How Local Number Portability Can Foster Competition," *Telecommunications*, February, 1998.

Kalinoglou, J., and J. R. Lilly, "Wireless LAN Technologies," *Faulkner Information Services*, March, 1998.

Kaufmann, A. J. S., "Helping to Build Careers," *Infoworld*, July 13, 1998.

Kay, A. C., "Computers, Networks, and Education," *The Computer in the 21st Century, Scientific American*, 1995.

Kennedy, M. D., "The Strategic Use of Telecommunications: Lessons Learned and the Path Ahead," *Telecommunications*, January, 1996.

Knight, F., "Growing Momentum for Global Communications," *Business Communications Review*, December, 1995.

Laino, J., "End-user Training," *Teleconnect*, June 1998.

Laino, J., "Reporting to Management," *Teleconnect*, July, 1998.

Laube, S., "Slow Down: Get Back to Management Basics," *Internet Week*, December 1, 1997.

Leibmann, L., "Too Much Information," *Network World*, June 15, 1998.

Lenat, D. B., "Artificial Intelligence," *Scientific American*, September, 1995.

Leong, K. C., "Help Wanted? Look Online," *Internet Week*, December 22, 1997.

Levine, H. D., et al., "The Telecommunications Act of 1996," *Datapro Reports—Managing Voice Networks*, April, 1998.

Lewis, B., "Recognizing Employees as Individuals Is a Good Way to Inspire Loyalty," *Infoworld*, October 27, 1997.

Lewis, B., "Used Strategically, Money Can Really Talk to Motivate and Reward Employees," *Infoworld*, October 13, 1997.

Lucky, R. W., "Computers and Society in the Future," *AT&T Technology*, Volume Seven, Number Two, 1992.

Malone, T. W., and J. F. Rockart, "Computers, Networks, and the Corporation," *The Computer in the 21st Century, Scientific American*, 1995.

Martin, J., *Local Area Networks*, Englewood Cliffs, NJ: Prentice-Hall, 1994.

Martin, J., *Telecommunications and the Computer*, Englewood Cliffs, NJ: Prentice-Hall, 1990.

Maes, P., "Intelligent Software," *Scientific American*, September, 1995.

Miller, S., "Videoconferencing," *Faulkner Technical Services*, February, 1998.

Mohan, S., "Leading an IT Team to Success," *Infoworld*, December 8, 1997.

Muller, N. J., *Desktop Encyclopedia of Telecommunications,* New York: McGraw Hill, 1998.

Muller, N. J., and R. P. Davidson, *The Guide to Frame Relay & Fast Packet Networking*, New York: Telecom Library, 1991.

Negroponte, N. P., "Products and Services for Computer Networks," *The Computer in the 21st Century, Scientific American*, 1995.

Newton, H., *Newton's Telecom Dictionary*, New York: Flatiron Publishing, Inc., 1997.

Network Technology Program—Graduate School of Industrial Administration, Carnegie Mellon University, January 16-February 6, 1997.

Olson, A., "Don't Be Last to Put People First," *Communications Week*, August 4, 1997.

Olson, A., "Taking IT Risks Can Bring Career Rewards," *Internet Week*, December 15, 1997.

Paul, L. G., "From Bad to Worse," *Network World*, May 25, 1998.

Patrowicz, L., "Rx for Burnout," *Network World*, October 20, 1997.

Patterson, D. A., "Microprocessors in 2020," *Scientific American*, September, 1995.

Pelton, J. N., "Telecommunications for the 21st Century," *Scientific American*, April, 1998.

Peterson, K., "Communications Managers' Survey," *Teleconnect*, October, 1996.

Peterson, K., "1997 Comms Managers Survey," *Teleconnect*, October, 1997.

Peterson, K., "What Can Voice Managers Offer Data Departments?" *Teleconnect*, July, 1997.

Prencipe, L., "Keeping to the Letter of the Law," *Network World*, August 4, 1997.

Prencipe, L., "The Real Cost of A Bad Hire," *Network World*, September 1, 1997.

Prencipe, L., "Use the 'Show Me' Interview," *Network World*, December 8, 1997.

Rennie, J., "The Uncertainties of Technological Change," *Scientific American*, September, 1995.

Rosenberg, J. R., *Dictionary of Computers, Data Processing & Telecommunications*, New York: John Wiley & Sons, Inc.

Siwolop, S., "The Skills Factor," *Information Week*, May 11, 1998.

Skvarla, C. A., "Wide Area Telecommunications Service (WATS) in the U.S.: Overview," *Datapro Reports—Communications Networking Services,* April, 1998.

Smith, M. H., "Virtual Private Network (VPN) Services: Overview," *Datapro Reports—Communications Networking Services,* April, 1998.

Soloman, A. H., "Telecommunications in the 1990s—Evolving Technologies (Part 1)," *Telecommunications,* January, 1990.

Soloman, A. H., "Telecommunications in the 1990s—Managing Networks to Serve User Needs (Part 2)," *Telecommunications,* February, 1990.

"Special Report: 1995 Salary Survey," *Datapro Worldwide IT Analyst,* July, 1996.

Sproull, L., and S. Kiesler, "Computers, Networks, and Work," *The Computer in the 21st Century, Scientific American,* 1995.

Steen, M., "Are You Getting A Fair Shake?" *Infoworld,* June 15, 1998.

Steen, M., "Finding and Keeping Good Support Staff," *Infoworld,* October 27, 1997.

Steen, M., "Job Satisfaction is More Than Money and Status," *Infoworld,* October 27, 1997.

Steen, M., "Networking's Visibility Increases," *Infoworld,* October 27, 1997.

Stevens, L., "Chill Out, Don't Burn Out," *Communications Week,* August 4, 1997.

Stevens, L., "Reassessing the Role of Recruiters," *Internet Week,* November 3, 1997.

Stix, G., "Domesticating Cyberspace," *The Computer in the 21st Century, Scientific American,* 1995.

"Structured and Unstructured Cable Plans," *Faulkner Technical Reports,* April, 1998.

Tadjer, R., "Empowered Staff: A Win-Win," *Internet Week,* September 15, 1997.

Tadjer, R., "Finding Work That Works," *Communications Week,* August 18, 1997.

"Telecom Manager Report," *Teleconnect,* November, 1990.

Tesler, L. G., "Networked Computing in the 1990s," *The Computer in the 21st Century, Scientific American,* 1995.

Trivett, D., "Basic Concepts of Communications," *Datapro Reports—Voice Networking Systems,* April, 1998.

Trowt-Bayard, T., *Videoconferencing: The Whole Picture,* New York: Flatiron Publishing Inc., 1997.

Turek, N., "IT Professionals Learn to Weigh Their Options," *Information Week,* June 29, 1998.

Waltz, M., "Closing the Door on Departures," *Network World,* April 6, 1998.

Ware, B. L., "Recognizing Discontent," *Network World*, August 18, 1997.

Weiser, M., "The Computer for the 21st Century," *The Computer in the 21st Century, Scientific American*, 1995.

Whitaker, J., and H. Winard, *The Information Age Dictionary*. Overland Park, KS: Intertec Publishing, 1992.

Wreden, N., "Staff Up Today for Tomorrow," Internet Week, September 22, 1997.

York, T., "Business Specialists Needed," *Infoworld*, April 6, 1998.

Ziegler, K., Jr., "The Enterprise Network—Who's in the Driver's Seat?" *InterNet*, June, 1992.

Zysman, G. I., "Wireless Networks," *Scientific American*, September, 1995.

About the Author

Robert A. Gable has been in the field of communications for 16 years, starting as a salesman for a major IXC in 1983. He became a senior telecommunications consultant with Ernst & Whinney (now Ernst & Young) and then worked for Kennametal Inc (a Fortune 1000 company), supporting their corporate telecommunications infrastructure both nationally and internationally. Mr. Gable is the author of two telecommuniactions books*: Inbound Call Centers: Design, Implementations and Management* and *Toll-Free Services: A Complete Guide to Design Implementation and Management.* His writing has also appeared in many industry publications such as the *Business Communications Review*, *Telecommunications Magazine*, and *Datapro.* He has also been a guest speaker at many national telecommunications conferences and is currently working for Hyperiorn Telecommunications. Mr. Gable lives with his wife and two children in Monroeville, Pennsylvania.

Index

Recent Titles in the Artech House Telecommunications Library

Vinton G. Cerf, Senior Series Editor

Access Networks: Technology and V5 Interfacing, Alex Gillespie

Achieving Global Information Networking, Eve L. Varma, Thierry Stephant, et al.

Advanced High-Frequency Radio Communications, Eric E. Johnson, Robert I. Desourdis, Jr., et al.

Advances in Telecommunications Networks, William S. Lee and Derrick C. Brown

Advances in Transport Network Technologies: Photonics Networks, ATM, and SDH, Ken-ichi Sato

Asynchronous Transfer Mode Networks: Performance Issues, Second Edition, Raif O. Onvural

ATM Switches, Edwin R. Coover

ATM Switching Systems, Thomas M. Chen and Stephen S. Liu

Broadband Network Analysis and Design, Daniel Minoli

Broadband Networking: ATM, SDH, and SONET, Mike Sexton and Andy Reid

Broadband Telecommunications Technology, Second Edition, Byeong Lee, Minho Kang, and Jonghee Lee

Client/Server Computing: Architecture, Applications, and Distributed Systems Management, Bruce Elbert and Bobby Martyna

Communication and Computing for Distributed Multimedia Systems, Guojun Lu

Communications Technology Guide for Business, Richard Downey, Seán Boland, and Phillip Walsh

Community Networks: Lessons from Blacksburg, Virginia, Andrew Cohill and Andrea Kavanaugh, editors

Computer Mediated Communications: Multimedia Applications, Rob Walters

Computer Telephony Integration, Second Edition, Rob Walters

Convolutional Coding: Fundamentals and Applications, Charles Lee

Desktop Encyclopedia of the Internet, Nathan J. Muller

Distributed Multimedia Through Broadband Communications Services, Daniel Minoli and Robert Keinath

Electronic Mail, Jacob Palme

Enterprise Networking: Fractional T1 to SONET, Frame Relay to BISDN, Daniel Minoli

FAX: Facsimile Technology and Systems, Third Edition, Kenneth R. McConnell, Dennis Bodson, and Stephen Urban

Guide to ATM Systems and Technology, Mohammad A. Rahman

Guide to Telecommunications Transmission Systems, Anton A. Huurdeman

A Guide to the TCP/IP Protocol Suite, Floyd Wilder

Information Superhighways Revisited: The Economics of Multimedia, Bruce Egan

International Telecommunications Management, Bruce R. Elbert

Internet E-mail: Protocols, Standards, and Implementation, Lawrence Hughes

Internetworking LANs: Operation, Design, and Management, Robert Davidson and Nathan Muller

Introduction to Satellite Communication, Second Edition, Bruce R. Elbert

Introduction to Telecommunications Network Engineering, Tarmo Anttalainen

Introduction to Telephones and Telephone Systems, Third Edition, A. Michael Noll

LAN, ATM, and LAN Emulation Technologies, Daniel Minoli and Anthony Alles

The Law and Regulation of Telecommunications Carriers, Henk Brands and Evan T. Leo

Marketing Telecommunications Services: New Approaches for a Changing Environment, Karen G. Strouse

Mutlimedia Communications Networks: Technologies and Services, Mallikarjun Tatipamula and Bhumip Khashnabish, Editors

Networking Strategies for Information Technology, Bruce Elbert

Packet Switching Evolution from Narrowband to Broadband ISDN, M. Smouts

Packet Video: Modeling and Signal Processing, Naohisa Ohta

Performance Evaluation of Communication Networks, Gary N. Higginbottom

Practical Computer Network Security, Mike Hendry

Practical Multiservice LANs: ATM and RF Broadband, Ernest O. Tunmann

Principles of Secure Communication Systems, Second Edition, Don J. Torrieri

Principles of Signaling for Cell Relay and Frame Relay, Daniel Minoli and George Dobrowski

Pulse Code Modulation Systems Design, William N. Waggener

Signaling in ATM Networks, Raif O. Onvural and Rao Cherukuri

Smart Cards, José Manuel Otón and José Luis Zoreda

Smart Card Security and Applications, Mike Hendry

SNMP-Based ATM Network Management, Heng Pan

Successful Business Strategies Using Telecommunications Services, Martin F. Bartholomew

Super-High-Definition Images: Beyond HDTV, Naohisa Ohta

Telecommunications Department Management, Robert A. Gable

Telecommunications Deregulation, James Shaw

Telemetry Systems Design, Frank Carden

Teletraffic Technologies in ATM Networks, Hiroshi Saito

Understanding Modern Telecommunications and the Information Superhighway, John G. Nellist and Elliott M. Gilbert

Understanding Networking Technology: Concepts, Terms, and Trends, Second Edition, Mark Norris

Understanding Token Ring: Protocols and Standards, James T. Carlo, Robert D. Love, Michael S. Siegel, and Kenneth T. Wilson

Videoconferencing and Videotelephony: Technology and Standards, Second Edition, Richard Schaphorst

Visual Telephony, Edward A. Daly and Kathleen J. Hansell

World-Class Telecommunications Service Development, Ellen P. Ward

For further information on these and other Artech House titles, including previously considered out-of-print books now available through our In-Print-Forever® (IPF®) program, contact:

Artech House
685 Canton Street
Norwood, MA 02062
Phone: 781-769-9750
Fax: 781-769-6334
e-mail: artech@artechhouse.com

Artech House
46 Gillingham Street
London SW1V 1AH UK
Phone: +44 (0)20 7596-8750
Fax: +44 (0)20 7630-0166
e-mail: artech-uk@artechhouse.com

Find us on the World Wide Web at:
www.artechhouse.com